The Oryx Guide to Natural History

The Oryx Guide to Natural History

The Earth and All Its Inhabitants

by
Patricia Barnes-Svarney
and Thomas E. Svarney

Oryx Press
1999

The rare Arabian Oryx is believed to have inspired the myth of the unicorn. This desert antelope became virtually extinct in the early 1960s. At that time, several groups of international conservationists arranged to have nine animals sent to the Phoenix Zoo to be the nucleus of a captive breeding herd. Today, the Oryx population is over 1,000, and over 500 have been returned to the Middle East.

© 1999 by Patricia Barnes-Svarney
and Thomas E. Svarney
Published by The Oryx Press
4041 North Central at Indian School Road
Phoenix, Arizona 85012-3397
www.oryxpress.com

All rights reserved. No part of this publication may be reproduced or transmitted
in any form or by any means, electronic or mechanical, including
photocopying, recording, or by any information storage and
retrieval system, without permission in writing
from The Oryx Press.

Published simultaneously in Canada
Printed and bound in the United States of America

∞ The paper used in this publication meets the minimum requirements of
American National Standard for Information Science—Permanence
of Paper for Printed Library Materials, ANSI Z39.48, 1984.

Library of Congress Cataloging-in-Publication Data

Barnes-Svarney, Patricia L.
    The oryx guide to natural history : the earth and all its
inhabitants / by Patricia Barnes-Svarney and Thomas E. Svarney.
      Includes bibliographical references.
      ISBN 1-57356-159-2 (alk. paper)
    1. Natural history. I. Svarney, Thomas. II. Title.
QH45.2.B37 1999
508—dc21   99-41783
                                                                       CIP

In memory of Eric Weiskopff—librarian and friend . . .

Contents

List of Tables   ix

Preface   xi

Chapter 1: Amphibians   1
Chapter 2: Animals   7
Chapter 3: Arthropods   17
Chapter 4: Atmosphere   21
Chapter 5: Bacteria and Viruses   29
Chapter 6: Birds   36
Chapter 7: Classification of Organisms   44
Chapter 8: Climate and Weather   50
Chapter 9: Cytology   58
Chapter 10: Dinosaurs   65
Chapter 11: Earth   73
Chapter 12: Earthquakes   83
Chapter 13: Evolution   91
Chapter 14: Extinction   99
Chapter 15: Fish   106
Chapter 16: Fossils   110
Chapter 17: Fungi   116
Chapter 18: Genetics   121
Chapter 19: Geologic Time Scale   127
Chapter 20: Humans   132
Chapter 21: Life   140
Chapter 22: Moon   145
Chapter 23: Mountains   150
Chapter 24: Oceans   155
Chapter 25: Plants   166
Chapter 26: Protista   173
Chapter 27: Reptiles   176
Chapter 28: Solar System   181
Chapter 29: Universe   190
Chapter 30: Volcanoes   196

Appendix A: Careers in Natural History   207

Appendix B: Natural History Sites on the World Wide Web   211

Glossary   217

References   221

Index   223

List of Tables

Chapter 1:	Common Types of Modern Amphibians—United States 4
Chapter 2:	Classification of the Modern Animal Kingdom I & II 12
	Classification of Mammals 16
Chapter 3:	Phylum Arthropoda 20
Chapter 5:	Modern Diseases from Bacteria 34
	Modern Diseases from Viruses 35
Chapter 6:	Common Modern Bird Species 38
Chapter 7:	Various Classifications Systems 48
	Human Taxonomy 48
	Tiger/Sweet Pea Taxonomy 49
Chapter 8:	Major Air Masses 53
	The Fujita and Pearson Tornado Scale 55
	Saffir-Simpson Damage Potential Scale for Hurricanes 56
Chapter 9:	Cellular Comparisons of Modern Animals, Plants, and Prokaryotes 62
Chapter 10:	Mesozoic Era Dinosaurs 70
Chapter 11:	The Continents 79
Chapter 12:	Ancient Earthquakes 88
	Modern Earthquakes 88
	Modified Mercalli Intensity Scale 88
	Richter Scale of Earthquake Magnitude 89
	Well-Known Earthquake Examples 89
Chapter 14:	Endangered Species 103
	Major Impact Craters around the World 105
Chapter 15:	Speed of Fish 109
Chapter 18:	Classical Genetics—Parents and Offspring 125
Chapter 19:	Geologic Time Scale 130
Chapter 22:	Composition of Lunar Rocks 149
	Types and Location of Lunar Rocks 149
	Characteristics of the Moon 149
Chapter 23:	Mountain Ranges of the World 153
	World's Highest Mountains 153
	Major Orogenies 153
Chapter 24:	Largest Oceans and Seas of the World 162
	Characteristics of Seawater 164
	Ocean Trenches 165
Chapter 26:	Kingdom Protoctista 175
Chapter 27:	Classification of Reptiles 179
Chapter 28:	Extrasolar Planet Discoveries 186
	Planetary Characteristics (each planet) 187–188
	Titius-Bode Law Calculations 189
Chapter 29:	Closest Stars Compared 194
	Temperature of Several Bright Stars Compared to Our Own Sun 194
Chapter 30:	Major and Notable Volcanic Eruptions 204
	Famous Volcanoes 204

Preface

The Earth has been in existence for about 4.55 billion years. If all of the topics in the natural world were addressed, the number of volumes of resulting books would be enormous. *The Oryx Guide to Natural History* offers the reader a condensed version of the most important topics—from the beginnings of the universe and our planet, to the plants and animals inhabiting our world.

The Oryx Guide to Natural History also traces the development of human understanding of our natural world. It examines the physical and biological changes that scientists believe have occurred over time, and in a modest way, it describes the evolution of natural history over time—both of the physical earth and the organisms that inhabit it—giving the reader a good understanding of why we think about our natural history the way we do.

Each of the 30 major topics contains the following:

- A brief introduction to the topic and its importance.
- An at-a-glance timeline of the topic. In some topics, the timeline lists the evolution of the organisms and the major discoveries made by scientists over time; other topics' timelines just mention the discoveries made by scientists over time.
- A detailed description of early and modern highlights within the topic—most of which are merely mentioned in the timeline. These sections detail some of the discoveries, expeditions, important publications, and major breakthroughs of each topic.
- Alphabetically sorted definitions of terms for a better understanding of the topic (words with definitions are set in bold type in the "History of" sections) with many of the definitions providing additional historical information.

In addition, the book presents pertinent tables, sidebars, and photographs. At the end of the book, two appendixes offer more information on types of natural history careers; and natural history sites on the World Wide Web for readers with an Internet connection; and major references.

This text is intended for those people interested in natural history (or science in particular), students needing a basic foundation on a topic in natural history, or people just looking for a good general reference book on this subject.

This book is not meant to answer all the questions about the Earth's natural history; it is only meant to be a compendium of the highlights. By dividing the book into major topics, we allow the reader a simple way of searching for information. First, for those people who want to know about a specific highlight or term within a certain topic, the reader can use the extensive index; for readers who want to know more about a certain topic, each entry can be read as a mini-book.

The information in this book was collected from a wide variety of sources, most listed in the reference section, and also from many studies published on the Internet. Like most books in science and science history, this work lists some dates and discoveries that are in debate—many of these have been debated for decades and will continue to be argued for decades to come. In addition, some discoveries have just occurred—and thus have not gone through peer review to verify their accuracy. We have made every effort to assure the correctness and reliability of the contents, but even as we wrote the text, new discoveries were being made. It seemed as if every week we were adding to the text—from the latest discoveries to changes in several theories.

We hope you enjoy this dip into 4.55 billion years of natural history. You will be surprised and amazed at what the natural world has had—and still has—to offer.

—*Patricia Barnes-Svarney and Thomas E. Svarney*

1. Amphibians

Introduction

Throughout most of the natural history of the Earth, amphibians have been part of the complex balance of life. In fact, amphibians were the first air-breathing animals to evolve onto land, and branches of the amphibians eventually gave rise to the reptiles. In other words, without the revolutionary transitional step of the amphibians from living in oceans to living on land, land animals would not exist. And the results were amazing, but not as much for the amphibians as other animals. Today, relatively few amphibians live on the Earth, the result of dead ends in evolution of various amphibian families, competition between the amphibians themselves and other animals—and of course, human encroachment on their habitats.

Timeline

(note: mya=million years ago)

Date	Event

Prehistoric Events

Date	Event
~360 mya	The first known amphibians evolve.
~345–280 mya	Amphibians become one of the dominant animals on the Earth, with the Carboniferous period often referred to as the "Age of Amphibians."
~180 mya	True frogs (frogs that resemble modern frogs) and toads may have evolved at this time.
~135 mya	Most amphibians become extinct, leaving no known fossils that link current amphibians with the ancient forms.

Modern Events and Discoveries

Date	Event
1661 A.D.	Italian physician and biologist Marcello Malpighi (1628–1694) uses a microscope to discover capillaries in the lungs of frogs.
1731	Swiss geologist Johann Scheuchzer (1672–1733) discovers a fossil he believes is the skeleton of a human, the *Homon diluvii testis*, or "man, a witness of the flood"; the skeleton is eventually proven to be that of an extinct giant salamander dating back 20 million years.
~1792	A controversy between Italian anatomist and physiologist Luigi Galvani (1737–1798) and Italian physicist Count Alessandro Giuseppe Antonio Anastasio Volta (1745–1827) over the twitching of frogs' legs (an electric current caused the animal's leg muscles to twitch) leads to an interest in investigating the electrical phenomena of animals; Galvani later announces his discovery of "animal electricity."
1917	The fossil remains of *Seymouria* (around 280 million years old), an organism showing both amphibian and reptilian characteristics, is discovered.
1932	A Danish scientific expedition finds the fossil of an *Ichthyostega* in Greenland—the oldest known fossils (approximately 360 million years old) that can be classified as amphibians.
1939	*Latimeria*, a living crossopterygian fish, is caught off the coast of South Africa; prior to this discovery, it was believed to have died out in the Cretaceous period after it gave rise to the amphibian line.
1950s	U.S. paleontologist Alfred Sherwood Romer (1894–1973) suggests that early fish moved onto land to escape drying pools during seasonal droughts, leading to the first amphibians.
1986	A complete frog is found fossilized in amber in the Dominican Republic; it is thought to be 35 to 40 million years old.

History of Amphibians

Early Amphibians

The phylogeny, or the evolutionary trail, of **amphibians** is a highly debated subject. It is thought that **caecilians**, or worm-like amphibians, played a crucial role in the evolution of the amphibians, in that similar organisms may have been the first to evolve. Many scientists believe that amphibians descended from a single group called Temnospondyls, a now-extinct amphibian; other scientists contend that only **frogs** are descendants of the Temnospondyls, while caecilians and **salamanders** are descended from the microsaurs, other now-extinct amphibians of the Carboniferous and early Permian periods. The debate will no doubt continue as amphibians fossil records are scarce. Scientists eventually hope to solve the phylogeny problem by comparing the biomolecular data from amphibian fossils with modern amphibians to unscramble the connections between frogs, **toads**, salamanders, **newts**, and caecilians.

We do know some things about early amphibians. In particular, they were much more fish-like than modern amphibians. And unlike today's amphibians that live in or near water, the majority of earlier amphibians lived only in the water. The progression to life on land seems to follow a general pattern. The first land plants were probably green algae that spread from the water during the Precambrian era; during the early Silurian period, vascular plants began to spread on land, creating tropical forests, and the early amphibians appeared less than a hundred million years after vascular plants.

As mentioned, the fossil record of amphibians is poor; thus, the true origin of amphibians and their subsequent evolution to modern amphibians is debated. The first amphibians are thought to have descended directly from early fish—but no one knows exactly what fish. It is believed that the first groups of fish that were most closely related to amphibians were the lungfish or the crossopterygians—both animals had fleshy fins with various bones and muscles inside (called lobe fins) and lungs. Most scientists think the link was a group of now-extinct crossopterygians called rhipidistians, which display similar front fin bones, skulls, and teeth to early amphibians. The limbs were no more than jointed lobed fins, and the head and tail were fish-like. The difference was that this early amphibian breathed air, something fish could not do.

But breathing was the least of an early amphibian's problems. It also had to develop a way of supporting itself and walking on land; in the water, the creature was virtually weightless. Walking was very different than swimming, so new muscles and bone structures evolved. Plus, new muscles and bone modifications were needed not only to hold up the animal's head, but also to hold up the belly, as the region could rupture from the weight of the internal organs.

Overall, scientists believe that amphibians initially appeared in the fossil record about 360 million years ago, during the late Devonian period. One of the earliest known amphibians on record is the *Ichthyostega* fossil found in Greenland. This 3-foot (1-meter) long animal had four limbs and a fin on its tail, and its backbone and rib cage were designed for supporting its weight on land. In many ways, it combined tetrapod (an amphibian-like creature thought by some scientists to be the first step toward amphibians) and fish features.

Scientists do not truly know why the first organisms moved from the water to land, but they have several theories. The classic theory, suggested by Alfred Sherwood Romer in the 1950s, states that the fish moved onto land to escape drying pools during seasonal droughts, but evidence of such extensive droughts has been limited. Another theory, and one that is more agreed upon, states that organisms moved to land to take advantage of the abundant food and oxygen supplies. There were more plants and animals (such as insects) to eat, and more oxygen in the atmosphere than in the water—all things these organisms needed to survive.

The ability to breathe air and walk on crude legs allowed the amphibians to spread to the land, although they were not able to stray too far from the water. This connection to water was due to two factors: First, most of the animals needed to lay their eggs in water. All amphibians in their larval stage needed water, too; for example, in the case of most frogs and toads, the animals start off as **tadpoles**—small, round-bodied young that need water to survive. Second, the animals could not let their bodies dry out from evaporation. (These characteristics are still true for most modern amphibians.)

Soon after this initial appearance, in evolutionary terms, some amphibians developed flattened bodies and heads and shorter limbs. The eye placement also changed to the top of the head, indicating that most early amphibians spent a great deal of life in shallow waters. Amphibians were the first vertebrates to have true legs, tongues, ears, and voice boxes.

The amphibians eventually played a major role in the Earth's natural history. During the Carboniferous period about 345 to 280 million years ago, they became one of the dominant animals on the Earth. In fact, the Carboniferous is often referred to as the "Age of Amphibians." After this time, amphibians began to decline in number of species, giving way to a new class of animals, the reptiles. By about 135 million years ago, most amphibians were extinct, and this extinction left scientists with a puzzle because no known fossils link modern amphibians with their ancient forms.

Although some of the first amphibian descendants gradually evolved into complete land animals, specific characteristics of other amphibians restricted their evolutionary advancement from water to land. Modern amphibians are essentially the "leftovers" of the groups that did not move completely from water environments to the land.

Modern Amphibians

Only a few major studies have been done on amphibians over the years, and they have certainly not attracted the attention that other species have—such as the dinosaurs. They are mostly categorized and studied in combination with other animals, such as reptiles. Recently, though, scientists have begun to pay more attention to amphibians, as they realized that changes in some amphibian populations (lower birth and survival rates and mutations) may be indicators of problems in our environment.

Modern amphibians (or Lissamphibians) are of the class Amphibia. They are quadruped vertebrates that include frogs, toads, salamanders, newts, and caecilians. The name comes from the Greek *amphi* meaning "both," and *bios* meaning "life" in reference to the fact that most amphibians live in and out of water. More than 4,000 species of amphibians are known on the Earth today, including about 360 species of salamanders, 3,800 species of frogs and toads, and 160 species of caecilians. They are found on all the continents except Antarctica. The majority of amphibians that live in colder climates hibernate in the winter; those in the arid climates often estivate (become inactive) during the hottest times of the year.

More recent studies of amphibians have concentrated on how amphibians fit into the natural scheme of our world. (See the sidebar in this chapter.) Even today, we are finding animals that were previously unknown in the amphibian world. For example, many new amphibians are being discovered, especially in rainforests around the world. In 1996, scientists may have found the smallest frog species in the Northern Hemisphere in a Cuban rainforest—a frog smaller than a dime, or about a third of an inch (about 1 centimeter) long. The new species, called the eleuth frog, after its genus, *Eleutherodactylus* (the formal Latin name is yet to be decided), produces one egg at a time, lays its eggs out of water, and hatches into developed frogs instead of tadpoles. Scientists believe these characteristics are survival techniques to avoid predation in ponds and streams. *See also* ANIMALS, CLASSIFICATION OF ORGANISMS, EVOLUTION, EXTINCTION, FISH, FOSSILS, GEOLOGIC TIME SCALE, LIFE, OCEANS, and REPTILES.

Topic Terms

amphibians—Amphibians are vertebrate animals. They are ectothermic, or cold-blooded, with their body temperatures staying about the same temperature as their surroundings. These amphibian features have not changed for hundreds of thousands of years.

With some rare exceptions, amphibians usually live in water during their early larval stages (the time between hatching and becoming an adult). After a certain period of time—which varies between individual amphibian species—a young amphibian metamorphoses, changing from the larval stage to a form that can live on land and in the water during its adult life. (The process of changing from a tadpole to a frog is also referred to as transformation.) The transformation from the larval to adult stage can take weeks to months—or even longer. For example, a species of bullfrog that measures about 6 to 7 inches (15–18 centimeters) in length as an adult may take three to four years to reach this size.

An amphibian's adult stage may be amphibious or totally aquatic. For example, most frogs metamorphose from the totally aquatic tadpole stage (with a compact body, short tail, but no legs) to the adult amphibious stage (with strong legs for jumping and no tail for living partially on land); others, such as certain salamanders, remain totally aquatic in their characteristics even after metamorphosing.

Most amphibians live in moist climates, although there are exceptions. Some frogs live only on land (but still must remain in a moist area), certain salamanders live only in water, and several types of amphibians are common in more arid areas. For example, to solve the problem of little water, species in these arid areas can store up to half their body weight of water in the bladder. To reproduce, most desert amphibians wait for puddles from sudden rains, then mate and lay eggs during that time.

Amphibians are related to reptiles but differ in many ways. Most amphibians go through a metamorphosis; reptiles do not go from the larval to adult stage, but look like miniature copies of their parents when born. In addition, unlike reptiles' waterproof skins, amphibians have moist, glandular skin with no scales; amphibians also lack claws on their toes.

Common Types of Modern Amphibians— United States

Frogs and Toads—*(Order: Anura or Salientia—a species without a tail and with muscular back legs used for jumping)*

　Narrow-mouthed Toads
　Spadefoot Toads
　True Toads (includes the American toad and common toad)
　Tailed Frogs
　Treefrogs (includes the spring peeper and barking treefrog)
　Cricket Frogs
　Chorus Frogs
　True Frogs (includes the bullfrog and green frog)

Caecilian—*(Order: Gymnophiona or Apoda—worm-like amphibians found mostly in the tropics, rarely in the United States)*

Salamander—*(Order: Urodela or Caudata—elongated animals with short legs and long tails)*

　Mudpuppies and Waterdogs
　Sirens
　Hellbenders
　Giant Salamanders
　Dusky Salamanders
　Brook Salamanders
　Red Salamanders
　Spring Salamanders
　Slender Salamanders
　Woodland Salamanders
　Climbing Salamanders
　Four-toed Salamanders
　Ensatina Salamanders
　Eastern Newts
　Pacific Newts

---

caecilians—Caecilians are of the order Gymnophiona or Apoda, and thus, are sometimes referred to as Gymnophionans or Apodans. They are legless, worm-like amphibians found everywhere around the world (except Australia and Madagascar), although they are usually rare outside the moist tropical regions. They are occasionally eyeless and have smooth skins; similar to earthworms, they burrow into the ground.

These creatures are the least studied order of living amphibians, although numerous studies are currently being conducted to determine the phylogeny of the caecilians and to understand their relationship to other amphibians. Living and fossil caecilians seem to indicate that the creatures played a crucial part in the development of early amphibians. Traditionally, the caecilians were thought to be a sister group of the other lissamphibians, but recent molecular studies show that the creatures may be closely related to just the salamanders.

frogs—Frogs are of the order Anura or Salientia. It is unknown when the earliest frogs appeared. Fossils that link modern amphibians to the ancient forms are rare, as are fossils that reveal the development of frogs. But scientists believe that frogs probably evolved about 180 million years ago.

There are thousands of types of modern frogs, with many more found every year, usually in the unexplored regions of the Amazon River basin in South America. The variation among members of the order is great. For example, some frogs have teeth in the upper jaw; one species has teeth in both jaws; others are without any teeth. The largest known frog is the Goliath frog from Cameroon, Africa; its body can be more than a foot (0.3 meter) long, and when extended it can be more than 2.5 feet (0.8 meter) long. For many years, the smallest frog known in the world was the banana frog from Cuba; it can fit on a person's thumbnail. This record may be eclipsed by the recently discovered eleuth frog.

Frogs develop from the tadpole (or larval) stage, with a compact body, short tail, and no legs, to the adult stage, with strong legs for jumping and no tail. In most frog species, tadpoles have gills to breathe underwater; when they become adults, gills are often replaced with lungs. Even then, these adults still also breathe through the skin.

Frogs are mostly twilight and nocturnal creatures; a minority of species are active during the day. Like many other amphibians, frogs eat algae and detritus (broken down plant and animal remains) as larvae. As adults, frogs are carnivores, consuming enormous quantities of insects and worms, usually with the help of their long tongues; larger frogs, such as the American bullfrog, will eat small fish, reptiles, and mammals—in other words, any creature small enough to swallow whole. Frogs are heavily dependent on their sense of vision, and occasionally smell, to catch prey; in fact, some species have bulging eyes that give the frog an almost 360-degree field of vision. Movement of the prey is often important, with some species not even reacting to prey unless it is moving.

Frog predators include alligators, snakes, owls, and even people. Frogs escape predators by leaping, some jumping distances up to 20 times their own body length in a single leap; following an erratic path as they jump; and hiding underwater or under aquatic vegetation.

Breeding season is usually in the spring and summer. During this time, the male frogs croak in an at-

This bullfrog, *Rana catesbeiana,* is a true frog, meaning it needs a permanent body of water as a habitat.

tempt to attract a female for breeding and to declare their territory. Several species puff up as they croak, causing the sound to become amplified, similar to how a stretched membrane of a drum works. Male frogs mate with the females by climbing on their backs and clinging using special adapted thumbs to hang on (the mating grasp is called amplexus). The male fertilizes the eggs as the female lays them in the water; each female lays thousands of eggs in a process called spawning—all at one time.

The eggs of most frogs (and toads) hatch into tadpoles, then later metamorphose into the adult phase. Not all species have the same method of reproduction. The eggs in some species hatch immediately into little froglets; a species of treefrog builds hanging nests for its eggs—the tadpoles dropping into the water below as they hatch.

Frogs, like toads, also molt at regular intervals. Most frogs undergo a series of ritual-like movements to remove the molting skin, which breaks off along seams where the new layer of skin is pushing through. In most cases, frogs will eat the skin as they peel it from their bodies.

newts—Newts come from a family of salamanders that lack the grooves found on the sides of most salamanders. Some newts use special tactics to defend themselves against predators. For example, the Pacific newt will warn off a potential predator by lifting its head and tail to expose brightly colored patches on its underbelly. If this tactic does not ward off a potential enemy, several glands on the newt's skin (seen as small bumps on its skin) secrete a irritating liquid if the predator tries to carry away the newt.

Newts have different overall stages of development. For example, the eastern newts and striped newts lay eggs in water in the spring, and their larval stage extends into the summer. In late summer, the newts lose their gills, turn red, and live on land; in this phase of their lives, they are called efts. After about one to three years, the efts turn color again, turning a spotted brown, and return to the water to mate, never again to return to land. Other species, such as the Pacific newts, do not go through an eft stage.

salamanders—Salamanders are tailed amphibians. They resemble lizards, but salamanders have thin, moist skin instead of scales and have four front toes instead of five. Similar to most amphibians, the majority of salamanders live on land near water; others retain their gills from the larval stage, living out their adult lives in water.

North America has the greatest number of salamander types in the world, numbering close to 85 species. The Great Smoky Mountains National Park in the southeastern United States is home to about 23 kinds of salamanders, more than can be found anywhere else in North America. The largest salamander, the Japanese giant salamander, is also the largest amphibian in the world, growing to more than 5 feet (1.5 meters) in length.

tadpoles—Young frogs and toads in the larval stage are called tadpoles. The tadpoles (often called "polliwogs") hatch from the egg clutch. Some tadpoles are so small that they can hardly be seen, while one of the largest ones, a type of bullfrog, measures 6 to 7 inches (15–18 centimeters) in length. Tadpoles are teardrop-shaped, with rounded bodies and gills outside the body. The growth stages vary in length between species. In general, a tadpole soon grows a tail and hind legs for swimming. Next, lungs and other organs begin to grow, as do the front legs. As they reach maturity (or metamorphose), the tadpole loses its gills. Finally, as an adult, the grown frog or toad absorbs its tail.

toads—Toads are amphibians that are directly related to frogs. Toads are of the order Anura or Salientia. This order has many families, but scientists classify only the toads in the family Bufonidae as true toads. It is not known when the earliest toads appeared, as few fossils link modern amphibians to the ancient forms; similar to frogs, though, it is thought toads evolved about 180 million years ago.

One of the largest known toads in the United States and the world (it is found in Australia, the Philippines, and southern Florida) is the giant, or marine, toad, measuring about 9 inches (23 centimeters); the smallest toads known in the world are the oak (southeastern United States) and Rose's (South Africa) toads, both of which measure about 1 inch (2.5 centimeters) long.

> ### Do Frogs Tell Us Something?
>
> Decades ago, miners brought canaries into coal mines to warn them of deadly carbon monoxide gas. They knew that canaries had a low tolerance to even a small amount of this gas. If a canary died, it meant that gas was in the mine—and that the miners should leave as soon as possible. In many instances, the death of a canary saved miners' lives.
>
> Some scientists now believe amphibians may be an indicator of the health of our planet. Frogs, for example, are sensitive to pollution because they live in two environments—land and water—and pollutants are easily absorbed through their skin. Because their skin is so thin, they are also sensitive to increases in such pollutants as lower atmospheric ozone (caused by emissions from industry and cars), and upper atmosphere ozone decreases (the "ozone hole").
>
> Because of this sensitivity, scientists are using populations and the physical appearance of frogs as indicators of possible environmental problems. What they have found is frightening: For decades, frogs were common in certain parts of the United States, but in the past two decades, it appears many species are disappearing. For example, Yosemite frogs from Yosemite National Park in California were once common. Today, even in the near pristine wilderness of the park's backcountry, frogs are rare. It is thought that pollutants from the nearby Central Valley—including smoke, smog, and pesticide residues—may be killing off the frogs. In another example, in 1995, students from the New Country School found deformed frogs in a wetland they were studying near Henderson, Minnesota. Many of the frogs had missing or extra legs or had deformed eyes or other parts. The next year was not much better: In 1996, deformed frogs were reported all over the state. Preliminary research showed that something in the water caused the deformities, but what it is still is not known—and it is not known if humans could be affected, too. A more recent study pointed to a frog parasite, but if so, scientists do not know the parasite's origin.
>
> Are frogs a true direct indicator of pollution? Scientists are still trying to decide. Even if they are not indicators, their disappearance will affect other species that frogs feed upon, or that feed upon the frogs. And some scientists believe that such a missing link in the food chain could have an dramatic effect on our natural world.

The African live-bearing toads are the only toads that do not lay eggs, but give birth to live young.

Similar to frogs, the majority of toads go through the tadpole (or larval) stage, with a compact body, short tail, but no legs; in the adult stage, the toad develops strong legs for jumping and loses its tail. Also similar to frogs, toad tadpoles breath underwater with gills, then lose the gills and develop lungs as adults.

Most toads seek food at twilight or at night, although some may be active during the day. They also eat just about the same as the frog: tiny algae and detritus as tadpoles, and insects, worms, and other animals as adults. The larger toads eat larger animals, such as fish, reptiles, and mammals.

Similar to frogs, toads' predators include alligators, snakes, owls, and people. Toads escape predators by leaping, following an erratic path as they jump. They also hide underwater, under aquatic vegetation, or under leaves on a moist ground to escape predators. Toads also have a pair of paratoid glands on top of their heads. These glands carry a poison that has an unpleasant taste to most predators (except the hognose snake); it also can make people ill or cause eye irritation.

Breeding season is usually in the spring and summer, and the procedure is much the same as the one the frog follows. Attracting a mate by making certain loud croaks, the males then mount the females' backs to mate and fertilize the eggs as they are laid. The eggs of most toads hatch into tadpoles, then later metamorphose into the adult phase. Not all species have the same method of reproduction, though. The eggs in some species hatch immediately into little toads, and the Surinam Toad carries the tadpoles on a spongy layer of skin on its back.

Toads and frogs do differ in many ways: Most frogs have smoother skin than toads, which almost always have warts. (Some frogs do have rougher skin, such as the barking treefrog of the southern United States.) Frogs live near more moist areas, whereas toads can live in drier areas; frogs lay their eggs in clumps, whereas toads lay their eggs in long strings (usually in double strands) in a pond; and frogs are generally more slim and faster than the sluggish, fat-bodied toads.

Some other physical differences include the following: Toads possess an oval, raised glandular area (parotid gland) behind the eye; frogs do not. Toads have L-shaped ridges (cranial crests) between and in back of the eyes; frogs do not. Many true frogs have long folds that lie from the eardrum (tympanum), along the side of the backs, and extend to the hip or groin area (dorsolateral folds). Treefrogs have digital discs at the ends of their fingers and toes; toads and most other frogs do not.

2. Animals

Introduction

Animals include many of the living organisms on the Earth; another name for animals is fauna. By way of definition, the animal kingdom, Animalia, encompasses all the multicellular eukaryotic (special celled) organisms that feed entirely on complex organic material for nutrition, have rapid motor response to stimulation, and have the capacity for spontaneous movement. The animal kingdom is indeed large, taking in the familiar divisions of vertebrates—the amphibians, reptiles, birds, mammals, sharks, and bony fish. It is not the intention of this chapter to list all the many members of the animal kingdom; instead it provides a brief synopsis of the more familiar creatures that have been, and are, important to the natural history of the Earth.

Timeline

(note: bya=billion years ago; mya=million years ago)

Date	Event
Prehistoric Events	
~3.75 bya	Life apparently evolves on the Earth as single-celled organisms; many scientists believe life evolved even earlier, but no fossil evidence has been found.
~1.2 bya	Some scientists believe, based on DNA sequences of certain genes in living animals, that the branching of several invertebrate phyla took place at this time.
~1 bya	Oxygen becomes a major gas in the Earth's atmosphere, helping life, including animals, to eventually develop on land.
just before 544 mya	Just before the Cambrian period the earliest sponges evolve in the oceans; these sponges are thought to be the most primitive multicellular animals.
~544 mya	At the start of the Cambrian period the ancestors of most modern animal groups suddenly appear in the oceans; scientists have dubbed this proliferation of species the Cambrian Explosion.
~520 mya	Representatives of most of the main groups (phyla) of animals have appeared on Earth in the oceans.
~440 mya	The first animals, probably tiny arthropods, crawl on land, and a second burst of growth and diversification takes place on Earth.
~360 mya	Amphibians, direct descendants of early fish, become the first air-breathing land vertebrates.
Modern Events and Discoveries	
~520 B.C.	Greek physician Alcmaeon of Croton (c. 535 B.C.–?) dissects animals; he distinguishes between veins and arteries, notes the optic nerve, and recognizes that the brain is the site of thought.
~335 B.C.	Greek philosopher and scientist Aristotle (384–322 B.C.) classifies animals; at this time, 500 species are known, and Aristotle divides them into eight classes.
~50 A.D.	Pliny the Elder (Gaius Plinius Secundus) (23–79 A.D.) of Italy writes *Historia naturalis* (*History of Nature*).
1551	Swiss naturalist Konrad von Gesner (also known as Conrad Gessner) (1516–1565) writes the first volume of *Historiae Animalium* (*The History of Animals*).
1596	Li Shi-Chen (circa late 1500s) writes the book *Ben-zao Gang-mu*, in which he describes more than 1,000 plants and 1,000 animals.
1599	Italian naturalist Ulisse Aldrovandi (1567–1605) publishes the first serious work in zoology, titled *Natural History*.
1664	In a book published posthumously, French philosopher and mathematician René Descartes (1596–1650) proposes (erroneously) that animals were purely mechanical beings.
1749	Comte de Georges-Louis Leclerc Buffon (1707–1788) writes the *General and Particular Natural History*, a 44-volume set of books explaining all that was known about the nature of animals and minerals at the time.

Date	Event
~1792	Italian anatomist Luigi Galvani (1737–1798) announces his discovery of "animal electricity."
1797	German scientist Alexander von Humboldt (1769–1859) describes the results of his 4,000 experiments in animal electricity, in *Experiments on the Excited Muscle and Nerve Fiber*.
1812	French naturalist and anatomist Georges Léopold Chrétien Frédéric Dagobert Cuvier (1769–1832) becomes involved in many facets of nature, including developing the first method of classifying mammals; this method establishes a system of zoological classification, dividing animals into four categories based on their structure; he also begins the field of comparative anatomy and becomes a paleontologist and taxonomist.
1832–1836	English naturalist Charles Robert Darwin (1809–1882) travels on the *HMS Beagle*, visiting many countries on his journey; he writes about the fossils, plants, and animals he sees.
1840	Darwin's book, *Zoology of the Voyage of the Beagle*, is published, describing the animals he saw or collected on his scientific voyage from 1832 to 1836.
1909	U.S. geologist Charles Doolittle Walcott (1850–1927) discovers the Burgess shale layer in Field, British Columbia, Canada, one of the most prolific layers of animal fossils from the Cambrian period ever found.
1940s	Ediacaran rocks, containing a confusing group of early, simple organisms, are found in the Ediacara Hills in South Australia; these primitive organisms are thought to represent an intermediate stage before the development of early animals.
1940s	A profusion of early Cambrian animal fossils are found in Siberia, in the former Soviet Union; found in sedimentary rock, the fossil assemblages are called "small shelly fossils."
1951	Nikolaas Tinbergen writes one of the first major books on animal instinct and behavior, titled *The Study of Instinct*.
1967	British biologist John B. Gurden clones the first vertebrate, a South African clawed frog.
1971	Sociobiologist Edward O. Wilson publishes his fundamental work on insect societies.
1984	Sheep are successfully cloned.
1987	The United States Patent and Trademark Office extends patent protection to all animals.
1995	Biologists determine that invertebrates and vertebrate animals share developmental genes.
1996	Although this topic is still highly debated, genes of living organisms suggest that the first animals emerged a billion years ago, much earlier than previously thought.
1997	Rock studies suggest that the Earth's outer shell, the crust, became unbalanced 535 million years ago, possibly triggering the Cambrian Explosion.

History of Animals

Early Animals

Life apparently began about 3.75 billion plus years ago on the Earth, in the form of single-celled organisms. For almost 2 billion years, the oceans were dominated by these simple organisms; their remains are found today in fossil stromatolites, which are layered structures of cells and debris in fossil rock form. Some of these early single-celled organisms developed photosynthesis, giving out oxygen as the waste product. Although it took approximately another billion years, the photosynthesis process increased the amount of oxygen while carbon dioxide levels decreased—leading to an oxygen-rich atmosphere that would eventually help life to develop on land.

Amazingly, five-sixths of the Earth's history had passed before **animals** became prevalent in the Earth's oceans. Just over 1.5 billion years ago, certain cells developed into more complex **eukaryotic cells**, the first complex cells that eventually led to all multicellular life. The first traces of such multicellular animals are found in rock just before the end of the Precambrian era. (The actual time that the first animals emerged is highly debated—including one study in 1996 that analyzed the genes of living organisms and suggested that the first animals may have emerged about 1 billion years ago.) The Precambrian era ended about 544 million years ago, marking the beginning of the Paleozoic era (and the beginning of the Cambrian period). At the beginning of the Cambrian period, a huge explosion of life occurred—a time known as the **Cambrian Explosion**.

Just before the Cambrian Explosion, a perplexing group of creatures appeared that scientists called the Ediacaran fauna—simple organisms that are difficult to categorize. Some of these first animals resembled segmented worms and jellyfish, and all were **invertebrates**.

One invertebrate group seems more easily categorized: the sponge, considered to be the first primitive multicellular animals.

All of the first animals on the Earth lived in the oceans, with the evolution to terrestrial animals taking many more millions of years. The first land organisms were algae, lichens, and bacteria that lived along the edges of shallow waters; next came the land plants. The first land animals were arthropods, such as scorpions and spiders, animals that came crawling out of the water about 440 million years ago. Many of these creatures have been found in Silurian period rock layers, usually in association with fossils of the oldest-known vascular land plants. About 360 million years ago, amphibians, direct descendants of early fish, became the first air-breathing land vertebrates. These amphibians were a vital part of the evolution of animals on land. Some of the first amphibian descendants gradually evolved into complete land animals—in particular the reptiles, about 300 million years ago. From the reptiles, the birds and modern reptiles evolved; and eventually some mammal-like reptiles developed into the true **mammals**. To compare, modern humans in their present form, as *Homo sapiens sapiens*, evolved only about 90,000 years ago.

The Study of Animals

The study of animals has had a long history, starting even before the advent of written accounts. Animals have always surrounded humans. We know they were an important part of our ancient past, especially in terms of survival, providing food, clothing, and even shelter. Early humans were omnivores, eating both plants and animals. Because of this diet, early humans learned to hunt the other animals for food. Evidence of such hunts have been found in caves, where ancient pictures of the hunts were drawn by early humans.

Efforts to define and classify animals were the first recorded animal studies. Aristotle was one of the first to classify animals around 335 B.C. in his book, *Historia animalium* (*History of Animals*). He named the 500 species known at that time (including 300 species of vertebrates described well enough for modern naturalists to identify the species); he also dissected about 50 of the animals. Around A.D. 50, scholar and natural historian Pliny the Elder (Gaius Plinius Secundus) of Italy brought out his *Historia naturalis* (*History of Nature*), a 37-volume work describing the known zoology, astronomy, and geography of his time. His volumes also mentioned most of the legendary monsters—including winged horses, unicorns, and mermaids—classifying them as real animals.

In the sixteenth through eighteenth centuries, scientists and naturalists published a plethora of books on the science of zoology, explaining the various animals found in their immediate regions. Swiss naturalist Konrad (also as Conrad) von Gesner wrote the first volume of *Historiae Animalium* (*The History of Animals*) in 1551, a four-volume set in which each book covered a portion of the animal kingdom known in his time. In 1596, Li Shi-Chen wrote *Ben-zao Gang-mu*, describing more than 1,000 plants and 1,000 animals known, plus about 8,000 medicinal uses for the plants and animals. The first book in zoology was published in 1599 by Italian naturalist Ulisse Aldrovandi, titled *Natural History*. By 1664, French philosopher and mathematician René Descartes proposed erroneously (in a book published posthumously) that animals were purely mechanical beings, with no "vital force" to make them different from other material objects. In 1749, Comte de Georges-Louis Leclerc Buffon wrote the *General and Particular Natural History*, the start of a 44-volume set of books explaining all that was known about the nature of animals and minerals at that time.

But it was scientists in the nineteenth century that truly paved the way for the modern study of animals. From 1832 through 1836, naturalist Charles Robert Darwin went on his famous voyage onboard the *HMS Beagle*, writing about fossils, plants, and animals he saw—which also helped him to develop his theory of evolution. (He eventually published *Zoology of the Voyage of the Beagle* in 1840, describing the animals from the voyage.) Naturalist and anatomist George Cuvier became involved in many facets of nature and described one of the first comprehensive systems of zoological classification, dividing animals into four categories: Vertebrata, Mollusca, Articulata, and Radiata, all based on their body structure. Cuvier was also considered the founder of the field of comparative anatomy and was an active paleontologist and taxonomist.

But even as these advances were made, little was known about the origins of animal species—or even how long animals had been on the Earth. By the late 1800s, no rocks older than the Cambrian period (544 million years ago) were found with animal fossils; while the rocks from the Cambrian period on exhibited the earliest animal fossil record of arthropods, mollusks, brachiopods, and other species. Charles Darwin concluded that the reason for the lack of animals earlier than the Cambrian period was difficult to explain—after all, his theory of evolution by natural selection predisposed a population to evolve, not to appear suddenly in an advanced form. Several theories were advanced to explain the sudden appearance of animals: (1) the older fossil-bearing rock eroded away; (2) the rock had metamorphosed, thus the fossils within the rock were crushed and folded; or (3) the animals evolved elsewhere, such as lakes and ponds, then eventually entered the oceans.

Eventually, by the middle of the twentieth century, two important discoveries were made. First, late Precambrian rocks (those earlier than the Cambrian period) were found that had evidence of complex life. The fossils within these sedimentary rock date from about 580 to 560 million years ago and include impressions of a variety of animals—but no evidence of hard parts. The Vendian period, or Ediacaran (after the Ediacara Hills in South Australia, where such fossils were first found by geologist R.C. Sprigg in the late 1940s), is a division within the Precambrian time—these fossils are now found from England to Namibia and from Russia to Newfoundland. The Ediacaran fauna were simple organisms that are difficult to categorize. Some of these first animals resembled segmented worms and jellyfish, and they were all invertebrates. Only one group from this period is understood very well: Sponges, considered to be the first primitive multicellular animals. In fact, in 1997, scientists studying fossil sponges in Mongolia discovered tiny glass spikes identical to the spicules of some modern sponges.

We now know that **Ediacaran fossils** were not the only evidence of animal life at the end of the Precambrian. Other animals left their mark, such as wavy or straight burrows made as the animals grazed on microbes on the surface; they also appeared to eat mud and silt, trying to sift out the organic particles and bacteria. But most of these animals had no hard parts—only soft parts that quickly decayed away, leaving only the trace fossils of their many activities.

The second discovery stemmed from the study of the first abundance of animal fossils found in rock: After billions of years of microscopic life, an apparent explosion of the numbers and types of animals occurred in the oceans. The early Cambrian seas were swarming with animals during the so-called Cambrian Explosion about 544 million years ago. The first evidence of the profusion of early Cambrian animals was discovered when the former Soviet Union explored geologic resources in Siberia after World War II. There, in the undisturbed sedimentary rock of the Cambrian period, were fossil assemblages of small skeletons, with a few larger than 0.5 inch (1 centimeter) long. These "small shelly fossils" (SSFs) were soon seen in spots all over the world: The Anabarites-Protohertzina layer contains the oldest fossils with hard parts (other than the earlier *Cloudina*, the oldest skeletal animal fossil, a tube-dwelling polyp); on top of the Anabarite-Portohertzina layer are a wide range of shells, tubes, and skeletal parts—and the first primitive mollusks, monoplacophorans (that move on a foot like a snail), and the rostroconchs (which probably gave rise to the bivalves such as clams and cockles).

Numerous theories have been advanced to explain the sudden proliferation of animals. Some scientists suggest animal life expanded into the vast unoccupied areas of the oceans; others believe that the existence of relatively few predators allowed the animals to thrive. One 1997 study suggested that the Earth's crust was in a state of imbalance about 535 million years ago, just before the Cambrian period. According to this theory, the Earth "lost its balance," listing to one side and causing the continents to quickly reposition themselves. Some scientists suggest that, although almost all life was microscopic prior to this event (except for some enigmatic soft-bodied organisms), the reorientation of the planet would change the climate and ocean currents—and could precipitate the explosion.

More recent studies of animals have concentrated mostly on the genetics and evolution of species. (See section on "genetics" and "evolution.")

To summarize the theories and evidence presented so far, the earliest invertebrate animals with hard parts began to appear in the oceans toward the end of the Cambrian period; the earliest **vertebrates**, as fish, evolved only a bit later in the late Cambrian. And by about 520 million years ago, representatives of most of the main groups, or phyla, of animals had appeared on Earth. In the past 3.75 billion plus years since life first began on the Earth, more than 2 billion species have lived on our planet, including animals and plants. Currently about 90 to 99.9 percent of all the species that ever lived on the Earth are now thought to be extinct.

Almost all species of animals have changed dramatically since the first animals stepped on land, adapting to changes in climate, continental movements, and massive extinctions. Today, animals range in size from no more than a few cells to organisms weighing many tons, such as blue whales and giant squids. The greatest number of animals inhabit the seas, with less numbers in fresh water, and even fewer on land. At this time, between 5 and 30 million different species of animals, plants, and other creatures (such as fungi and bacteria) live on the Earth, although most scientists place the number at about 9 or 10 million species. Vertebrates and flowering plants have the fewest number of species; insects account for the greatest numbers—but the true numbers of most species of organisms are unknown. The reason for the uncertainty is that we know we have not discovered all the organisms on the Earth—including all the animal species. *See also* AMPHIBIANS, ARTHROPODS, BIRDS, CLASSIFICATION OF ORGANISMS, CYTOLOGY, DINOSAURS, EVOLUTION, EXTINCTION, FOSSILS, GENETICS, GEOLOGIC TIME SCALE, LIFE, PLANTS, and REPTILES.

Topic Terms

animals—All animals are members of the Kingdom Animalia, also called Metazoa. This kingdom does not contain the prokaryotes (in some classifications, prokaryotes are included in the Kingdom Mondera, which includes bacteria and blue-green algae) or the protists (in some classifications, protists are included in the Kingdom Protista, which includes unicellular eukaryotic organisms). All members of the Kingdom Animalia are multicellular; all are heterotrophs (they rely directly or indirectly on other organisms for their nourishment), with most ingesting food and digesting it internally. And they include both terrestrial and marine animals.

Animals cells differ greatly from plant cells, as they lack the rigid walls of the plant cells. The bodies of all animals, except sponges, are made up of these cells, and the cells are organized into tissues, each specialized to some degree to perform specific functions within the animal's body. In most animals, the cells are further specialized and organized into specific organs within the body, such as kidneys or livers. Also contrary to plants and other organisms, most animals are capable of complex and relatively rapid movement.

The majority of animals reproduce sexually, by means of differentiated eggs and sperm. Most are diploid (the cells of adults contain two copies of the genetic material), and the progression from egg to adult is characterized by distinctive stages of development. In particular, in most animals, a zygote forms, the product of the first few division of cells following fertilization; a blastula is then created, a hollow ball of cells formed by the developing zygote; and a gastrula forms when the blastula folds in on itself to form a double-walled structure with an opening to the outside (blastopore).

Research continues on the evolutionary relationships of the major groups of animals, and thus, the classification of animals is not agreed upon universally. The following lists two diverse classifications—the first, an older system, and the second, a more recent classification scheme (in general listed here from more primitive to advanced):

Classification of the Modern Animal Kingdom I *

Kingdom: Animals (Animalia)
  Phylum: Porifera
  Phylum: Cnidaria
  Phylum: Platyhelminthes
  Phylum: Nematodes
  Phylum: Rotifers
  Phylum: Bryozoa
  Phylum: Brachiopods
  Phylum: Phoronida
  Phylum: Annelids
  Phylum: Mollusks
    Class: Chitons
    Class: Bivalves
    Class: Scaphopoda
    Class: Gastropods
    Class: Cephalopods
  Phylum: Arthropods
    Class: Horseshoe crabs
    Class: Crustaceans
    Class: Arachnids
    Class: Insects
    Class: Millipedes and centipedes
  Phylum: Echinoderms
  Phylum: Hemichordata
  Phylum: Cordates
    Subphylum: Tunicates
    Subphylum: Lancelets
    Subphylum: Vertebrates
      Class: Agnatha (lampreys)
      Class: Sharks and rays
      Class: Bony fish
      Class: Amphibians
      Class: Reptiles
      Class: Birds
      Class: Mammals
        Subclass: Monotremes (often classed as an order)
        Subclass: Marsupials (often classed as an order)
        Subclass: Placentals
          Order: Insectivores
          Order: Flying lemurs
          Order: Bats
          Order: Primates (including humans)
          Order: Edentates
          Order: Pangolins
          Order: Lagomorphs
          Order: Rodents
          Order: Cetaceans
          Order: Carnivores
          Order: Seals and walruses
          Order: Aardvark
          Order: Elephants
          Order: Hyraxes
          Order: Sirenians
          Order: Odd-toed ungulates
          Order: Even-toed ungulates

* This is only one classification structure; there are several others, as not all scientists agree with the classification of certain animals.

Classification of the Modern Animals II **

Kingdom: Mesozoa
  Phylum: Mesozoa
Kingdom: Parazoa
  Phylum: Porifera
  Phylum: Placozoa
Kingdom: Eumetazoa
Kingdom: Radiata
  Phylum: Cnidaria
  Phylum: Ctenophora
Kingdom: Bilateria
Kingdom: Protostomia
Kingdom: Acoelomates
  Phylum: Platyhelminthes
  Phylum: Nemertea
Kingdom: Pseudocoelomates
  Phylum: Rotifera
  Phylum: Gastrotricha
  Phylum: Kinorhyncha
  Phylum: Gnathostomulida
  Phylum: Nematoda
  Phylum: Priapulida
  Phylum: Nematomorpha
  Phylum: Acanthocephala
  Phylum: Entoprocta
  Phylum: Loricifera
Kingdom: Eucoelomates
  Phylum: Mollusca
  Phylum: Annelida
  Phylum: Arthropoda
  Phylum: Echiurida
  Phylum: Sipuncula
  Phylum: Tardigrada
  Phylum: Pentastomida
  Phylum: Onychophora
  Phylum: Pogonophora
Kingdom: Deuterostomia
  Phylum: Phoronida
  Phylum: Ectoprocta
  Phylum: Brachiopoda
  Phylum: Echinodermata
  Phylum: Chaetognatha
  Phylum: Hemichordata
  Phylum: Chordata

** This system was outlined C.P. Hickman and L.S. Roberts, *Animal Diversity,* New York: McGraw-Hill, 1994.

Cambrian Explosion—The Cambrian Explosion refers to an approximate time period, about 544 million years ago, in which an explosion of animal abundance took place in the oceans. The Cambrian Explosion, named after the Cambrian period of geologic time (between 544 and 510 million years ago) is best exemplified in the Burgess shale layer found in Canada. This formation showed a period of intense evolutionary diversity, including strange arthropods, trilobites, and worm-like animals; the majority of animals found in the Burgess shale (about 90 percent) became extinct. The reasons for the intense period of growth is unknown. Some scientists suggest animal life expanded into the vast unoccupied areas of the oceans; other believe there were few predators for most species, allowing the animals to become profuse; and still others suggest the bizarre and diverse animals were actually experiments in animal evolution and design.

chordates—The name is from the phylum Chordata, usually divided into three subphyla, listed under their common names as the tunicates, lancelets, and vertebrates. Chordata organisms' overall characteristics include an internal skeleton, a specific alignment of the nervous system, and, within the alimentary tract, a chamber whose walls include what are often called gill slits. (These openings are only in the embryos of terrestrial species.) One of the first known chordates—and thus, perhaps one of the first members of our own phylum of the chordates—was probably the *Pikaia*, a worm-like animal found in the Burgess shale layer of Canada. It is often difficult to estimate the importance of this phylum, as humans are one of the included species. Thus, the vertebrate subset of the chordates includes the most advanced and highly developed animals—at least from a scientific viewpoint; from a more commercial or practical point of view, animals in the other vertebrate subphylums have also supplied humans with food, fur, leather, feathers, wool, beasts of burden, domesticated animals, and sundry products.

Ediacaran fossils—Ediacaran fossils are Precambrian fossils found within the sedimentary rock (from about 580 to 560 million years ago) that include impressions of a variety of animals, but no hard parts. The Ediacaran, or Vendian, fossils were first found in the late 1940s, but they took several decades to interpret.

Their origins and true nature are still highly debated. By the 1980s, some paleontologists proposed that the Ediacaran fossils were true jellyfish and would have been preserved in the beach sand, probably from being stranded on the beach (for example, after a violent storm). But other features of these animals caused the scientists to believe the Ediacaran fauna were a separate development, not related to the modern animal phyla. Still other scientists are now suggesting that the animals were an intermediate stage of simple animals; in this case, the fauna didn't work—perhaps in terms of the cell membranes or nuclei within the cells—and the "experiment" was soon discarded. Thus, the actual start of animal life in the early oceans is still a matter of conjecture and theory.

Some evidence shows that the Ediacaran fossils seem to disappear some distance in the rock strata, just below the first appearance of the Cambrian period animals. Scientists believe this distribution may indicate that the Ediacaran faunas experienced a mass extinction—either gradually or quickly dying out before the end of the Precambrian era. Other scientists believe they have found evidence of the Ediacaran animals in certain modern animals—showing the animals did not die out, but evolved into a different lineage altogether.

eukaryotic cells—Eukaryotic cells are the more complex cells usually associated with certain plants, animals, protists, and fungi; more primitive cells are called the prokaryotic cells. The eukaryotic cells have a nucleus enclosed within a nuclear membrane; inside are chromosomes, the long, filament-like structures composed of nucleic acids and proteins that carry the units of inheritance, called genes. The cells also carry mitochondria, Golgi bodies, endoplastmic reticulum, and most of the other organelles associated with complex cells.

invertebrates—Invertebrates are animals with no backbone; actually, most invertebrates have hard, external skeletons or use pressure in their body fluids to support themselves and move. Even though most people are familiar with the vertebrate animals (especially the more well-known mammals), few people realize that invertebrates encompass over 90 percent of all living animals. In fact, only one of the about 25 animal phyla, the Chordata, contains vertebrates. (And even some invertebrates are listed under this group.) The diversity—size, shape, color, physical features, and behavior—of invertebrates is amazing. Some have not changed for hundreds of millions of years, such as the dragonflies and horseshoe crabs; while other invertebrates are currently in the process of rapid evolution, such as the fruit fly.

The most successful group of invertebrates are the arthropods, or joint-legged animals. The majority of these creatures have hard, outer skeletons; their jointed legs allow various species to run, jump, swim, fly, sting,

Do You Feel Like Hibernating in Winter?

Some of us like to curl up in the middle of a cold winter afternoon, pull up the cover, and take a short nap. It's not that we are hibernating—we are just trying to get warm for a while. Plus, keeping the cold at bay often takes up energy. Thus, we often snooze in the cold afternoons of winter.

It may seem as if many animals are hibernating because they sleep so much, but sleep alone is not considered to be hibernation. Many animals spend most of their time resting, to conserve energy, save calories, digest their meals, stay camouflaged, warm up, or tend to their surrounding family or help with other families. For example, lions spend up to 75 percent of their time resting to allow their heavy meals to digest. Panda young have been known to sleep up to 20 hours per day. Even some of the smaller animals, bees and ants, will work only about 20 percent of the time; hummingbirds, the smallest of birds, perch on a branch about 80 percent of their days and 100 percent of their nights. Sloths fit their name, lazing about 15 hours per day—long enough in one place so that two algae species inhabit their fur and claws. None of this sleeping and resting is considered hibernating, though.

Some species take resting a bit farther. For example, some plants "go dormant," mostly in the winter; while certain insects contain a special "antifreeze" in their systems, or use a reduced metabolic rate, that allows them to survive in cold weather. The Arctic woolly bear caterpillars spend most of their lives in "suspended animation," coming to life about a few weeks each summer to feed and grow. They not only are trying to survive the cold, but also the summer parasites that can kill about five times more woolly bear caterpillars than winter does. Thus, the long "suspended animation" time helps these caterpillars survive.

Certain animals, such as frogs, snakes, salamanders, and many types of mammals, take resting to the limit in a process called hibernation. Not every hibernating animal is a true hibernator in the winter. For example, bears are often what people think about when the word "hibernation" is mentioned. In reality, they drowse through the winter, living off their body's fat supplies—energy stored by the animal gorging itself before winter. The bear's temperature doesn't lower like true hibernators, allowing the animal to act quickly if they are threatened. Even more amazing, virtually no mammals in the Arctic hibernate, because of the risk of freezing to death. Grizzly bears come the closest to upsetting these two hibernation observations, as they do hide and go into a deep sleep in the winter.

For animals, surviving a winter often means spending months underground, cooling down their body to temperatures far below what would be comfortable to most humans. For example, dormice and insectivorous bats lower their body temperature as they sleep—sometimes near 40° F (4° C) or lower. The true definition of hibernation is a state of suspended animation when a mammals slows its metabolism, lowering its temperature from about 90° to 32° F (32.2° to 0° C); its pulse and breathing rates are hardly discernible. For example, the hedgehog during hibernation breathes once every six minutes.

What causes true hibernators to sleep for such long periods? The reason probably has to do with surviving the long winter when foodstuff is not as readily available—a technique that many animals have developed for survival. As for the time for hibernation, not everyone agrees on what sets it off: Some scientists believe that the fat deposits, not the cold, trigger the hibernation; and some believe that some genetic predisposition lets them know when to hibernate.

Modern studies of hibernation have cleared up a few mysteries. One study done by several scientists at Stanford University in the late 1990s examined what prompts animals to begin and end hibernation. These scientists believe that the brain's hypothalamic suprachiasmatic nucleus (SCN) regulates hibernation, along with governing the body's daily rhythms. The experiment used several golden-mantled ground squirrels, animals that normally hibernate for five to seven months a year. By taking away the SCN, they were able to understand how the animals respond to hibernation. For 2.5 years, the team compared the responses of squirrels with SCNs with those without the SCNs—showing that the animals without the SCNs would arouse themselves more during a normal hibernation period. The animal's abilities to take in food were also disturbed, showing how important such a small center of the brain can affect a natural rhythm.

---

signal, and do sundry other activities. Modern invertebrates vary in size from the single-celled animals, such as amoebae and parameciums, to the largest living invertebrate, the Atlantic giant squid. This animal can weigh up to 2 tons, with tentacles almost 50 feet (15 meters) long, and has the largest eyes ever to evolve, measuring about 16 inches (40 centimeters) in diameter.

mammals—Mammals are an important part of the Earth's natural history, and one of their species (humans) dominates the planet. (But contrary to popular belief, they are not the most abundant.) Since they evolved hundreds of millions of years ago, mammals have played a major part in the development of organisms on the planet. One reason is that most mammals are important parts of the terrestrial and marine food chains; another reason is that they also contribute to keeping other organisms in balance within the food chain as they consume food.

Mammals themselves are part of a larger tetrapod group called the synapsids, a branch from the early reptiles that came into existence hundreds of millions of years ago. But not all synapsids were mammals. For example, a nonmammalian member of the synapsids was the *Thrinaxodon*.

All modern mammals share three characteristics not found in other animals. They produce milk using modified sweat glands (mammary glands), have hair, and have three middle ear bones. Mammals feed milk—a substance rich in fats and proteins —to their newborn young. The mammary glands, which take a variety of shapes, are usually located on the ventral (front) surface of females along an area from the chest region to the groin. They vary in number from two (one right, one left, as in humans) to a dozen or more.

Mammals also have hair with various characteristics and for different reasons. Adults of some species tend to lose most of their hair, but hair is present during some phase of the ontogeny of all species. Mammalian hair, made of a protein called keratin (a protein that also makes up the human fingernails and toenails), serves at least four functions: Hair slows the rate of heat exchange with the surrounding environment, allowing the animal to insulate itself against extreme climate conditions. The animal's specialized hairs (such as whiskers, or "vibrissae," that are often surrounded by muscles controlling the hair's position) often include a sensory function, allowing the animal to tell when it has contacted an object.

Color of animal hairs often has a direct affect on the animal, serving it in many capacities. Hair color can often serve as camouflage, as color and pattern of the hair hide or announce the animal's presence. One good example is the conspicuous color pattern of a skunk as a warning to predators. Hair color can also announce the presence of danger, such as the white underside of the tail of a white-tailed deer. Color may also communicate social information, such as the differences in sex—for example, the different colors of male and female capuchin monkeys. Finally, hair provides some protection, either simply by providing an additional protective layer against abrasion or sunburn, or by presenting a threat to predators or animals entering a fought-after territory—such as the spines of a porcupine, or the erect hair on the back of a wolf.

The final mammal characteristic is the inner ear's chain of three bones—the malleus, incus, and stapes—that allow mammals to hear sounds. Two of these, the malleus and incus, are derived from bones involved in jaw articulation in most other vertebrates.

Another lesser characteristic found in most mammals are highly differentiated teeth. In general, for all mammals, teeth are replaced just once during an individual's life (diphyodonty) with the first set of teeth being called the "milk teeth." Most mammals also have a lower jaw made up of a single bone (the dentary), four-chambered hearts, a secondary palate separating air and food passages in the mouth, a muscular diaphragm separating thoracic and abdominal cavities, a highly developed brain, endothermy and homeothermy, separate sexes with the sex of an embryo being determined by the presence of a Y or two X chromosomes, and internal fertilization.

The class Mammalia includes around 5,000 species placed in 26 orders. (Systematists do not yet agree on the exact number or on how some orders are related to others.) Mammals can be found in all continents and seas. Partly because of their high metabolic rates, they often play an ecological role that seems disproportionately large compared to their numerical abundance, mainly because they have to eat more to maintain their higher metabolic rate.

There are three major groups of living mammals and one extinct group, the Multituberculata—animals that were both widespread and successful, but gradually became extinct. The first of three groups of living mammals is the placental, or *Eutheria*, mammals of which humans (under the order Primates) are a member. Another example of a placental mammal was the quagga. It is generally thought to have been related to horses and zebras, and was a yellowish-brown zebra-like animal with stripes only on its head, neck, and forebody. The reason for the past tense is simple: This animal, once a native to the desert areas of the African continent, was hunted to extinction in the 1880s.

The other two groups are the marsupials, or Metatheria, that include all of the pouched animals, such as opossums, kangaroos, and Tasmanian devils. The last group are the less diverse monotremes, or Monotremata. They are warm-blooded, have hair, and produce milk just like other mammals—but they lay eggs and do not birth live like marsupials and placentals.

Here is a representation of a recent mammal classification. As with all the classification systems, we still have much to learn about how to connect the many mammal species, and scientists continue to put together a better picture of mammal development.

> **Classification of Mammals**
>
> Subclass: Prototheria
>   Order: Monotremata—Monotremes: platypus and echidnas
> Subclass: Metatheria (marsupials)
>   Order: Didelphimorphia
>   Order: Paucituberculata
>   Order: Microbiotheria
>   Order: Dasyuromorphia
>   Order: Peramelemorphia
>   Order: Notoryctemorphia
>   Order: Diprotodontia
> Subclass: Eutheria (placentals)
>   Order: Insectivora—Insectivores: shrews, moles, hedgehogs, and tenrecs
>   Order: Macroscelidea—elephant shrews
>   Order: Scandentia—tree shrews
>   Order: Dermoptera—colugos
>   Order: Chiroptera—bats
>   Order: Primates—primates
>   Order: Xenarthra–edentates: sloths, armadillos and anteaters
>   Order: Pholidota—pangolins
>   Order: Lagomorpha—rabbits and pikas
>   Order: Rodentia—rodents
>   Order: Cetacea—whales, dolphins, and porpoises
>   Order: Carnivora—carnivores
>   Order: Tubulidentata—aardvark
>   Order: Proboscidea—elephants
>   Order: Hyracoidea—hyraxes
>   Order: Sirenia—dugongs and manatees
>   Order: Perissodactyla—horses, rhinos, tapirs
>   Order: Artiodactyla—antelope, giraffe, camels, pigs, and hippos

vertebrates—Vertebrates are animals with a backbone. The reason for the backbone, or vertebrae, is to support the back, to hold up the body, and to protect the spinal cord. The subphylum Vertebrates, under the phylum Chordata, includes three classes of fish (Agnatha, or lampreys and hagfish; Chondrichthyes, or sharks and rays; and Osteichthyes, or bony fish), amphibians, reptiles, birds, and mammals.

The first vertebrate known was the conodont ("cone teeth"), a 2-inch (5-centimeter) eel with a finned tail; it lived about 515 million years ago. This animal had gripping teeth that were made of bone cells, enamel, and a supporting layer of mineralized cartilage. This bone is the proof scientists use to verify that the conodonts were vertebrates, as the development of bone, no matter where it is located, is considered the sign of a vertebrate. Conodonts died out about 200 million years ago, when the dinosaurs ruled the Earth.

3. Arthropods

Introduction

Arthropods include an incredibly diverse group of animals such as insects, crustaceans, spiders, scorpions, and centipedes. There are far more species of arthropods than in all other phyla on the Earth combined—and the total count of undescribed species in the largest assemblage of arthropods, the insects, probably numbers in the tens of millions. Members of this phylum have been responsible for the most devastating plagues and famines humans have known. Yet other species of arthropods are essential for our existence, directly or indirectly providing us with food, clothing, medicines, and protection from harmful organisms.

Timeline

(note: mya=million years ago)

Date	Event

Prehistoric Events

Date	Event
~440–420 mya	The first animals, probably tiny arthropods, crawl on land, and a second burst of growth and diversification takes place on the Earth. (The first burst occurred about 544 million years ago, called the Cambrian Explosion.)
~320 mya	Insects undergo rapid evolution, diversifying greatly by this time.

Modern Events and Discoveries

Date	Event
~1610	Italian mathematician and astronomer Galileo Galilei (1564–1642) uses the newly invented microscope (discovered independently by Zacharias Janssen and Hans Lippershey) to study insect anatomy.
1668	Italian biologist, physician, and poet Francesco Redi (1627–1697) concludes that flies were not spontaneously produced on meat but are hatched from eggs deposited on the food; but he still does not deny spontaneous generation existed.
1710	English naturalist John Ray's (1627–1705) work, *Historia insectorum*, is published posthumously.
1734	French entomologist René Antoine Ferchault de Réaumur (1683–1757) starts the first book of six volumes on the history of insects, titled *The History of Insects*; it was finished in 1742.
1737	Hermann Boerhaave (1668–1738), a Dutch physician, posthumously prints Jan Swammerdam's (1637–1680) *Bible of Nature* (*Biblia naturae*); the book, originally published in 1658, is Swammerdam's records of the dissection of insects under a microscope.
1775	Danish entomologist Johann Christian Fabricius (1745–1808) classifies insects based on their mouth structure, not their wings.
1825	English naturalist Henry Walter Bates (1825–1892) proposes his theory of insect mimicry.
1892	U.S. bacteriologist Theobald Smith (1859–1934) discovers Texas cattle fever spreads through ticks; it is the first arthropod known to spread disease.
~1943	Austrian zoologist, entomologist, and ethnologist Karl von Frisch (1886–1982) publishes a study on the dance of bees, recording the detailed movements of the insects; he determines the bees are actually communicating with each other.
1994	Scientists suggest that insects may have evolved wings by first skimming on water.
1995	A possible new phylum of animals is found residing on lobsters; they do not harm the lobsters but live on the creatures.
1997	A protein is isolated in insects that acts as a potent antifreeze, which may explain how certain insect survive in colder regions of the world.

History of Arthropods

Early Arthropods

Arthropods are a group of animals whose name means "joint-legged," a characteristic that sets them apart from all the other invertebrates. They have hard **exoskeletons** and jointed legs; being invertebrates, they have no backbone.

Early arthropods were probably the first animals to crawl on land. Some of the first were probably **arachnids**—in particular, spiders and scorpions. The earliest known fossils of arachnids have been found in Silurian-period rock layers (from 440 to 410 million years ago), usually in association with the oldest-known vascular plants. (The first true land plants first appeared about 420 million years ago.) Some scientists suggest that the animals crawled on land to seek out these abundant food sources.

By about 320 million years ago, around the Pennsylvanian (or Upper Carboniferous) period, insects had gone through a rapid evolution. They also became greatly diversified, both in terms of the number of species and in terms of their body sizes and shapes; for example, some dragonflies had wingspans of 30 inches (76 centimeters) and certain cockroaches were more than 4 inches (10 centimeters) long.

Modern arthropods may be much smaller, but overall, they have changed little physically from their ancient ancestors. Today, the phylum Arthropoda includes more species than any other phylum, especially because of the class Insecta, or the **insects**.

Arthropod Studies

One of the earlier ideas about arthropods, in particular the insects, was that they appeared instantly. The impetus for this was simple. To ancient peoples, the appearance of flies would seem to be a spontaneous occurrence. After all, the animals would be grubs one minute, and seemingly the next, they would change into flies—a kind of "magical" transformation. It was not until 1669 that Dutch naturalist Jan Swammerdam contradicted the idea of instant metamorphosis of insects in his treatise on insects. He also discovered the reproductive parts of insects, giving a better explanation of the proliferation of these animals.

Other insect studies followed in the eighteenth century, such as French entomologist René Antoine Ferchault de Réaumur's six-volume history of insects. Later in the century, Danish entomologist Johann Christian Fabricius classified insects based on the structure of their mouth organs, not their wings. In the nineteenth century, various scientists proposed theories about arthropod development, mostly based on the study of insects, such as British naturalist Henry Walter Bates's idea of insect mimicry—which says certain species sometimes resemble one another for reasons of defense. This helped to verify Charles Darwin and Alfred Wallace's theory of evolution.

By the twentieth century, a plethora of studies were done on insects. For example, in the early 1900s, Austrian zoologist, entomologist, and ethnologist Karl von Frisch discovered details about honeybees—in particular, the precise pattern of movements done by the forager (worker) honeybees when they returned to the hive. These dances communicate to the other bees the direction and distance of a food source. Although the actual interpretation of the dance has been debated, it sparked more entomological studies to explain how behavior of insects can communicate a great deal more than previously thought.

Today, studies deal with all sorts of behavior of the arthropods. For example, scientists study spiders' webs and how they are used to communicate between spiders and other species; how killer bees, or Africanized honeybees, are migrating and interbreeding; how certain arthropods can be used in agriculture to keep away other pests; how and why fireflies produce light; and how arthropods are diversified in the rainforests of the Earth. *See also* ANIMALS, CLASSIFICATION OF ORGANISMS, EVOLUTION, FOSSILS, FUNGI, GEOLOGIC TIME SCALE, LIFE, and PLANTS.

Topic Terms

arachnids—Arachnids are a major class of arthropod and include spiders, scorpions, mites, and ticks; their numbers approach 100,000 species. The arachnids have four pair of jointed appendages—or in total, eight legs. The anterior appendages are often used for manipulating, piercing, and sucking out the fluid of their prey; on some arachnids, the first pair of appendages often contain poison glands. The second pair is usually used for sensing. (On some arachnids, they can even detect certain chemicals.) The other four appendages are usually used for walking. Their body usually has two parts; they have a hard outer skeleton, and their jaws usually bear fangs.

Spiders are the hunters of the class and have eight pairs of "eyes," with their heads held separate from the rest of their body. Scorpions are mostly ground-dwellers, and prefer dark corners to hide in. They are some of the larger arachnids, with lobster-like pincers and a flexible abdomen with 12 segments. Mites and ticks are also arachnids and have small, round, circu-

Arthropods' (Somewhat Undeserved) Bad Reputations

Most of the arthropods have a place in human's past history and are even a part of many mythologies—mainly because of their detrimental effects on humans and human surroundings, such as agriculture. For example, the arachnid name (spiders, scorpions, mites, and ticks) not only evoke squirminess in most people, but it also has a bad name in many legends, such as the woman who was changed into a spider by the goddess Athena for challenging her to a weaving contest. Or the spider of West African lore, who is a trickster, much in the tradition of the coyote in Native American tales.

Humans seem to fear arthropods around the world for many reasons. Besides their abundance, one of the chief reasons has to do with their behavior and their features. Spiders, members of the class *Arachnida*, for example, are feared for their role as land-living hunters. In fact, spiders are truly the hunters of all the invertebrates, sneaking up on their prey, running down their prey, or spinning webs to catch it. The fangs of these hunters are what people remember—especially in species such as the tarantula, the world's largest spider. This spider is hairy, with fierce-looking mouthparts; it often rears up on its hind legs if it is disturbed and rubs its appendages together, creating a hissing sound. It also can shoot out barbed and pointed hairs, which can cause the skin to burn and itch for days.

Scorpions also evoke fear. They are mostly ground-dwellers and prefer dark corners to hide in. They are some of the larger arachnids, with lobster-like pincers, and they have a venom-laden tail that they use to defend themselves and disable their prey. In fact, the Arizona scorpion's sting is often fatal to humans. But not all scorpions are dangerous; the whip scorpion is perfectly harmless and the "whip" on its tail has no discernible use.

People dislike mites and ticks because they are often parasitic. A recent surge in the interest in ticks is because of a disease that seems to be spreading in the eastern part of the United States—Lyme disease. In the past few decades, this debilitating disease—a bacteria spread from the deer tick to humans—has caused many problems (although there is now a vaccination against the disease).

Thus, it is true that many arthropods are not the most comfortable of creatures, but they do have a place in our natural world. In many cases, they are merely menacing-looking, but do not attack people unless provoked, like the "cutest" of mammal animals. In fact, the arachnids include some of the best architects in the animal world. In particular, spiders' silken webs are spun from digit-shaped glands in the abdomen, catching insects and thus, benefiting humans; certain insects eat other insects in the gardens and agricultural fields; and the crustaceans, put in with other more annoying animals in the arthropod phylum, have an even better reputation—the crabs, lobsters, prawns, shrimp, and crayfish supplying food for humans.

Of course, we can't leave out the insects. If someone were to write about each insect in the world, the books would cover volumes, as over 500,000 species have been described so far—and with maybe hundreds of thousands or even millions yet to be discovered. As for insects' reputations, everyone has their own story—from the huge American cockroaches (will they really be the only survivors of a nuclear holocaust?) to the plagues of locust and munching caterpillars. But they do have their place in the food chain.

---

lar bodies; their heads project out from their bodies, and most are parasitic.

arthropods—The phylum of Arthropods have a number of shared characteristics. Arthropods have segmented bodies, both externally and internally. Some segments are fused to form specialized body regions called tagmata, and include the head, thorax, and abdomen. (The process and condition of fusion is called tagmosis.) Arthropods also usually have an exoskeleton. The one bad thing about the arthropod exoskeleton is that it restricts the growth, weight, and size of the animals. The only way for an arthropod to grow is to molt; thus, throughout their lifetimes, most of these animals molt by shedding their outer layer, expanding in size, and then laying down a new, hard, outer covering to which the muscles are reattached. In other arthropods, the organism will undergo a rapid transformation, or metamorphism in just one molt—for example, a caterpillar to a butterfly. With insects, other types of metamorphism are possible: The complete metamorphosis is when the insect goes though several distinct stages of growth to reach adulthood. This group includes ants, moths, butterflies, termites, wasps, and beetles; the incomplete metamorphosis occurs when the insect does not go through all the stages of complete metamorphosis, such as grasshoppers, louses, and crickets.

In the primitive arthropods, each body segment had a pair of jointed appendages; in all modern arthropods, many of these appendages have been dramatically modified or even lost. The appendages they do have move using a complex muscular system, divided into smooth and striated muscles, similar to chordates, including humans. Most of the arthropods have two compound eyes (composed of facets that produce images that resemble an overall mosaic of the observed object) and one to several simple eyes (ocelli); either or both kinds of eyes may be fewer or altogether absent in some groups. Their body cavity is open, with the space loosely filled with tissue, sinuses, and blood. They usually fertilize the female internally, but certain groups fertilize outside the body; most of them lay eggs.

Phylum Arthropoda *

Subphylum: Chelicerata
- Class: Merostomata (horseshoe crabs, eurypterids)
- Class: Pycnogonida (sea spiders)
- Class: Arachnida (spiders, ticks, mites, scorpions)

Subphylum: Crustacea
- Class: Remipedia
- Class: Cephalocarida
- Class: Branchiopoda (fairy shrimp, water fleas)
- Class: Maxillopoda (ostracods, copepods, barnacles)
- Class: Malacostraca (isopods, amphipods, krill, crabs, shrimp)

Subphylum: Uniramia
- Class: Chilopoda (centipedes)
- Class: Diplopoda (millipedes)
- Class: Insecta

* The systematic relationships of arthropod groups is not fully understood, which is not surprising given the size and diversity of the phylum. This classification is from C.P. Hickman and L.S. Roberts, *Animal Diversity,* New York: McGraw-Hill, 1994; R.C. Brusca and G.J. Brusca, *Invertebrates,* Sunderland, MA: Sinauer Associates, Inc., 1990; Pearse, V., et al., *Living Invertebrates,* Boston: Blackwell Scientific Publications, and the Pacific Grove, CA: Boxwood Press, 1987.

---

crustaceans—Another subphylum (or class in some classifications) within the phylum Arthropoda is the Crustacea (crustaceans); they number more than 25,000 species. The crustaceans are a numerous and variable group of animals, including lobsters, shrimp, crabs, barnacles, crayfish, woodlice, pill bugs, and water fleas. The majority are aquatic (marine and freshwater), but others, such as the woodlice and pill bugs, are terrestrial. Their overall similarities are few: In most cases they vary in size (from small planktonic forms to large lobsters and crabs), appendages (size, shape, and use), and numbers of legs. They all seem to have a body covered with a hard shell, and two pairs of feelers, or something similar, in front of their mouth.

Some of the most important crustaceans are similar to the water fleas. These animals' bodies are surrounded by a two-part shell, and they swim using long hairy antennae. This group includes the copepods, major members of the oceanic plankton, and one of the most numerous creatures on the Earth—not to mention an important player in the world's food chain.

exoskeleton—The exoskeleton of an arthropod is actually a hard outer skeleton. It is primarily made up of the protein chitin (a flexible polysaccharide). In lesser amounts, the exoskeletons include lipids, other proteins, and calcium carbonate. Most of the animals usually grow by molting their exoskeletons (shedding the outer layer and growing a new one) in a process called ecdysis.

insects—The insects, or the class Insecta, are the largest class of living organisms. They all have four characteristics in common: (1) three pairs of jointed legs at some stage of their lifecycle; (2) a hard outer skeleton, the exoskeleton; (3) a body divided into three parts; and (4) one or two pairs of wings.

Insects number over 500,000 described species; although some scientists estimate that around one to 10 million living species of insects inhabit the Earth, most in the unexplored areas of the tropical rainforests. Each year, about 7,000 new insect species are described; many are lost, too, as more rainforest habitats are destroyed. Overall, insects have infiltrated almost every habitat in the world—land, sea, air, polar regions, underground, and under the sea. They range in size from almost microscopic spring-tails to the 6-inch (15-centimeter) Goliath beetles.

In terms of effect, it is thought that the insect class has the most widespread influence on humans, especially in terms of health. In particular, they are often responsible for direct injury to a human, or indirect, by destroying crops and spoiling stored foodstuff. They also often carry disease, not only to humans, but to domesticated animals. Because of their great mobility, they can spread disease rapidly. So far, the major diseases spread by insects include typhus, malaria, yellow fever, bacillary dysentery, and tick-borne diseases, such as Lyme disease.

metamorphosis—In certain arthropods, especially insects, organisms undergo metamorphoses. There are several types of metamorphism. In the complete metamorphosis, the insect goes through several distinct stages of growth to reach adulthood, such as a catepillar to butterfly. This group includes ants, moths, butterflies, termites, wasps, and beetles. The incomplete metamorphosis occurs when the insect does not go through all the stages of complete metamorphosis; examples include grasshoppers, louses, and crickets.

4. Atmosphere

Introduction

An atmosphere is a collection of gases around a planet or satellite. Besides Earth, several other bodies in the solar system have an atmosphere, including Venus, Mars, Jupiter, Saturn, Neptune, Uranus, and many of the outer planets' satellites. But these atmospheres appear to be too hostile to support life, currently making Earth the only known planet that contains abundant flora and fauna. Our atmosphere is one of the layers of the Earth's biosphere, a thin blanket that keeps the flora and fauna alive. It has always had an important connection to the natural history of the planet—especially to all organisms that need its many gases to survive.

Timeline

(bya=billion years ago; mya=million years ago)

Date	Event

Prehistoric Events

Date	Event
~4.3 bya	Carbon dioxide makes up 54 percent of the Earth's atmosphere; oxygen is almost nonexistent.
~2 bya	Plants begin to produce oxygen through the process of photosynthesis (although the idea of when oxygen truly began to accumultate has recently been challenged).
~2 bya	Oxygen rises to 1 percent of the atmospheric composition; carbon dioxide decreases from 54 percent to about 4 percent, the excess is taken up by rocks and plants; the protective ozone layer also develops at this time
~600 mya	Oxygen levels continue to rise as more life spreads around the planet; levels reach 21 percent oxygen, similar to the modern percentage.

Modern Events and Discoveries

Date	Event
1643	Italian physicist and mathematician Evangelista Torricelli (1608–1647) invents the mercury barometer.
1646	French mathematician and physicist Blaise Pascal (1623–1662) confirms that the atmospheric air has weight.
1650	German scientist Bernhard Varen (Bernhardus Varenius, 1622–1650) proposes that the air in the equatorial regions is thinned by the Sun's heat, and in response, the cold, heavier air of the polar regions flows toward the equator.
1684	Englishman Martin Lister (1638–1712) mistakenly believes that the trade winds are actually the "breath" of the sargasso weed; it goes along with Aristotle's idea that winds are caused by exhalations.
1686	English scientist Edmond Halley (1656–1743) publishes the first meteorological chart, indicating the prevailing winds over the tropical ocean, and attempts to explain causes of the trade winds and monsoons.
1753	George Hadley (1685–1768), English lawyer and climatologist, describes the Hadley cells (localized areas of circulation) of the Earth's atmosphere in detail.
1770	English chemist and minister Joseph Priestly (1733–1804) discovers oxygen and shows it is consumed by animals and produced by plants.
1775	Priestley determines that "dew" forms when hydrogen explodes with oxygen.
1823	English meteorologist John Frederic Daniell (1790–1845) presents the first study of the atmosphere and trade winds.
1831	U.S. meteorologist William Redfield (1789–1857) determines the rotary motion of storms, noting that winds in storms move counterclockwise around a center that is being pushed by the prevailing winds.
1835	French physicist Gustave-Gaspard Coriolis (1792–1843) describes the apparent deflection of global winds caused by the Earth's rotation, an effect that now carries his name.

Date	Event
1850	U.S. oceanographer and meteorologist Matthew Maury (1806–1873) draws a plot of surface winds all around the world, including the prevailing winds and the doldrums.
1856	William Ferrel (1817–1891), U.S. meteorologist, describes the middle-latitude global wind cells, known as the Ferrel cells; he determines that winds are deflected to the right in the Northern Hemisphere and to the left in the Southern Hemisphere.
1863	Irish physicist John Tyndall (1820–1893) discovers the greenhouse effect.
1902	Léon-Philippe Teisserenc de Bort (1855–1913), French meteorologist, discovers that the troposphere and stratosphere are two distinct layers of the Earth's atmosphere.
1902	English scientists Oliver Heaviside (1850–1925) and Arthur Kennelly (1861–1939) discover the layer of the atmosphere that reflects radiation at radio wavelengths; the Heaviside-Kennelly layer is named after both scientists.
1913	French physicist Charles Fabry (1867–1945) discovers the existence of an ozone layer in the upper atmosphere.
1935	English mathematician and geophysicist Sydney Chapman (1888–1970) determines the Moon's gravitational effect on Earth's atmosphere.
1957	Scientists detect a "hole" in the ozone layer over Antarctica during the International Geophysical Year; it would take until 1980 to recognize the true extent of this ozone hole.
1990	Most industrialized countries agree to stop producing chlorofluorocarbons (CFCs) to curtail the depletion of ozone; the measure remains controversial, with many scientists questioning the actual relationship between CFCs and the ozone hole.
1993	A paleontologist finds what appears to be the remains of photosynthetic bacteria in 3.5-billion-year-old rock, which would mean that oxygen began to accumulate much earlier than scientists previously believed; this finding has yet to be confirmed.
1998	The Antarctic ozone hole reaches record size, growing to about 10.5 million square miles (27.3 million square kilometers), more than a half-million square miles larger than two years previously.

History of the Atmosphere

Early Atmosphere

No one knows the true origin of the Earth's **atmosphere**. Some scientists believe the atmosphere originated as gases from part of the solar nebula collected; others think that the gases also could have been produced from the planet's early and very dynamic volcanic activity, or even a combination of the two processes. The early Earth probably had a thicker atmosphere, too, but the young, active Sun boiled away the lighter materials—elements still found today around the more distant gas giant planets.

By about 4.3 billion years ago, carbon dioxide made up about 54 percent of the Earth's atmosphere, with oxygen being a very small part of the composition. About 2 billion years ago, plants in the oceans began to produce oxygen by photosynthesis, which involved taking in carbon dioxide and releasing oxygen; around that time, the **ozone** layer began to form. (This date was recently debated, as a paleontologist in 1993 believes he has found what appears to be remains of photosynthetic bacteria in 3.5-billion-year-old rock, but it has yet to be proven.) Either way, by 2 billion years ago, approximately 1 percent of the atmosphere was oxygen and only 4 percent carbon dioxide—the originally high levels of carbon dioxide reduced by the intake of plants and taken up by carbonate rocks.

This sudden, geologically speaking, outpouring of oxygen helped to build up the ozone layer. As the oxygen levels increased, the number of ocean animals also increased. And once the protective ozone layer was in place, it allowed the marine plants and animals to safely spread onto land—as it protected the organisms from the harmful ultraviolet rays of the Sun. By about 600 million years ago, atmospheric oxygen continued to increase as volcanoes and climate changes buried a great deal of plant material—plants that would have absorbed oxygen from the atmosphere if they had decomposed in the open. And as the atmosphere grew, so did its interaction with the **water** of the oceans—creating the interactions present in the atmosphere that create climate and weather.

Today, the planet's atmospheric composition has changed dramatically from its early history. The Earth's modern atmosphere measures 21 percent oxygen, 78 percent nitrogen, and only 0.036 percent carbon dioxide.

Atmospheric Studies

The study of the weather has long been a concern of all cultures. After all, humans have always been surrounded by the weather and the strange occurrences it brought. Most of humans' early conclusions about the weather were based on superstition and myth. But gradually the studies by philosopher-scientists such as Benjamin Franklin and Thomas Jefferson began to shed light on weather patterns in the eighteenth century.

But when it comes to the atmosphere, and atmospheric physics, it wasn't until the nineteenth century that any outstanding atmospheric studies were published (with the exception of George Hadley). And the true development of atmospheric research didn't begin until after World War II.

Despite this early ignorance of the atmosphere in general, by the 1700s, scientists realized that the overall flow of Earth's atmosphere could be broken down into localized areas of circulation. They discovered that the engine for this circulation is the heating of the atmosphere by the Sun: As rays hit the equator, large amounts of warm, humid air rise in the tropics, creating a band of low pressure. This air rises to the troposphere where it spreads out north and south toward the poles.

These circulating air masses in the atmosphere, called **cells**, produce the prevailing global winds: From the equator to about 30 degrees north and south latitudes are the Hadley cells, named after English lawyer and climatologist George Hadley, who described them in 1753. The cells from about 30 degrees to 60 degrees north and south latitudes are the Ferrel cells, named after U.S. meteorologist William Ferrel, who described them in 1856. Finally, the polar cells range from 60 degrees latitude to the respective pole, in each hemisphere.

The **trade winds**, another atmospheric phenomenon, were first studied and confirmed by John Frederic Daniell in 1823; the global prevailing winds are also affected by the rotation of the Earth, a phenomenon discovered by French physicist Gustave-Gaspard Coriolis in 1835. The rotating Earth deflects the winds to the right in the Northern Hemisphere and to the left in the Southern Hemisphere by means of the **Coriolis effect**. Even the idea of the **greenhouse effect** was discovered in the nineteenth century (1863), by Irish physicist John Tyndall.

Despite this progress, much information about the atmosphere remained a mystery. Earth's many **atmospheric layers** were not discovered until the twentieth century. For example, the Appleton layer is a region in the upper ionosphere extending between approximately 93 miles (150 kilometers) and 186 miles (300 kilometers) above the surface of the Earth. It was named after the English physicist Sir Edward Appleton (1892–1965), who won a Nobel Prize in physics for 1947 for his investigations of the upper atmosphere, including the discovery of this layer named after him.

After World War II, atmospheric science advanced at a rapid pace—often due to the inventions and techniques discovered in the war. For example, high-speed computers, aircraft, high-altitude balloons, rockets, and satellites used in atmospheric science were all progeny of the advances during the war. In addition, sensors, such as radar, lasers, and sundry other instruments, were also perfect for studying the atmosphere.

By 1960, atmospheric science took a major step in the United States when the National Center for Atmospheric Research (NCAR) was established in Boulder, Colorado. NCAR was to be operated by a consortium of universities with federal support from the National Science Foundation—and is still considered one of the top atmospheric science centers in the world today.

The main focus of modern atmospheric science is monitoring the major changes in the atmosphere, and the patterns that cause that change—now and in the future. One of the major concerns over the past few decades has been the depletion of the ozone layer above Antarctica, and more recently, above the Arctic. By about 1980, scientists had detected a "hole" in the ozone layer over Antarctica; since then it has grown with the Southern Hemisphere's spring and shrunk in the fall. (See **ozone hole**.) But overall, it seems to have become larger—close to one and a half times the area of the United States. (Although some recent readings show little growth in the hole). Scientists began to suspect that chlorofluorocarbons (CFCs) released into the atmosphere by industry and appliances such as air conditioners were the cause of the hole, although this view is still somewhat debated. By 1990, most industrialized countries agreed to stop producing CFCs to curtail the depletion of ozone. The concern is understandable: The layer of ozone gas protects all creatures—including humans—from the ultraviolet radiation of the Sun. *See also* CLIMATE AND WEATHER, EARTH, EXTINCTION, GEOLOGIC TIME SCALE, OCEANS, SOLAR SYSTEM, and VOLCANOES.

Topic Terms

atmosphere—Heat regulation is the process by which the Earth's temperature range is moderated by the atmosphere—one reason why the planet supports life as we know it. The average global surface temperature is about 55° F (13°C). The Earth's temperature has been about the same since life evolved more than 1 billion years ago. For example, even during the ice ages, or periods when great ice sheets covered the polar regions of the planet, the tropics remained mostly unaffected; thus, the average temperatures around the Earth only dropped by about 9° F (5° C).

Without the atmosphere, the incoming solar radiation by day would heat the Earth to high temperatures; at night, this heat would escape and temperatures would plummet—close to the temperatures in space—or well below zero. The atmosphere dampens these extremes. During the day, some of the incoming solar radiation is reflected back toward space by clouds that also absorb some of this radiation; thus, only a percentage of the total incoming solar radiation reaches the Earth's surface. Some of this radiation is reflected back into space by snowy and desert areas, as these lighter areas have a high albedo (reflective power); the Earth's land and water absorb what is left, reradiating it as infrared radiation, also known as heat.

Overall, the atmosphere doesn't absorb most of the Sun's radiation. Therefore, the Earth is mainly heated from below—by long-wave terrestrial radiation; the Sun's rays heat the Earth's surface, and the heat is reradiated. This is why the planet's atmosphere decreases in temperature with increasing altitude.

atmospheric layers—The Earth's atmosphere has several major layers. They include the following:

troposphere—The troposphere is the lowest layer of the Earth's atmosphere, starting at the surface and ending at the tropopause. All of our weather occurs within the troposphere, with the air in constant motion, flowing horizontally and vertically. The height of this layer varies with the season: In the summer, it can extend to about 5 miles (8 kilometers) above the poles, and about 11 miles (18 kilometers) above the equator; in the winter, the air is thinner and is just a little less thick. In the troposphere, the temperature decreases with altitude.

tropopause—The tropopause is the division between the troposphere and stratosphere about 7 miles (11 kilometers) above the Earth's surface. It is where the weather is no longer a factor, and the temperature stops falling with height.

stratosphere—The stratosphere is a layer of the Earth's atmosphere beginning at the tropopause and ending at the stratopause. It averages from a height of about 7 miles (11 kilometers) above the Earth in the mid-latitudes throughout the year, to about 30 miles (50 kilometers) above the surface at the stratopause. The thin air of this layer is mostly rarefied, with an almost steady temperature of around –58° F (–50° C) to about 12 miles (20 kilometers); after that the air gradually rises in temperature with height. Supersonic aircraft reach such heights.

ozone layer—The ozone layer is found in the stratosphere and is a collection of oxygen molecules (three molecules of oxygen [O_3] instead of the two molecules [O_2] we breathe). The ozone layer intercepts the Sun's damaging ultraviolet rays, and thus, permits life to exist on Earth. The ozone is present at different thicknesses around Earth—on the average at a height of approximately 9 miles (15 kilometers) to 25 miles (40 kilometers) above the Earth's surface.

stratopause—The stratopause is the boundary between the stratosphere and beginning of the mesosphere (and the start of the ionosphere) at a height of about 30 miles (50 kilometers) above the Earth's surface. It has a relatively high temperature, about 32° F(0° C) as this layer absorbs the Sun's ultraviolet radiation.

ionosphere—The ionosphere is the region where atmospheric particles become ionized (the atoms become ions) by ultraviolet radiation from the Sun or other radiation. It extends from a height of about 30 to 50 miles (50 to 80 kilometers) to about 250 to 370 miles (400 to 600 kilometers). How this process works is that X-rays and ultraviolet radiation from the Sun cause the nitrogen and oxygen molecules in the air to ionize, producing free electrons, and because the air is thin at this height, the particles remain ionized. Radio waves transmitted into the ionosphere are partially absorbed and partially refracted (bent) back toward

The Earth's Atmospheric Layers

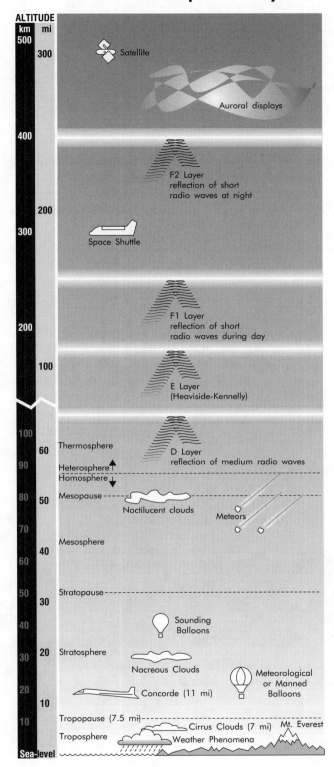

the Earth by these ionized particles. The amount of refraction depends on the frequency of the transmitted radio waves; very high frequencies, for example, are not refracted at all. The ionosphere is not uniform throughout and contains several more layers within the ionosphere layer itself. Although they can vary with time of day, latitude, and season, overall, the layers include the following:

first layer—from the stratopause to the mesopause is a layer of the ionosphere that allows the reflection of long radio waves.

D-layer—Just above the mesopause to just in the middle of the thermosphere, at about 62 miles (100 kilometers) in height, is the D-layer, in which the reflection of medium radio waves occurs.

Heaviside-Kennelly (or E) layer—This layer reflects medium radio waves.

F-1 layer—Reflection of short radio waves during the day.

Appleton (or F) layer—The Appleton layer—discovered by English physicist Sir Edward Appleton—is a region in the upper ionosphere that extends between approximately 93 miles (150 kilometers) and 186 miles (300 kilometers) above the surface of the Earth. A specific band of frequencies was found to be refracted by the Appleton layer in the atmosphere, and this bending makes it possible to transmit radio waves over a longer range than would be possible for the same waves transmitted along the ground. This discovery made long-range communication possible before communication satellites. The best refraction takes place at night (when the electron concentration in lower layers falls away because of the absence of sunlight)—which is one reason why radios receive more stations at night.

F-2 layer—This layer reflects short radio waves at night; it starts at about 250 miles (400 kilometers) above the Earth's surface.

mesosphere—The mesosphere lies between the stratopause and the bottom of the thermosphere; it ranges from about 30 miles (50 kilometers) in height, to about 50 miles (80 kilometers). The temperatures again begin to fall in the mesosphere, decreasing to about –148° F (–100° C) at the top of the layer.

mesopause—The mesopause is the boundary between the mesosphere and thermosphere, and is found at about a height of 50 miles (80 kilometers) above the Earth's surface.

thermosphere—The thermosphere is the last layer and is considered the outermost part of the Earth's atmosphere (although certain types of space weather can influence the atmosphere at this height). It extends from the mesopause at about 50 miles (80 kilometers) above the Earth's surface and into outer space. The temperatures in this layer increase with height, reaching several thousand degrees.

cells—The overall flow of the Earth's atmosphere can be broken down into localized areas of circulation, or cells. The engine for this circulation is the heating of the atmosphere by the Sun. The first cells, called the Hadley cells in the Northern and Southern Hemispheres, start as the Sun's rays heat the air around the equator. This heating results in the rising of large amounts of warm, humid air in the tropics, creating a band of low pressure. The air continues to rise into the troposphere where it spreads out north in the Northern Hemisphere and south in the Southern Hemisphere, toward the respective poles. As the air travels, it begins to cool and sinks back toward the surface at around the 30 degrees north and south latitudes, completing the cycle for the first cell.

The second cells, the Ferrel cells, form because not all the air in the Hadley cells flows back toward the equator—rather, some of it flows toward the poles along the surface. At about 60 degrees latitude, north and south, this relatively warmer air runs into the cooled air from the poles, causing it to rise. Some of this air reverses course at the higher altitudes and heads back toward 30 degrees latitude, where it cools and sinks. Again, similar to the situation at 30 degrees latitude, some of the rising air at 60 degrees north and south latitude continues to flow toward the poles, where it cools and sinks to the surface. It then returns toward 60 degrees latitude, creating the final cells, called the Polar cells (sometimes called the polar Hadley cells).

Coriolis effect—The Coriolis effect (or force) is a concept that explains the movement of objects in a rotating system—including the global winds within the Hadley, Ferrel, and Polar cells in the Northern and Southern Hemispheres. The true definition for this "force" is the deflection of a moving body caused by unequal rates of rotation between two rotating systems. In the Northern Hemisphere, the winds apparently curve to the east (or "to the right"), and in the Southern Hemisphere, the winds curve to the west (or "to the left"). Between the cells and the Coriolis effect, six major wind bands are created across the planet: The polar easterlies, westerlies, and the trade winds, one each in the Northern and Southern Hemispheres.

greenhouse effect—The so-called greenhouse gases produce what is known as the greenhouse effect. Greenhouse gases absorb some radiated heat from the ground that passes through the atmosphere, which then reradiates the heat back to Earth, keeping the planet warm like a giant blanket. The gases include carbon dioxide, water vapor, chlorofluorocarbons (CFCs), nitrous oxides, ozone, methane, and other trace gases. Together, these gases are critical to the heat regulation of the planet.

Many of the greenhouse gases are produced by human activities. For example, carbon dioxide is emitted as a byproduct of fossil fuel use, or grassland and rainforest burning; CFCs are emitted by propellants in aerosols, coolants in older refrigerators, and gases in foam plastics; water vapor is produced by irrigation and the burning of fuels; lower atmospheric ozone (or ozone produced close to the Earth's surface, as opposed to the ozone layer higher in the atmosphere) is mainly produced by cars, buses, and trucks; and nitrous oxides come from animal wastes, nitrate fertilizers, and the burning of grasslands and fossil fuels. Whether such activities have increased the amount of greenhouse gases in the atmosphere, and whether they will lead to a rise in the Earth's average temperature are still topics that are being hotly debated. In many ways, we will not know the true answer in our lifetime, as the amount of time scientists have been keeping records is so small—only about three centuries—while climate change can take hundreds to thousands of years.

magnetism—Earth has a large magnetic field, which is used by navigators to determine the proper direction; the magnetic field also interacts with particles from the Sun to create colorful auroras. Birds, and

This aurora is caused by the reaction of the Sun's particles with the Earth's magnetic field. The image was taken from the space shuttle. *(Photo courtesy of NASA)*

> ### The Other Atmospheres—and What We Know
>
> Not all planets and satellites of the solar system contain atmospheres. Mercury; the Moon; the Martian moons, Phobos and Deimos; and most of the smaller moons around the outer planets do not have atmospheres. However, Venus, Earth, Mars, Jupiter, Saturn, Uranus, Neptune, and Pluto all have atmospheres, ranging from thick to thin. The atmospheres of the outer planets are much thicker and exert more pressure than on Earth. For example, just above Jupiter's central core, the pressure is about 2 million times greater than Earth's pressure at sea level. Few satellites have an atmosphere because their small sizes do not allow them to retain gases; Saturn's moon Titan has the thickest atmosphere.
>
> But we've watched one planet very carefully—mainly because of one startling characteristic, the greenhouse effect. Venus is entirely covered with a thick layer of carbon dioxide gas. The excessive carbon dioxide creates a runaway greenhouse effect, with temperatures at the surface of the planet reaching close to 900° F (482° C). Earth itself could have the same conditions—if our place in the cosmos had been any different.
>
> Earth's early atmosphere was probably composed mainly of water vapor, carbon dioxide and monoxide, nitrogen, hydrogen, and other gases released by volcanoes, with carbon dioxide as the major constituent. It was a hot planet and devoid of life as we know it—or if any life were around, it was so small as to not survive as fossils in the rock layers we find today.
>
> Some scientists believe that our early atmosphere was similar to the atmosphere on Venus today: a great deal of carbon dioxide, mostly from gases spewed out by volcanic activity, gases left over from the formation of the solar nebula, and even some gases from comets and asteroids striking the forming planet. Venus probably also had a great deal of carbon dioxide in its early atmosphere, the gas produced not only by possible early volcanic activity, but also by the intense heat of the Sun scorching carbonate rocks and causing the release of carbon dioxide gas. On Earth, this atmospheric condition changed; but on Venus, the planet could not rid itself of the carbon dioxide.
>
> The result was a condition we now refer to as the greenhouse effect—the average global temperatures rising as the greenhouse gases, including carbon dioxide, trap the Sun's energy and keeps it within the atmosphere. But things changed for Earth. The carbon dioxide in the atmosphere gradually dwindled over millions of years, as life on Earth eventually produced oxygen, and excess carbon dioxide was used for photosynthesis—with even more carbon dioxide taken up by carbonate rocks.

perhaps other animals, also use the Earth's magnetic field for navigation.

ozone—Ozone, or three molecules of oxygen bonded together instead of the usual two (O_3, as compared to the O_2 we breath), forms when sunlight strikes certain particles in the atmosphere. In the lower layer of the atmosphere called the troposphere, ozone is produced when particles emitted by cars and industries form a blue-tinged gas of ozone (less than one part per million of this gas is poisonous to humans, which is why many larger cities release daily ozone levels as part of their pollution reporting).

At the ozone layer itself, only a small amount of ozone is present, equivalent (if it were uniform) to a thin layer less than 0.25 inch (6 millimeters) thick. Because the layer is so thin, it is very sensitive to increases in solar activity, volcanic eruptions, and pollutants, particularly chlorofluorocarbons (CFCs), all of which seem to cause the ozone molecules to break down.

The ozone layer shields the Earth's surface from most of the Sun's harmful ultraviolet rays. The layer is thought to have developed about 2.5 or 3 billion years ago; its development was thought to have allowed plants and animals to evolve on land, as the ozone protected the fragile organisms from the harmful rays of the Sun. Any weakening of this protection through the depletion of this gas increases the possibility of damage to plants and animals.

ozone hole—Scientists have discovered an area with smaller than normal amounts of ozone, known as an "ozone hole," in the atmosphere above Antarctica. This hole, first discovered during the International Geophysical Year (IGY) from 1957 to 1958, expands and contracts with the seasons, reaching a maximum with the Antarctic spring (September through November). In the Northern Hemisphere, there is also evidence of ozone depletion, although this area has not been as extensively examined as Antarctica.

By 1980, the true extent of the hole was discovered; it expanded continuously in size due to increased ozone depletion during the 1980s, causing some scientists to theorize that increased CFC levels were to blame. This speculation led to the Montreal Protocol of 1987 that, with subsequent revisions, was aimed at reducing and/or eliminating the use of CFCs throughout the world, thus halting or reversing the rate of ozone depletion. But the true role of CFCs in the ozone depletion process has not been agreed upon.

Many scientists believe the ozone hole is still increasing in size, caused by artificial pollutants that reach the upper atmosphere; other scientists believe that the ozone hole size changes are normal. They maintain that because we only have a short period of collected data about it (in fact, recent measurements indicate that the hole has not become any larger), we don't know what normal fluctuations entail.

But if the average size of the ozone hole is increasing for any reason, it is estimated that many sensitive organisms in the oceans (especially the upper ocean water planktons, an important part of the food chain) and on land will be affected; the incidence of skin cancer, eye cataracts, and other health problems in humans may also rise. It is estimated that a 10-percent decrease in ozone leads to a 20-percent increase in ultraviolet radiation reaching the surface.

trade winds—These are prevailing easterly winds found between 30 degrees north and 30 degrees south. They are formed as a result of the returning southerly air in the Hadley cells interacting with the rotation of the Earth, or the Coriolis effect. Historically, these winds helped sailing ships traveling in an easterly directions—not to mention weather systems.

water—Water was, is, and always will be one of the most important elements on the Earth's surface. It constitutes a global volume of about 330 million cubic miles (1,400 million cubic kilometers), and its distribution varies widely. It drives nature and is an integral part of life and the formation—and erosion—of the Earth's land and seascape.

Approximately 97.2 percent of all water on Earth is found in the oceans; of the remaining 2.8 percent, 2.15 percent is in ice sheets, and 0.62 percent in groundwater. The remaining 0.03 percent consists of 0.011 percent in saltwater lakes and inland seas; 0.013 percent in rivers, freshwater lakes, and wetlands; 0.005 percent in the soil; and 0.001 percent in the atmosphere.

Water moves around the Earth via a water cycle. Simply put, the cycle includes evaporation from the oceans and other waterways, condensation to clouds, precipitation, evaporation from the surface, evapotranspiration from the surface and vegetation, and water seeping into the ground (percolation, infiltration, and groundwater flow), then back to the oceans to start the cycle again.

One important part of the Earth's water cycle is groundwater—and in particular, the aquifers that give most people their potable water supplies. Also called an underground lake, an aquifer is underground water confined by impermeable layers of rock. As water percolates deep underground through porous rock (which water is able to travel through), it can eventually hit an impermeable rock layer, which water cannot travel through. If the water has percolated between two layers of impermeable rock, it can form a confined aquifer, with the water under pressure between the two rock layers. When a well is sunk at such a location, water shoots to the surface in an artesian well. If the impermeable rock has faults, the artesian springs follow the cracks in the rock to the surface; desert oases often form near these springs.

Some aquifers are quite large; for example, the Ogallala Aquifer, under the central United States, covers 156,000 square miles (404,000 square kilometers). Although the water in many larger aquifers has been there for thousands of years or longer, many of the main aquifers are being depleted by humans. For example, today, more than 150,000 wells tap into the Ogallala Aquifer, which is rapidly becoming depleted. The problem with draining this, and many other aquifers, is that the climate has changed since the aquifer was formed—and the climate is now too dry to replenish the lost groundwater.

5. Bacteria and Viruses

Introduction

Although they are invisible to the naked eye, bacteria and viruses have had a major effect on the natural history of the Earth, especially their interactions with other organisms. Bacteria are the most abundant lifeforms on the Earth, and they are all around us—inside organisms, soil, water, and the ocean depths. Viruses also have a place in the natural history of the planet, causing most of the diseases that affect organisms and existing on the border between the living and the nonliving.

Timeline

(note: bya=billion years ago)

Date	Event

Prehistoric Events

Date	Event
~3.5 bya	Bacteria evolve on Earth; the only evidence so far for this date are the oldest fossils known, which are bacteria-like organisms.

Modern Events and Discoveries

Date	Event
1546	Italian scientist Girolamo Fracastoro (c.1478–1553) believes that infections and epidemics arise in certain ways, including his idea that diseases are transmitted from person to person through imperceptible organisms; although he conducted no experiments to reach his conclusions, his theories are the precursor to the germ theory of the nineteenth century.
mid-1600s	Dutch microscopist and zoologist Anton van Leeuwenhoek (1632–1723) invents the single-lens and double-convex microscopes; his inventions allow him to discover protozoa (1677) and bacteria.
1683	Van Leeuwenhoek discovers bacteria in the human mouth.
1717	Mary Montagu (1689–1762), an English author, records a smallpox inoculation procedure from Turkey; as part of the procedure, the smallpox matter is inserted into the vein with a needle, or a thread soaked in the smallpox fluid is drawn through a scratch in the patient's arm.
1773–1786	Danish microscopist Otto Friedrich Müller (1730–1784) taxonomically separates bacteria from protozoa.
1796	English physician Edward Jenner (1749–1823) is the founder of virology, inoculating a small boy with cowpox—the first inoculation against a virus.
1809	French chef, inventor, and bacteriologist Nicolas François Appert (c.1750–1841) demonstrates a procedure for preservation of foods by canning with glass jars.
1835	Italian scientist Agostino Bassi (1773–1856) shows a disease of silkworms is caused by a fungus; the discovery gives impetus to the germ theory of disease; because of his work on insects, he is also called the father of insect pathology.
1843	U.S. physician and author Oliver Wendell Holmes (1809–1894) proposes that puerperal, or childbed, fever is contagious; his suggestion to wash and wear clean clothes is scoffed at by his peers—but eventually it will be proven as the correct way to eliminate germs.
1854	English physician John Snow (1813–1858) verifies that disease is often spread by water, not air, based on his observations of the London cholera epidemic; he argues that the cholera is spread by a living, water-born cell or germ that could be reduced by boiling water or washing soiled linens; his suggestions slowed down the spread of the disease.
mid-1800s	Swiss-born U.S. bacteriologist Theodor Albrecht Edwin Klebs (1834–1913) is the first person to link various bacteria to disease, establishing this link in endocarditis, tuberculosis, typhoid, syphilis, and other diseases.
1857	French chemist and microbiologist Louis Pasteur (1822–1895) demonstrates that lactic acid fermentation is carried out by living bacteria.

Date	Event
1872	German botanist and bacteriologist Ferdinand Julius Cohn (1828–1898) coins the term "bacterium" and founds the field of bacteriology.
1877	German bacteriologist Robert Koch (1843–1910) develops a way to obtain pure cultures of individual bacterium.
~1880	Koch and Pasteur help to further the development of the germ theory of disease.
1881	Pasteur develops an immunization against anthrax by using the weakened bacteria that causes the disease; he calls the process vaccination, in recognition of Edward Jenner's use of cowpox, or *vaccinia*.
1882	Koch demonstrates that tuberculosis was an infectious disease caused by bacteria, not by heredity; he also establishes Koch's laws, which prove a given disease is caused by a specific microorganism.
1883	Klebs and German bacteriologist Friedrich Löffler (1852–1915) discover the deadly diphtheria bacillus, called the *Klebs- Löffler bacillus*, the old name for *Corynebacterium diphtheriae*.
1885	Pasteur develops a vaccine for rabies (hydrophobia) and saves the life of Joseph Meister, a young boy who had been bitten by a rabid dog.
1888	French bacteriologist Pierre Roux (1853–1933) discovers that the diphtheria bacteria does not cause the disease—it is caused by a toxin from the bacteria.
1890	German physician, bacteriologist, and chemist Paul Ehrlich (1854–1915) standardizes the diphtheria antitoxin; he becomes the founder of the field of immunology.
1890	German immunologist Emil Adolf von Behring (1854–1917) discovers antibodies.
1898	Loeffler and German bacteriologist Paul Frosch (1860–1928) find evidence that the cause of foot-and-mouth disease in livestock was an infectious particle smaller than bacteria; it is the first clue to the nature of viruses.
1915	U.S. bacteriologist Frederick William Twort (1877–1950) finds that certain viruses prey on bacteria, today called phages; this discovery was made independently by Canadian Félix-Hubert D'Hérelle in 1916.
1926	U.S. bacteriologist Thomas Rivera notes the difference between bacteria and viruses; virology becomes a major field of study.
1928	Scottish bacteriologist Sir Alexander Fleming (1881–1955) discovers penicillin.
1935	U.S. biologist Wendall Meredith Stanley (1904–1971) isolates a virus in crystal form, thus proving it is protein in nature.
1941	Russian-born U.S. biochemist Selman Abraham Waksman (1888–1973) coins the term "antibiotic" to describe compounds produced by microorganisms that kill bacteria.
1952	U.S. virologist Jonas Edward Salk (1914–1998) invents the vaccine for polio, consisting of the dead virus.
1957	Polish-American microbiologist Albert Bruce Sabin (1906–1993) develops an oral vaccine for polio using living, but weakened, polio organisms.
1970s	Oil-eating bacteria is developed at General Electric; it is also the subject of a 1980 Supreme Court ruling that states "a live, human-made microorganism is patentable subject matter."
1977	Smallpox is eradicated worldwide.
1981	Acquired immunodeficiency syndrome (AIDS) from the human immunodeficiency virus is first reported in the United States by the United States Centers for Disease Control and Prevention.

Date	Event
1986	The United States Department of Agriculture grants the first license to market a living organism; the organism was produced by genetic engineering—a virus to vaccinate against a herpes disease in swine.
1987	U.S. plant pathologist Steve Lindow and Greek plant pathologist Nickolas Panopoulos develop a genetically altered bacterium, a mutant of the common parasite, *Pseudomonas syringae*; it is also the first time a human-made microbe is released into the environment.
1996	An outbreak of *E. coli* bacteria in Canada and in north and northwestern United States occurs; it is traced to a brand of apple juice, and it would be one of many outbreaks in North America caused by a variety of reasons, mostly involving badly handled foods.

History of Bacteria and Viruses

Early Bacteria and Viruses

Early **bacteria** and **viruses** were probably similar to their modern counterparts. In fact, bacteria are thought to have been some of the first living organisms on Earth. In particular, the oldest fossils known, about 3.5 billion years old, are of bacteria-like organisms.

The problem with viruses is that, unlike bacteria, they have left no fossil record. Some scientists believe, though, that some viruses may have left evidence of their existence in an indirect way. For example, many scientists theorize that viruses may have been responsible for certain extinctions that occurred in the Earth's long natural history. It was once thought that the extinction of the dinosaurs about 65 million years ago was due to the outbreak of viral diseases. Now scientists believe this idea seems unlikely. One virus could have killed off the dinosaurs and similar species, but what about the other animals—from ammonites and formanifera, to ichthyosaurs and numerous other animals—that became extinct, too?

Studies of Bacteria and Viruses

One of the most important reasons for discovering more about bacteria and viruses stemmed from the diseases they spread. The beginning of scientists' understanding the relationship between microorganisms and diseases stretches back into the eighteenth century. Although the scientists could not understand the true reasons behind the diseases, they did develop several measures to counteract some diseases at this time. For example, in 1717 (and even a few years before) several reports of smallpox inoculation procedures—one by Englishwoman Mary Montagu—showed that some people knew exposing the person to the smallpox itself would cure the disease. In 1721, however, the procedure met with controversy: After American scientist Cotton Mather heard about successful inoculations from a slave that had arrived from Africa, American Zabdiel Boylston decided to try to stop an epidemic of smallpox. The results were positive, but the controversy still existed about how to stop these diseases.

The first true—and more accepted—inoculation would occur in 1796, undertaken by English physician Edward Jenner. The story is famous, not only for the chance Jenner took, but for the results. To eliminate smallpox in a young boy, Jenner infected the boy with cowpox. His idea was to use one virus to immunize against another virus—it was hoped that this procedure would make the person immune to the offending virus. Jenner was also the first to coin the word "virus," was the founder of virology, and a pioneer in **vaccination**. His ideas, albeit modified, are still used to defend against diseases caused by viruses.

By the nineteenth century, researchers made major breakthroughs in the studies of viruses and bacteria, and their effects on other organisms. Scientists determined that the microorganisms did exist—although their exact natures would not be discovered until the twentieth century.

By the middle of the nineteenth century, scientists had determined that doctors were spreading germs from one patient to the other. Historically, scientists had believed that, overall, disease spread through bad air (miasmal theory) or imbalances in the body (humoral theory of disease). In 1843, U.S. physician and author Oliver Wendell Holmes proposed that doctors and nurses should wash hands between patients, which lowered the number of child deaths. Other physicians, such as Ignaz Semmelweiss in 1847, would also recommend that doctors wash their hands between deliveries (washing the hands washes away much of the offending bacteria). But years passed before the idea became acceptable practice.

One of the first scientists to link bacteria to disease was Edwin Klebs, who established this connection in such diseases as endocarditis, tuberculosis, and syphilis. In 1878, he was the first to transmit a disease—syphilis—from humans to monkeys; thus, he could be considered the first to determine that certain animals could be used to study diseases in humans. In 1881, he discovered the typhoid bacillus. And along with German bacteriologist Friedrich Löffler, he discovered the deadly diphtheria bacillus in 1883, called the Klebs-Löffler bacillus. Löffler would not only go on to discover that some animals were immune to the disease but also other diseases associated with bacteria.

By the middle of the century, Louis Pasteur influenced others to experiment with certain antiseptics, as he had shown that antiseptics would stop the spread of disease in silkworms. He urged French army surgeons to sterilize their instruments during the Franco-Prussian War of 1870, and eventually he developed a vaccine for anthrax by a discovery he made by accident. Bacteria could be weakened, and in their weakened state, they failed to cause disease but still caused the body's immune system to create antibodies—the beginning of several vaccines he would develop.

Also in the late nineteenth century, Robert Koch, the famous German founder of modern bacteriology, determined four rules that must be met to prove that a bacterium was the cause of a particular disease. The rules, which are still valid in many ways today, include the following: (1) The microorganism must be found in great numbers in all diseased animals, but not in healthy animals; (2) the offending microorganism must be isolated from a diseased animal and grown outside the body in a pure culture; (3) if the isolated microorganism is injected into healthy animals, it must produce the same disease; and (4) the suspected microorganism must be recovered from experimental hosts (the ones used in rule number 3), isolated, compared to the first microorganism, and verified as identical.

Koch and Pasteur were also major advocates of the germ theory of disease that was slowly developing as they and others made discoveries about how diseases spread or even discoveries of bacterium that created certain diseases. The original germ theory applied only to bacteria, but as Pasteur was putting together his treatise on the germ theory in 1880, Charles Laveran had discovered that a protist (sometimes called a protozoan) caused **malaria**, not a bacterium. This discovery meant that not all diseases were caused by bacteria—a finding that would be shown to be correct in the next century as scientists learned more about viruses, fungi, and protists. But the germ theory was entrenched, and still is today: Most of our noncommunicable diseases (cancer, diabetes) are often suspected to be from some type of "germ."

Historically, it took until the beginning of the twentieth century to distinguish between bacteria and viruses. This discovery, made in 1926 by Thomas Rivera, would divide the studies of bacteriology and virology, and it led to new discoveries about how both microorganisms truly work in our environment. And similar to when they were first discovered, the fight against harmful viruses and bacteria continues to today—not only how to fight diseases transmitted by both microorganisms, but also how to use the microorganisms for human benefits.

As research into bacteria and viruses continued, other vaccines soon followed—ones that had a tremendous impact on our lives. For example, in 1952, Jonas Edward Salk invented the vaccine for polio (a major crippling disease at the beginning of the twentieth century), consisting of the dead virus. In 1957, Albert Sabin developed an oral vaccine for polio using living, but weakened, polio organisms. Sabin's vaccine was actually longer-lasting and replaced the Salk vaccine introduced several years before.

In recent years, one harmful virus in particular has been given a great deal of media coverage: the human immunodeficiency virus (HIV) that produces **acquired immunodeficiency syndrome** (AIDS). In addition, agents recently designed in the laboratory and isolated from natural sources are being used to fight certain viral infections. These agents are not antibiotics: That term is used in reference to bacteria. So far, none of the antiviral agents have killed off the virus, but many of the drugs so far developed have inhibited the replication of certain viruses.

Advances in other areas of science have furthered the study of bacteria and viruses in recent years. Genetics, for example, have played an important role in our exploration of bacteria and viruses. In particular, bacteria have been genetically altered to help humans—and even the environment. For example, in 1987 plant pathologists Steve Lindow from the United States and Nickolas Panopoulos from Greece developed a genetically altered bacterium, a mutant of the

> ### Deadly Bacteria and Viruses
>
> Before people knew the value of effective sanitation and the discovery of antibiotics, a series of bacterial diseases swept through many cultures in the form of epidemics. Here are a few of the more well-known problems with bacteria and viruses:
>
> - A bubonic plague strikes Europe in A.D. 541 and lasts for three years; it reaches China by A.D. 610.
> - By 1331, a bubonic plague in China begins; it reaches western Eurasia by 1346 and northern Europe in 1347, lasting four years.
> - The first case of influenza, caused by a virus, is reported in France in 1414; in 1918, an influenza epidemic would kill 20 million people as it spread around the world from northern France.
> - Smallpox is brought to the Americas around 1518.
> - Smallpox decimates the Aztec people in 1520, as the culture had no childhood immunity to the disease. The smallpox would continue to spread, reaching Peru by 1525. The disease eventually kills off large numbers of Native Americans.
> - Louis Pasteur develops a vaccine for rabies in 1885, an infectious disease caused by a virus. It is often transferred to humans via the bite of an infected warm-blooded animal, usually by its saliva.
> - Lyme disease is first reported in 1975, in Old Lyme, Connecticut. It is caused by a bacterium called *Borrelia burgdorferi*, carried by the deer tick.
>
> What about today—do any viral diseases seem to stand out? AIDS is, of course, one of our main concerns. But others, too, seem to crop up now and then. One case in particular began in 1995, when it was thought that malaria was making a comeback in the United States. This disease, transferred by mosquitoes, was once endemic (prevalent) in this country; in more recent decades, the disease has only been reported in travelers visiting or coming from countries where malaria was endemic. But in Houston, Texas, in that year, three patients were hospitalized with malaria symptoms—and the malaria parasites. They were treated and released, but records showed an additional 21 infected people. Fortunately, scientists found the reason: All 21 had traveled to countries in which malaria was endemic.
>
> And when it comes to bacteria, there are a few concerns—in particular, is it possible that antibiotics will not work some day? It may be true. Many of the antibiotics used to eliminate bacteria don't work for certain people anymore, as tougher strains of certain diseases, such as tuberculosis, have increased in the past decade or so. The reason for these antibiotic-resistant bacteria is simply evolutionary characteristics. In any population—including bacteria—some genes will undergo mutations; some of the mutations include bacteria that carry the ability to survive antibiotics. Over the years, this process can create a large pool of bacteria that are resistant to one or more antibiotics.
>
> And to add to the evolutionary pressure, bacteria of different species often swap genes with each other (called conjugation), which allows the trait of drug resistance to become enhanced. But most scientists agree: It is not that we couldn't develop more antibiotics to resist bacterial diseases—the drug companies simply have not thought it would make a profit to bring these new antibiotics to the market, until now.

common parasite *Pseudomonas syringae*. The bacterium was designed to give plants the ability to withstand temperatures below zero—or to retard frost on the plants. This bacterium was important for another reason: It was the first time scientists approved a human-made microbe to be released into the environment. And in the 1970s, a biochemist at General Electric developed oil-eating bacteria—a microorganism used to eat oil at spills.

Viruses are also used in genetics. The main reason is that viruses can transfer genetic material between different species of hosts. In addition, they carry out a natural form of genetic engineering, as they often incorporate genetic material from their host, transferring the information to a new host (called transduction). This recently has been looked at as a possible evolutionary mechanism—a way of "evolving" more highly disease-resistant organisms in the future.

Another example of genetics being used in the study of bacteria and viruses includes a more direct method of vaccination that shows a great deal of promise—genetic vaccines. In many ways, it is genetic manipulation: The vaccine, actually genetically altered DNA-coated projectiles, carries an encoded gene that produces a certain virus protein. This vaccine is directly injected into the cells of living organisms. Once inside the cell, it replicates the protein, creating an immune response—and thus, antibodies against the certain viruses. The details of such a process are currently being worked on, but are not fully in practice. *See also* ANIMALS, CLASSIFICATION OF ORGANISMS, CYTOLOGY, EVOLUTION, FOSSILS, GENETICS, GEOLOGIC TIME SCALE, LIFE, OCEANS, PLANTS, and VOLCANOES.

Topic Terms

acquired immunodeficiency syndrome—Acquired immunodeficiency syndrome, or AIDS, is caused by the human immunodeficiency virus (HIV) that attacks the T-cell lymphocytes of the immune system. This attack on the T-cells (cells that protect the body and strengthen the immune system) causes the body to be vulnerable to serious illnesses that would usually be repelled by a healthy immune system. It is thought to be contracted by the transfer of blood or other body fluids from an infected person to a noninfected person. Because of this outbreak, blood for transfusions has been tested for HIV for the past two decades; people at risk for this disease include those engaging in unprotected sex and those sharing unsterilized drug needles.

bacteria—Bacteria (the plural term is bacteria; the singular term is bacterium) are the most abundant lifeforms on the planet. They are single-celled lifeforms, invisible to the naked eye, and exist all around us. Found inside living organisms; in soils, rivers, ponds, lakes, and oceans; and deep on the ocean floor, bacteria are of immense importance in our natural world because they are extremely adaptable and have the capacity for growth and reproduction.

Even though bacteria cause diseases among animals (including humans) and plants, these microorganisms also play an important and necessary part in our lives. For example, some antibiotics come from certain bacteria (actinomycetes); yogurt and sourdough bread use bacteria, giving them a tangy taste; and certain bacteria live symbiotically in our digestive tract, allowing us to digest food.

Certain bacteria are also important to our environment. For example, humans have long created compost piles for use on the land. In compost bins nitrogen is created by organic wastes; this nitrogen feeds certain bacteria that help break down the dead organic matter still further. More recently, anaerobic bacteria have been added to all landfill sites where they thrive in the compacted, air-free layers of waste, digesting the organic matter and releasing methane and carbon dioxide. (It took years before it was realized that these gases had to be contained so as not to contribute to possible global warming.)

Some of the common diseases caused by bacteria are listed below. Some are highly contagious in childhood, such as diphtheria and whooping cough (pertussis); others can be contracted at any time, such as strep throat, tuberculosis, and pinkeye; other diseases are caused by coming in contact with a certain bacteria from an "outside" source, such as the deer tick carrying Lyme disease or the bacteria that causes gangrene.

Modern Diseases from Bacteria	
Anthrax	Osteomyelitis
Botulism	Pinkeye
Cat Scratch Fever	Pneumonia
Cholera	Salmonella
Diphtheria	Shigella
E. coli	Staphylococcus
Gangrene	Strep Throat
Giardiasis	Syphilis
Gingivitis	Tetanus
Legionnaires' Disease	Tooth Decay
Leprosy	Trench Fever
Lyme Disease	Tuberculosis
Mad Cow Disease	Typhoid Fever
Meningitis	Whooping Cough

malaria—Malaria, once thought to be caused by a bacterium, is a rather complicated disease. In this case, a protozoan parasite called *Plasmodium* causes the problem. The organism is transmitted by female anopheline mosquitoes that feed on human blood after mating. Initially, if a mosquito bites an infected person, the *Plasmodium* are taken into the mosquito and begin to reproduce in the gut of the insect, eventually releasing new *Plasmodium* cells; these cells migrate to the mosquito's salivary gland and are then passed on to the next human it bites. Once in the human bloodstream, the new *Plasmodium* cells (called sporozoites) infect liver cells and eventually red blood cells. This infection results in liver failure and anemia, a reflection of the growth of the *Plasmodium* cells in the body. Antimalaria drugs kill off the parasite, but until the early 1930s, when they were first developed, the only treatment was quinine, which had been in use for more than 300 years.

vaccination—A vaccination is a procedure to immunize the recipient from a disease caused by a virus or bacterium; it is from the word *vaccinia*, Latin for "cowpox" (in reference to the first inoculation by Edward Jenner, in which he used cowpox to protect a young boy from smallpox). In general, a person is inoculated by injecting a mild dose of the disease-causing virus or bacteria into the bloodstream; the dose is usually purified proteins or a weakened or dead form of the disease-causing virus or bacteria. The dose causes the body's immune system to build up antibodies that will recognize and attack a particular infection, but won't actually cause the infection. Some common vaccinations include smallpox, chicken pox, and influenza. Currently, many vaccines need to be developed, including those to fight against the human immunodeficiency virus (HIV).

viruses—Viruses are genetic entities that exists somewhere between a living and nonliving state; thus they have always been confusing "organisms" to scientists. They affect every living thing—from plants and animals to bacteria—with 100 or more viruses having the potential to infect a single species. Viruses differ quite a bit from one another. Some viruses infect only humans (smallpox); some infect humans and one or two additional kinds of animals (influenza); and some infect only a particular kind of plant (tobacco mosaic virus) or bacteria (lambda bacteriophage that infects *E. coli*).

Viruses exist only to make more viruses; and with the possible exception of bacterial viruses (viruses that kill harmful bacteria), all viruses seem to be harmful, as they reproduce by killing the host cell. The smallest viruses are the spherical-shaped satellite viruses, which measure only 18 nanometers (18 billionths of a meter) across. To compare, the herpes virus measures 100 nanometers; the average bacterium is 1,000 nanometers wide; and a typical plant cell is 500 to 5,000 times larger than the satellite viruses. These satellite viruses have such small protein shells that they cannot hold the necessary genetic material to replicate on their own—thus, they are actually parasites of parasites, clinging to other viruses in regular cells and pirating other viruses' genetic material.

In general, a virus depends on the host cell it infects to reproduce. Without the host cell, a virus exists as a protein coat (capsid) that is often enclosed within a membrane. At this stage, the protein coat surrounds either DNA or RNA, the codes for the virus, and the virus is "inert." (If a virus is dormant inside a host cell, it is said to be in a lysogenic phase.) The virus is often "stimulated" when it comes into contact with a host cell (i.e., when an activated virus enters a lytic phase). The virus inserts its genetic material into the host, taking over the cell's function; the cell then changes, producing more material—viral protein and genetic material—but for the virus, not itself. As new viruses are formed, they self-assemble, then kill the host cell by bursting it—and continue on to infect other cells.

It is difficult to get rid of viruses. One reason is the virus's main characteristic: It is a minute parasitic organism that reproduces only inside the cell of its host. The virus replicates itself by invading the host cells, thus taking over the cell's mechanisms for DNA replication; as the virus "breaks out," it causes disease. Historically, viruses were discovered but not understood as far back as the invention of the microscope—around the turn of the seventeenth century. But it took another century before any true knowledge of viruses—and what was present within these entities—was uncovered.

Some of the more familiar human diseases associated with viruses are listed below. Many of these have vaccines, such as chickenpox, malaria, and smallpox. Others are still major diseases, such as AIDS/HIV, that do not yet have a vaccination. (Note: Even some types of cancer—though definitely not all—have been linked to viruses.)

Modern Diseases from Viruses	
AIDS/HIV	Influenza
Chicken pox	Leukemia
Dengue Fever	Measles
Ebola	Mononucleosis
Elephantiasis	Mumps
Encephalitis	Plague
Fifth Disease	Poliomyelitis
Flu	Rabies
Foot and Mouth	Rubella
Hanta Virus	Shingles
Hepatitis A	Smallpox
Hepatitis B	Warts (certain types)
Hepatitis C	Yellow Fever
Herpes	

6. Birds

Introduction

Birds are one of the most diverse, adaptable, and interesting animals on Earth. They are found everywhere and have adapted to a wide range of climates and habitats. Their origins are fiercely debated among paleontologists; some feel that birds are the only living descendants of the dinosaurs, while others are just as vehement that they are not. Whatever their past, they seem to have been present on Earth, in one form or another, for approximately 150 million years, since the late Jurassic period. Today, birds thrive on diets ranging from seeds to mammals, and everything in between; they are colorful, with beautiful songs and calls. But the main reason for our fascination with birds is their ability to fly through the atmosphere (with a few exceptions).

Timeline

(note: mya=million years ago)

Date	Event
Prehistoric Events	
~270 mya	The reptiles, having evolved from the amphibians, rapidly spread, developing into thousands of species; of the four main lines, the diapsid line evolves into the archosaurs, or "ancient reptiles."
~250 mya	A great extinction event occurs, killing off nearly 96 percent of all living species; the resulting ecological niches that opened allow for the evolution of the reptiles.
~150 mya	One of the earliest birds, *Archaeopteryx lithographica*, evolves at this time.
~65 mya	A mass extinction occurs, leading to the disappearance of the dinosaurs from the known fossil record; somehow, birds survive, diversifying and increasing in number, and all the modern families of birds evolve.
~1.8 mya	By the ice age years of the Quaternary period, modern bird families diversify into the species we now see today all around the world.
Modern Events and Discoveries	
1680	The last flightless dodo bird dies on Mauritius, an island in the Indian ocean; the species is exterminated by Dutch settlers who arrived on the island in 1598.
1713	English naturalist John Ray's (1627–1705) work, *Synopsis of Birds*, is published posthumously.
1803	John C. Otto of the United States is the first person known to conduct studies by banding birds.
1827	French-American ornithologist, naturalist, and artist John James Audubon (1785–1851) publishes his volumes of ornithology, *Birds of America*—one of the first bird guides published..
mid 1800s	Naturalist and writer John Burroughs (1837–1921) writes his first book, called *Wake Robin*, his most popular book about birds.
1835	English naturalist Charles Robert Darwin (1809–1882) studies the closely related finches (today called Darwin's finches) in the Galápagos Islands during his travels on the scientific vessel *H.M.S. Beagle*; the observations would prove to be instrumental in his evolutionary theories.
1855	German scientist Alexander von Middendorff (1815–1894) suggests that birds were capable of detecting the Earth's magnetic field and using it to navigate.
1861	The first *Archaeopteryx* is found in a German quarry, the first known fossil of a bird; it is also the earliest known bird, living some 150 million years ago.
1905	The National Audubon Society, a nonprofit organization, is founded; it begins as the National Association of Audubon Societies for the Protection of Wild Birds and Other Animals, Inc.

Date	Event
1934	U.S. naturalist and artist Roger Tory Peterson (1908–1996) publishes the first field guide, *Field Guide to the Birds*, which would become one of the best-known birding guides in the world.
1952	Gustav Kramer demonstrates Sun-compass orientation in birds.
1957	German ornithologist Franz Sauer suggests certain birds orient themselves using the stars.
1969	U.S. paleontologist John Ostrom suggests dinosaurs may have been endothermic (warm-blooded)—thus, more active and similar to birds
1972	The U.S. government restricts the use of the pesticide DDT, which is shown to adversely affect the environment, especially the eggshells of certain birds.
1985	A semipalmated sandpiper banded and released in Massachusetts is shot four days later in Guyana 2,800 miles (4,500 kilometers) away, thus setting a record as the longest distance ever known for flight over water by a shorebird.

History of Birds

Early Birds

According to present knowledge, **birds** arose from theropod dinosaurs at some point in the Jurassic. Birds, together with the rest of the dinosaurs, the crocodilians, and their relatives, are classified together in a group called Archosauria. One of the earliest birds is thought to have been the *Archaeopteryx lithographica*, an animal that lived about 150 million years ago. It may have evolved from smaller carnivorous dinosaurs, although this is still debated.

During the late Mesozoic era, several lines of birds eventually became extinct; these birds shared the world with early **Neornithes** (recent birds). The early birds known as the ichthyornithiformes were able to fly and bore teeth in their jaws; their lifestyles, and probably their structures, were similar to modern sea gulls. The hesperornithiformes also had teeth, but their wings were very small, thus, they were probably flightless (although some fossils indicate that some lines of the hesperornithiformes could fly). They could swim, though, and had powerful swimming feet, probably similar to modern cormorants. One of the most debated fossil birds is the enantiornithes, birds that first appeared in the lower Cretaceous period; they were either toothed or toothless, and their sizes ranged from sparrow-size to those with feet-wide wingspans. Because of the lack of fossils, their true origins and ancestry is highly debated.

After the Cretaceous period came the Cenozoic era; in the first period, the Tertiary, all the modern families of birds evolved. By the ice age years of the Quaternary period, modern bird families diversified into the species we now see today all around the world.

Bird Studies (Ornithology)

Similar to all the natural sciences, the early study of birds was couched in myth and superstition. Many of the larger animals were thought to have special powers, which also gave rise to some typical sayings, such as "wise, old owls," or "eagle-eye."

Although birds received some scientific attention in early centuries, ornithology, the study of birds, flourished in the nineteenth century. Several of the more popular ornithologists, such as John James Audubon, fostered the idea of bird identification. **Bird anatomy** also came into play in determining avian details. Charles Darwin, in his famous expedition on the *H.M.S. Beagle*, studied one part of birds' anatomy—

Birdbanding is one of the best ways for scientists to keep track of birds. This songbird, a catbird, has a narrow band on its leg.

bird beak development. He used the beaks of the various finches in the Galapagos Islands to develop parts of his evolutionary theories.

Although the existence of bird fossils was proposed in 1825, the discovery that confirmed their existence and led to the hunt for more bird fossils, was that of the *Archaeopteryx lithographica*. This bird lived about 150 million years ago and is considered one of the earliest birds. An almost complete skeleton—called the "London specimen"—was found in 1861 in a quarry in Germany. It was the basis for a continuing debate between supporters and detractors of Charles Darwin's newly published theory of evolution. A third skeleton was discovered in 1877 (the "Berlin specimen"); subsequent finds over the years, including the latest in 1992, bring the current total to seven.

The biggest controversy surrounding modern birds is their origin. Some scientists believe *Archaeopteryx lithographica* was the transition between dinosaurs and modern birds—mainly because the fossilized skeletons show dinosaur (or reptilian) characteristics, such as bony tails, teeth, and claws on the fingers, along with bird-like characteristics, such as feathers, wishbones, and beaks. The idea of birds as a kind of dinosaur was revived in 1969 by paleontologist John Ostrom, who suggested that dinosaurs may have been endothermic (warm blooded)—thus, more active and similar to birds. At this time, too, scientists were able to better study dinosaur physiology (cells and tissues) using advanced technology. They noticed the physiological similarities and differences between bird fossils, other animal fossils, and modern birds themselves. It is hoped that further studies such as these will eventually lead to the actual—and agreed upon—bird lineage.

Today, many scientists believe birds and dinosaurs evolved separately from a common reptilian ancestor, while others believe that modern birds are direct descendants of the dinosaurs. But so far, no one has yet found positive, acceptable fossil evidence to support or disprove either one of these ideas.

Other bird studies concentrate on modern birds—especially **bird behavior** and **bird flight**. Bird behavior has fascinated ornithologists for centuries, and today, scientists are attempting to understand courtship behavior, eating habits, birds' needs for various habitats, the speed of bird flight, support and steering (wings), and how birds control takeoffs and landings.

One of the more intriguing studies in bird flight has dealt with bird migration and orientation. In the early twentieth century, scientists suggested that birds use the Sun as a guide; since then, several birds have been shown to have a "Sun compass." In 1957, Franz Sauer, a German ornithologist, found that certain birds orient themselves using the stars. In the early 1960s, some scientists tried to prove that bird migration was influenced by the Moon, but so far, no such evidence has been discovered. Scientists have also used radar to determine how migrations take place—and some researchers believe that many birds migrate using learned landmarks. Still other studies, especially those started by Alfred von Middendorff in 1855, suggested that birds are capable of detecting the Earth's magnetic field—a debate that has continued for more than 100 years.

Modern birds are classified as part of the Animal kingdom, subphylum Vertebrata, and have their own class known as Aves. This class has 22 generalized groups; each group has one or more orders. In turn, each order has a specific number of families, and each family contains numerous species—showing that birds are very diverse. *See also* ANIMALS, CLASSIFICATION OF ORGANISMS, DINOSAURS, EVOLUTION, EXTINCTION, GEOLOGIC TIME SCALE, LIFE, and REPTILES.

Topic Terms

birds—There are thousands of modern bird species. Some of the species of living birds, from the more primitive to the advanced species, are listed below. (Two notes: Bird names can be capitalized or set in all lowercase; in general, most science books list common bird names in all lowercase letters. This listing is only one classification—there are others.)

Common Modern Bird Species			
Group	*Order*	*Family*	*Number of Species*
Ratites	ostriches	ostriches	1
	rheas	rheas	2
	cassowaries	cassowaries	3
	emus	emus	1
	kiwis	kiwis	3
Tinamous	tinamous	tinamous	47
Grebes	grebes	grebes	20
Loons	loons	loons	5
Penguins	penguins	penguins	18
Tube-Nosed Marine Birds	tube-noses	albatrosses	13
		diving-petrels	4
		shearwaters & petrels	66
		storm-petrels	21
Pelicans & Relatives	pelicans & relatives	anhingas	4
		boobies & gannets	9
		cormorants	33

Group	Order	Family	Number of Species
		frigatebirds	5
		pelicans	8
		tropic birds	3
Long-Legged Wading Birds	storks & relatives	hamerkop	1
		herons	64
		shoebill	1
		storks	17
		ibises & spoonbills	33
	flamingos	flamingos	6
Waterfowl	waterfowl	ducks, geese & swans	147
		screamers	3
Birds of Prey	birds of prey	falcons	60
		hawks, eagles, kites & osprey	218
		new world vultures	7
		secretary-bird	1
Game Birds	game birds	curassows & guans	44
		grouse, quail & pheasants	212
		megapodes	12
Cranes, Rails & Relatives	cranes, rails & relatives	12 including:	
		bustards	24
		button-quails	14
		cranes	15
		kagu	1
		limpkin	1
		mesites	3
		rails & coots	141
		seriemas	2
		sunbittern	1
		sungrebes	3
		trumpeters	3
Shorebirds, Gulls, Auks, & Relatives	shorebirds, gulls, auks, & relatives	auks, murres, & puffins	23
		avocets & stilts	13
		coursers	16
		crab-plover	1
		ibisbill	1
		jacanas	8
		oystercatchers	7
		painted-snipes	2
		plovers	64
		sandpipers	86
		sheathbills	2
		skuas, gulls, terns, & skimmers	98
		thick-knees	9
Pigeons & Doves	pigeons & doves	pigeons & doves	304
		sandgrouse	16
Parrots	parrots	cockatoos	18
		lories	55
		parrots	269
Cuckoos & Relatives	cuckoos & relatives	cuckoos	130
		hoatzin	1
		toruacos	19
Owls	owls	barn-owls	12
		typical owls	133

Group	Order	Family	Number of Species
Nightjars & Relatives	nightjars & relatives	frogmouths	13
		nightjars	77
		oilbirds	1
		owlet-nightjars	8
		potoos	5
Swifts & Hummingbirds	swifts & hummingbirds	hummingbirds	320
		swifts	82
		crested-swifts	4
Colies	colies	colies	6
Kingfishers & Woodpeckers	kingfishers & relatives	bee-eaters	24
		cuckoo-rollers	1
		ground-rollers	5
		hoopoes	1
		hornbills	45
		kingfishers	92
		motmots	9
		rollers	11
		todies	5
		wood-hoopoes	8
	trogons	trogons	37
	woodpeckers & relatives	barbets	81
		honeyguides	14
		jacamars	17
		puffbirds	34
		toucans	33
		woodpeckers	200
Perching Birds	perching birds	73, including:	
		buntings, grosbeaks, tanagers, wood-warblers, & relatives	795
		crows & jays	105
		finches	122
		thrushes & relatives	1350

bird anatomy—Bird anatomy includes circulatory, respiratory, skeletal, and digestive systems; beaks; and feathers, which are all specialized for this species. The following are some of the major systems:

beaks/mandibles/bills—Although they vary in size, color, and shape, bird bills play an important role in avian survival. Not only are these "instruments" used to gather and break up food, but birds also use their beaks to scratch, preen, collect (such as nesting material), and threaten or peck their rivals. The bill usually has two parts—an upper and lower mandible separated by the mouth; the nostrils are located on the upper part of the bill (with exceptions like the kiwi, whose nostrils are at the tip of the bill), which allows the animals to breathe without opening the mouth.

circulatory system—The circulatory system takes in oxygen and other materials, which are then rapidly distributed to the organs and muscles of a bird's body, while carbon dioxide and waste products are removed. This system, like that of other animals, uses a blood supply as the transport medium. The blood is pumped throughout the body by the heart, which is a relatively large, four-chambered organ capable of generating high pressure with rapid contractions—in other words, a much faster heartrate than humans.

digestive system—The digestive system of birds begins with the specialized beak; for example, seed-cracking beaks are found on seed-eating birds, and flesh-tearing beaks are found on birds of prey and vultures. From there, the digestion also differs depending on the species. For example, owls eat food such as mice, swallowing the entire animal; the food then passes to a glandular stomach that digests the food chemically. The gizzard, a muscular stomach, is also important to most birds' digestive tracts. This muscle actually grinds the food, often with the aid of course sand or pebbles, serving as the "internal teeth" of the animals. The indigestible parts of the food are often regurgitated by many birds in the form of pellets. Overall, digestion in birds is rapid and highly efficient, in order to maintain the animals' high internal temperatures—usually in the range of 101° to 112° F (38° to 44° C).

feathers—Feathers are the unique structures that insulate birds in the cold, allow them to fly, and provide protective coloration. Because reptiles and birds are related to each other, feathers are thought by some scientists to have evolved from the reptilian-type scales. Modern bird feathers are light, strong, and flexible, and composed of mostly keratinous, or protein, material. Feathers are not all the same, and the specialized feathers found on specific parts of a bird's body have definite purposes. The contour feathers are found on the outer body, wings, and tail. The wing and tail feathers are normally large and stiff, to enable flight. Wing feathers can be classified into primaries, secondaries, and tertials; the long flight feathers on the tail are called rectrices, or steering feathers. Another type of contour feather is the auriculars that grow around the ears. Semiplume feathers are hidden beneath the contour feathers and provide insulation, flexibility, and, in waterbirds, buoyancy; they also fill out the outline of the body. Down feathers are small, fluffy, soft feathers found beneath the contour feathers that provide insulation for the bird. Filoplumes are hair-like feathers associated with contour feathers. Powder feathers do not resemble feathers at all, but appear to be a powdery substance thought to provide protection from moisture. Different species of birds have varying amounts of these types of feathers. In addition, some species have very specialized feathers, such as the bristles found on the toes of barn owls and the crest on the blue jay.

Molting is the periodic shedding of old feathers to be replaced by new ones. Each feather is a dead structure incapable of further growth and must be completely replaced. All adult birds molt at least once during the year; some molt two, or even three, times. The underlying reasons for molting are many. The tips of feathers become worn and broken during flights through trees and bushes. Wear also occurs when contacting the ground during takeoffs and landings, and while walking around. There may be fading of the feathers' colors as well. To stay at the peak of flight efficiency, new feathers must be grown to replace the damaged ones.

The prebasic molt is a complete molt; that is, all of the feathers on the bird are replaced during this molt. However, all of the feathers are not replaced at precisely the same time. Each bird species has a definite sequence, or pattern, by which this molt happens. For example, in the passerines, or perching birds, the primary feathers of the wing molt first, starting at the middle of the wing and working toward the tips. When the molt of these feathers is near completion, the secondaries of the wing begin molting, beginning at both ends and working toward the middle. There are also partial molts, such as the prealternate, in which the head, body, and tail feathers are replaced. Most birds molt the wing and tail feathers a few at a time and equally from both sides of the body. These feathers are critical for flight, and molting enables the bird to continue flying. Some aquatic birds, however, lose all of their flight feathers at the same time. They remain flightless until the new feathers grow in.

respiratory system—The respiratory system of the bird takes in oxygen for energy and removes carbon dioxide. This system is different than that of mammals. Birds' lungs do not inflate and deflate with each breath; rather, a constant volume of air is always present. In addition, the air flows through the lungs to adjacent air sacs present in the bird's body. These adaptations contribute to the bird's level flight and a higher energy level.

skeletal system—The bird's skeletal system is the supporting framework of the entire bird. It is made up of hard calcareous bones (made of calcium), and tough, somewhat springy cartilaginous bones. As with most animals, the bones protect the bird's vital organs and parts, and are attachment places for the muscular system. Birds have light and strong skeletons; in particular, the slender, hollow, air-filled bones give most of these animals their ability to fly. And each part of the skeleton has a particular specialization, allowing the animals to fly, walk, swim, or run. For example, the breast muscles that drive a bird's wings evolved in parallel with the increase in size of the sternum (breastbone) of all modern flying birds.

bird behavior—Of all the bird behaviors, people most notice the calls, songs, and nesting (during the breeding season) of most of the common birds. Here are some of the explanations behind these behaviors:

calls—Calls are brief, simple sounds uttered by birds for many purposes. They are usually of one to five notes, in contrast to the longer and more complex songs. For example, young birds use location calls to let their parents know where they are hidden; grown birds use assembly calls to gather a scattered flock, migration calls to keep contact within a group, and warning calls to signal the presence of danger.

songs—Songs are elaborate vocal signals, usually associated with birds. It is difficult to know what bird songs were used by birds millions of years ago. Based on what we know about birds today, scientists speculate that bird songs were used much in the same way then as they are today—to claim territory and for mating, usually by the male (although some birds, such as certain species of woodpeckers, will "drum" on posts with their beaks to claim their territory—a crude sort of "song"). The songs are mostly simple and repetitive; and although some songs are partially learned, many of the more complex bird songs are instinctive. For example, a song sparrow raised by a canary will sing a song sparrow melody. The variation of the songs is seemingly just as varied as the many songbird species. Examples of bird songs include the melodies of the cardinal, the hooting of an owl, and the machine-like sounds of nighthawks.

nesting—A bird's nest is its "home," the place in which the bird raises its family. When most people think of nests, the usual bowl-shaped structure typical of the robin comes to mind. In reality, nests can be a multitude of sizes and shapes and made of grasses, twigs, rootlets, mosses, and mud; nests can also be on rock ledges, bare rocks, depressions in the ground or on floors of abandoned buildings, in old nests, in dead trunks of trees, in marshes, or in burrows. The most important factor in choosing location has to do with the chosen environment—in particular, how close the animal is to its food supply.

bird flight—Flight is the ability of birds to move through the atmosphere of our planet using wings and feathers. It also includes certain related traits, including the ability to navigate and migrate.

Other than bats and certain insects, birds are the only animals capable of true flight on Earth. (The ability to glide does not count.) A bird's wing structure has bones connected by joints. The primary feathers extend from the middle of the wing (near the wrist joint) to the wing tip. The secondaries extend from the shoulder joint to the middle of the wing. The bird has complete control over the area from the wrist joint outward, similar to the way humans can control their hands and fingers. This combination of feathers, skeletal structure, and flexibility allows each wing to have a dual use: The wings of birds combine the function of a propeller and a lifting surface, or airfoil. Using the action from the wrist, the primary feathers are moved, forward and down, then upward and back during each cycle, or flap. This process provides the power, or force, needed to move through the air. The inner feathers, or secondaries, are moved by the action of the shoulder joint into an airfoil, or lifting surface, configuration. This process provides the lift needed to stay in the air.

Bird flight is also important when discussing bird migration, navigation, and orientation. Migration occurs (for most birds) twice a year, with most species of birds flying great distances to and from their ancestral breeding grounds—from cold to warm areas. In the Northern Hemisphere, these trips generally take the birds from warmer, southern areas where they spend the winter, back to the general areas where they were born. There, as spring comes to these northern areas, the birds mate, lay eggs, and raise their young. The onset of shortening days and colder weather triggers hormonal changes (although the real reason is yet unknown) that compel the birds to migrate in the opposite direction—to the south, where food will be more plentiful.

Navigation (the ability to maintain a direction independent of landmarks) and orientation (understanding direction) are also important to birds, as a bird in flight must be able to navigate not only for migration, but also for hazards such as trees, branches, and predators. It is thought that some birds use various techniques, including orientation by the Sun and stars. (So far no evidence has been found that shows the Moon plays a role.) They migrate using landmarks, such as coastlines, and possibly the Earth's magnetic field—although birds probably use many other clues that have yet to be discovered.

Neornithes—The Neornithes is a subclass of birds (Aves) that includes about 9,000 species of recent (or living) birds; they are classified based on their palate (inside the mouth) anatomy. They are divided into four suborders: the Odontognathae (fossil); the Palaeognathae, divided into two subgroups, the ratites (such as ostrich, rhea, emu, and other large, flightless birds; although the smaller kiwi is included) and tinamous (such as the South American tinamous); Impennes (penguins); and Neognathae (all other living birds, from hummingbirds to plovers). Other classification systems differ, including one that divides the Neornithes only into the Palaeognathae and the Neognathae (which would include all other living birds).

The Bird and Dinosaur Evolution Controversy

Where did birds come from? Paleontologists and anatomists have been asking—and arguing about—this question for years. The debate has two major sides: Some scientists believe birds are actually direct descendants of the ultimate reptiles, the dinosaurs; while others believe that birds *are* dinosaurs.

The similarity between birds and reptiles was noticed as far back as 400 years ago, but it wasn't until the great fossil discoveries of the 1800s that someone pointed out the possible connection. This similarity between birds and dinosaurs was discovered and summarized by English biologist Thomas Henry Huxley, who was a staunch advocate of evolution. He found numerous commonalties between modern birds, the 150-million-year-old *Archaeopteryx lithographica* fossil that had been recently discovered, and the fossils of theropod (meat-eating) dinosaurs. In light of Charles Darwin's then new theory of evolution, the bird-type fossil was compelling evidence for a link between dinosaurs and birds.

But in the 1920s, ornithologist Gerhard Heilmann theorized that birds were not descended from dinosaurs, using the argument that birds had wishbones and dinosaurs did not. He believed birds and dinosaurs evolved separately from a common, reptilian ancestor.

For the next 40 years, the common ancestor theory predominated until, in the 1960s, paleontologist John Ostrom of Yale University found a dinosaur fossil he named *Deinonychus antirrhopus*. Ostrom compared the bones of *Deinonychus* and *Archaeopteryx* and found them to be extremely similar. He linked both to other fossils in a subset of the theropods called coelurosaurian dinosaurs. Similarities between birds and coelurosaurians include a fused bone in their wrists which gives the "hand" the rigidity necessary to support wing structure, a pelvic structure which is set farther back on the body, elongated forelimbs, clawed "hands," light hollow bones, reversed first toes, large eye sockets, and S-curved necks. Since this discovery, the theory of birds evolving from dinosaurs has been predominant, although the common ancestor theory is still very much alive; both points of view have been substantiated by recent fossil finds, fueling the debate over the ancestry of modern birds.

Since Ostrom's theory was announced, many new bird-like fossils have been found around the world. In northern China, fossils show that birds similar to our modern species were present almost 140 million years ago; they include magpie-sized *Confuciusornis* and the sparrow-size *Liaoningornis*, both of which had beaks, not teeth. The species seems too advanced to have evolved from dinosaurs or *Archaeopteryx*—which is seen by some to be an evolutionary dead-end. Instead, several scientists feel that birds and dinosaurs share many features because of convergent evolution, not because they are closely related.

And not everyone is convinced about the bird-dinosaur connection. Some paleontologists think lizard-like archosaurs—an ancient group of reptiles—were the real ancestors of birds (and dinosaurs), giving rise to *Confuciusornis*, *Liaoningornis*, and *Archaeopteryx*, and along separate lines, the dinosaurs. Another recent fossil find, *Sinosauropteryx*, a theropod dinosaur with what appears to be feathers, also shows the outlines of the abdominal cavity. This cavity, according to John Ruben, a zoologist at Oregon State University, shows a clear separation between the lungs and other internal organs, indicating a crocodilian-type lung—unlike that of birds, and making it unlikely that this type of dinosaur evolved into modern birds. It is more likely, says Ruben, that both bird lungs and crocodilian-type lungs evolved independently from a common ancestor. Ann Burke and Alan Feduccia of the University of North Carolina have studied bird embryos and note a difference between the "fingers" on the front limbs of birds and theropod dinosaurs. Though both start out with five fingers and end up with three during embryo growth, birds lose the equivalents of thumbs and pinkies, while dinosaurs lose the fourth and fifth digits.

All of this data has been challenged by proponents of the bird-dinosaur link, which has been greatly strengthened by recent fossil finds. One of the most significant has come from Argentina, where the fossil remains of a 7-foot (2.2-meter) long, carnivorous dinosaur have been found—the most

The Bird and Dinosaur Evolution Controversy (cont'd.)

bird-like dinosaur fossil yet discovered. Named *Unenlagia comaheunsis*, "half-bird from northwest Patagonia," the 90-million-year-old dinosaur had a bird-like pelvis and arms that could flap like wings. It existed, says Fernando Novas of the Museo Argentino de Ciencias Nauturales in Buenos Aires, at the same time as birds but was a living member of the missing dinosaur line from which birds evolved.

Another find comes from the Pyrenees mountains of northern Spain—a fossil of a 135-million-year-old baby bird that hatched approximately five million years after the first appearance of *Archaeopteryx*. Similar to modern birds, the 4- to 5-inch (10- to 13- centimeter) long animal had wings, feathers, and tiny holes in its immature bones. But it also had a head and neck similar to carnivorous dinosaurs, with sharp teeth and powerful neck muscles for chewing. Another fossil found in Mongolia was of an *Oviraptor* dinosaur, which apparently curled over its eggs, similar to the nesting behavior of chickens. And, in the Red Deer River Badlands of Canada, the remains of a carnivorous ornithomimid dinosaur were found that included traces of keratin around the front of its skull. Keratin, the material found in hair and fingernails, is also found in the beaks of birds. This fossil was the first carnivorous dinosaur that showed evidence of a beak. Although it is an ornithomimid, and not a coelurosaurian, it shows that dinosaurs could make the transition from teeth to a beak-like structure. This wealth of recent evidence has led some paleontologists to speculate birds are not just very much like dinosaurs, or that birds and dinosaurs both evolved from some common ancestor—but birds are truly dinosaurs. To these scientists, birds are dinosaurs with adaptations, such as beaks, light weight, and agility, that enabled them to survive the catastrophe that wiped out other dinosaurs 60 million years ago and to evolve further. In essence, these scientists believe that the dinosaurs didn't go extinct—their descendants are all around us today.

7. Classification of Organisms

Introduction

The classification of organisms is not necessarily an important part of the Earth's natural history—but it is one of the best methods of better understanding all the organisms on our planet. It is, and will probably be, next to impossible to categorize all the species on our planet. Not only do tens of species go extinct every week, but many species remain hidden, living in remote areas or buried deep under soil. In the meantime, scientists are trying to get some idea overall of the organisms covering our planet—and the only manageable way to achieve this is by classifying them.

Timeline

Date	Event
Modern Events and Discoveries	
~1220	Scottish scientist Michael Scot (c.1175–1234) translates Aristotle's *History of Animals*, *Parts of Animals*, and *Generation of Animals* into Latin.
~1250	German scientist Albertus Magnus (c.1200–1280) introduces the work and methods of Aristotle to Europe by reading the translations of Scot; Magnus' further works classify European plants and vegetables into basic types.
1539	German scientist Hieronymus Tragus (Jerome Bock, 1498–1554) arranges plants by relation or resemblance, the first known attempt at a natural classification of plants.
1551	Swiss naturalist Konrad (also as Conrad) von Gesner (1516–1565) writes the first volume of *Historiae Animalium* (*The History of Animals*), helping to found modern descriptive zoology; it also organizes each known animal species into a classification system.
1554	Italian naturalist Ulisse Aldrovandi (1522–1604) proposes a systematic study of plant classification in his publication *Herbarium*.
1583	Italian physician and botanist Andrea Cesalpino (also Andreas Caesalpinus, 1519–1603) suggests a plant classification in his publication *De plantis*; it classifies plants according to roots and fruit organs.
1623	Swiss biologist Gaspard Bauhin (1560–1624) introduces the two-name system of classifying plants, one for genus, one for species, in his book *Pinax theatri botanici*; it is called binomial classification.
1660	Italian histologist Marcello Malpighi (1628–1694) tries to classify all living organisms on one scale, based on the relationship between the relative size of the organism's respiratory system and the level of the being; plants are at the bottom, and humans are at the top of the scale.
1694	German botanist Rudolph Jakob Camerarius (1665–1721) elaborates on Cesalpino's plant classification; he classifies them based on the nature and number of stamens and pistils in the flower.
1735	Swedish botanist Carl von Linné (Carolus Linnaeus) (1707–1778) introduces the first classification system to keep track of organisms on the Earth in his book, *Systema naturae*.
1737	Von Linné explains his method of systematic botany in the book *Genera plantorum* (*Genera of Plants*); he classifies 18,000 species of plants in the book.
1749	Comte de Georges-Louis Leclerc Buffon's (1707–1788) book, *General and Particular Natural History*, is published, written in his native French and subsequently translated; he also interprets the modern definition of the term "species."
1753	Von Linné uses binary nomenclature in botany in his book, *Species plantarum* (*Species of Plants*).
1788	English naturalist Gilbert White (1720–1793) writes *The Natural History and Antiquities of Selborne*, describing relationships between plants and animals.

Date	Event
1802	French philosopher Jean Baptiste Lamarck (1744–1829) divides the invertebrates into 10 classes and the vertebrates into four classes—fish, reptiles, birds, and mammals; his divisions are based on the nervous systems of the organisms and arranged in a linear fashion.
1806	French chemist Antoine de Fourcroy (1755–1809) classifies organic substances based on whether they are of vegetable or animal origin.
1816–1822	Lamarck modifies his early scale of organism classification, as he did not discover the evolutionary link between the invertebrate and vertebrate classes; this time he uses a branched arrangement, similar to the more modern "family tree" idea.
1864	German biologist Ernst Heinrich Haeckel (Häckel, 1834–1919) outlines the essential elements of modern zoological classification.
1895	The third meeting of the International Zoological Congress meets, appointing an International Commission on Zoological Nomenclature; it still classifies animals and interprets rules and procedures for naming them.
late 1900s	Cladistics becomes important in the study of organism classification.

History of Classifications

Early Classifications

Although humans had been observing animals for ages, some of the first more complex **classification** systems were developed in the thirteenth century. Around 1220, Michael Scot translated Aristotle's *History of Animals*, *Parts of Animals*, and *Generation of Animals* into Latin. Albertus Magnus read the translations and then, based on the volumes, introduced the methods of Aristotle to Europe. His treatises classified plants and vegetables into basic types and described the many structures and functions of parts of plants, including the root, stem, seed, flower, and leaf.

A long list of scientists, both in zoology and botany, proceeded to develop their own systems of classifying organisms. For example, in 1554, Ulisse Aldrovandi proposed a systematic study of plant classification in his *Herbarium*; Andrea Cesalpino also concentrated on plants in his *De plantis* (1583), classifying plants according to roots and fruit organs; and in 1694, Rudolph Camerarius elaborated on the plant classification of Andrea Cesalpino, basing his system on the nature and number of stamens and pistils of flowers. Gaspard Bauhin was the first to actually use binomial (two-term) classification based on plants—genus and species—although the divisions were not accepted as scientific nomenclature for several years.

But it was not until 1735 that Swedish naturalist Carl von Linné developed one of the most extensive classifications of organisms—one that is still in use today. When Von Linné was a young student, a problem plagued scientists: There was no agreed upon system for classifying living things. The reason this presented a problem was that more and more animals and plants were being discovered, and to keep track of them all some system had to be developed. Von Linné first looked at all living organisms and assigned those that were similar (by physical looks mostly) to one of several classes. Those with more detailed similarities within the class were divided into orders; then into genera; and finally, into species.

Von Linné's most important contribution to classification was to give every organism a unique Latin scientific name—the organism's genus and species in descriptive Latin terms. Called binary notation (or nomenclature), it is still used today. For example, the greater horseshoe-nosed bat's scientific name is *Rhinolophus ferrumequinum*, from *rhinos* (Greek for "nose"), *lophos* (Greek for "crest"), *ferrum* (Latin for "iron"), and *equinum* (Latin for "horse"). By using Latin (and often Greek) names, scientists across the world were able to communicate without confusion. For example, animals may have had different common names from country to county, but in the classification scheme, the Latin name was known across language barriers.

Not everyone agreed with von Linné's classification system. Comte de Georges-Louis Leclerc Buffon, for instance, did not believe that a system based on external characteristics of the organism would work. In de Buffon's mind, the best way to develop a classification system was to use the reproductive history of the organisms instead. For example, to determine if a deer is a different species than a zebra, de Buffon stated that if and only if the matings between a deer and zebra did not produce offspring or even sterile offspring would they be classified as a different species. Although this method is useful for determining individual species, it does not help group animals or plants that are similar and have similar ancestors but cannot breed. Needless to say, de Buffon's classification method was not adapted by the scientific community, but the modern idea that species are a group of organisms that can breed and create fertile offspring is based on de Buffon's ideas.

Around the same time as von Linné's work, other scientists were closely examining the Earth's organisms. Gilbert White, a clergyman known as the first English naturalist, simply observed nature and wrote about his discoveries. His book, *The Natural History and Antiquities of Selborne*, was one of the best early books on the Earth's natural history. His detailed analysis of the relationships between plants and animals was far ahead of its time, and was eventually used as a basis for future modern classifications schemes.

In 1889, the First International Zoological Congress met in Paris to agree upon a system for systematically naming the animals around the world. By 1895, after the third meeting of the International Zoological Congress, an International Commission on Zoological Nomenclature was appointed. The commission still oversees the classification of animals and interprets rules and procedures for naming them. The work of the commission is aided by regional organizations. In the United States, for example, the Committee on Classification and Nomenclature of the American Ornithologists' Union—a special committee of experts on the **taxonomy** of birds—assigns the names, both scientific and common, for North American birds.

Modern Classifications

In the past few decades, scientists have found many more organisms—and in trying to keep track of the animals, plants, and other organisms, they have turned to classifications based on evolution. Overall, not everyone agrees on a single classification scheme. (No doubt the reader will notice the different classification listings for the same animals throughout the text). For example, even the largest divisions of classification, the number of kingdoms, are debated. Some scientists divide their classification system into only four kingdoms and do not include the Kingdom Protoctista. (They classify these animals with the Kingdom Fungi.) Other scientists believe there are actually five kingdoms: the animal, plant, fungi, protists, and monerans. And still others divide the classification into anywhere between six to 20 kingdoms!

Today, classification is only part of the much larger field of phylogenetic systematics—reconstructing the pattern of events leading to the distribution and diversity of life. In general, classification is mainly the creation of names for groups, while systematics goes beyond this idea to develop new theories behind the mechanisms for evolution. In other words, systematics attempts to interpret the interrelationships between the evolution of living things—to interpret how life has diversified and changed over time. Many modern taxonomists are turning to classification systems based on phylogenetic systematics. Because such systematics are meaningful, not arbitrary, and are based on the evolutionary history of life, they are able to interpret and integrate the properties of newly discovered or poorly known organisms. Systematics are especially useful in studying the patterns of relationships among taxa.

Currently, scientists are using **cladistics**, a particular method of hypothesizing relationships among organisms based on their common evolutionary history. The basic idea is that animals that share an evolutionary history are "closely related" and should be grouped together. In particular, the groups are divided into those organisms that share unique features that were not present in distant **ancestors**.

And although cladistics sounds easy in theory, in practice the analysis is difficult and time consuming. For this reason, cladistic listings have not been completed for many facets of biology. And there is a great deal of controversy in certain interpretations within the cladistics method. In other words, analyses have been done, but not to all organisms on the Earth and not everyone agrees on those that have been done. While the cladistics method holds promise as a more agreed upon system, a great deal of work remains to be done in the classification of animals. *See also* AMPHIBIANS, ANIMALS, ARTHROPODS, BACTERIA AND VIRUSES, BIRDS, CYTOLOGY, DINOSAURS, EVOLUTION, EXTINCTION, FISH, FOSSILS, FUNGI, GENETICS, GEOLOGIC TIME SCALE, HUMANS, LIFE, PLANTS, PROTISTA, and REPTILES.

Topic Terms

ancestors—To explain what an ancestor is in biological terms, it helps to think of one of the visual results of cladistics, a cladogram, which is a branching diagram somewhat similar to a family tree. A family tree traces back our ancestry; for example, in your family, your father's parents would be the ancestors that gave rise to your father; from there, the family tree would list your father and mother as your parents. The tree branches from one person to the next—backward and forward in time—a genealogical roadmap of our past. Cladograms, like family trees, tell the pattern of ancestry and descent. But unlike family trees, ancestors in cladistics ideally result in only two descendent species. Also unlike family trees, new species form from the splitting of old species, not the joining of couples. In formation of the two descendent species, the ancestor "dies off" after the splitting takes place.

As humans, we all have ancestors, as do other species, and we can trace our lineage back to one set of ancestors. Simply put as a cladogram, sometime in the past an ancestral species (father/mother) of *Homo sapiens sapiens* walked the Earth. This ancestor went extinct (died), but left descendant species (children).

cladistics—Cladistics is a method that allows scientists to hypothesize relationships among organisms by examining their shared characteristics and evolutionary history; it is now accepted by most scientists as the best method to understand such relationships and to classify animals. The main idea behind cladistics is that animals which share common traits and a common evolutionary history are "closely related" and should be grouped together. Overall, the groups are separated out by unique features that were not present in distant ancestors.

But cladistics is really not that easy. It is not enough for organisms to share characteristics. In fact, two organisms may have many similar characteristics, but they will not be considered members of the same group. For example, consider a jellyfish, starfish, and human: Which two would be most closely related in cladistics? The jellyfish and starfish are both aquatic invertebrates with radial symmetry—but they do not belong to the same group. This is be-

The Species 2000 Programme

One way that scientists around the world are trying to keep up with all the species on our planet is called the Species 2000 Programme, a uniform and validated quality index of names of all known species for use as a practical tool. By using a combination of statistical and scientific data, the organizers—a "Federation" of database organizations working closely with users, taxonomists, and sponsoring agencies—hope that the project will produce four major results: (1) an electronic baseline species list for use in inventorying projects worldwide; (2) the index as an Internet gateway to species databases worldwide; (3) a reference system for comparison between inventories; and (4) a comprehensive worldwide catalog for checking the status, classification, and naming of species.

The overall Species 2000 Programme was established by the International Union of Biological Sciences (IUBS), in cooperation with the Committee on Data for Science and Technology (CODATA) and the International Union of Microbiological Societies (IUMS) in September 1994.

But the project really began in March 1996, as senior representatives of 18 taxonomic databases met in Manila, Philippines, to give the formal go-ahead to the Species 2000 Programme. The workshop was sponsored by the United Nations Environment Programme (UNEP) and the Global Environment Facility (GEF). Some of the taxonomic groups being addressed by the organizations starting the Species 2000 Programme were viruses, bacteria, corals, mollusks, crustacea, diptera, ichneumon wasps, moths and butterflies, curculionid beetles, fish, birds, mammals, fungi, cacti, palms, legumes, umbellifers, and fossil plants.

Why put such a list together? One major reason is the lack of data. According to the worldwide sponsors, they hope to provide the names of organisms that are the key to biodiversity communications—in other words, provide access to the accumulated knowledge of all life on Earth. Currently, despite the obvious value of a catalog, no comprehensive indexing system yet exists for the 1.75 million animals, plants, fungi, and microorganisms named by science. In fact, it is estimated that the existing global species databases may presently account for only some 40 percent of the total known species.

According to the treatise on Species 2000, this lack of a widely accessible index, with built in mechanisms for maintenance and updating, is a significant constraint on all nations wishing to fulfill their obligations under the United Nations' Convention on Biological Diversity. The Species 2000 Programme will enable users worldwide to verify the scientific name, status, and classification of any known species, with the information sent to the Clearing House Mechanism under the United Nations Convention on Biological Diversity run by several groups around the globe.

The result of this ambitious undertaking will be an extensive listing of all the earth's organisms, and operation of a dynamic Common Access System on the Internet through which users can locate a species by name across an array of online taxonomic databases. They also hope to produce a stable species index called the Species 2000 Annual Checklist, available on the Internet and on CD-ROM, to be updated once a year. The lists will also uncover some of the gaps we now have in species listings and will contain a system of links to each species' entry that will allow the user to find other information about that species (such as museum or herbarium links). For more information, link to (in North America) <http://www.atcc.org/sp2000/>.

cause cladistics also includes the evolutionary relationships of the organisms. In fact, in cladistic analysis, the starfish and human would be more closely related—as it is not just shared characteristics (features of the organism) that scientists use to determine a group, but the presence of shared derived characteristics.

Cladistics is based on three main assumptions: First, any group of organisms are related by descent from a common ancestor—a general assumption made for all evolutionary biology. It essentially means that life arose on Earth only once, and therefore all organisms are related in some way or other. Because of this assumption, we can take any collection of organisms and determine a meaningful pattern of relationships, provided we have the right kind of information. Again, the assumption is that all the diversity of life on Earth has been produced through the reproduction of existing organisms.

The next assumption is that organisms do split into other groups—one of the most controversial assumptions in cladistics. It means that new kinds of organisms can arise when existing species or populations divide into exactly two groups. But not everyone agrees: Some biologists believe multiple new lineages can evolve from a single original population at the same time (or close enough to seem to be the same time), but no one knows if it actually occurs. Another caveat is the possibility of interbreeding between distinct groups. But to this point, no method—cladistics or otherwise—accounts for the possibility of interbreeding.

Finally, the most important point of cladistics is that the characteristics of organisms change over time. Without theses changes, we would not be able to recognize different lineages or groups. Cladists call the original state of the characteristic "plesiomorphic" and the changed state "apomorphic." (The words "primitive" and "derived" were once used, but cladists avoid such terms as they were abused in the past.)

classification—Dozens of classifications are used in science—all differing greatly. No real right or wrong classification has emerged as of yet; at least, no classification is agreed upon by all the scientific community. This is probably why systematics is becoming the more prevalent method used to understand present and past organisms that have roamed our natural world.

The following list shows three different classification systems, based only on Kingdoms:

Four-Kingdom Classification

Kingdom Plantae	Plants
Kingdom Animalia	Animals
Kingdom Fungi	Multicelled organisms, mostly parasitic
Kingdom Monera	Bacteria

Five-Kingdom Classification

Kingdom Plantae	Plants
Kingdom Animalia	Animals
Kingdom Fungi	Multicelled organisms, mostly parasitic
Kingdom Monera	Archaebacteria, bacteria, photosynthetic blue-green algae
Kingdom Protoctista	Neither animal, plant, fungi, or prokaryotes

Six-Kingdom Classification

Kingdom Plantae	Plants
Kingdom Animalia	Animals
Kingdom Fungi	Multicelled organisms, mostly parasitic, usually from spores
Kingdom Prokaryotae	Certain bacteria and blue-green algae
Kingdom Protoctista	Single- and multi-celled eukaryotes, including some fungi
Kingdom Archaebacteria	Bacteria that produce methane

taxonomy—Taxonomy, or systematics, is one way to organize organisms—a hierarchical system of taxonomic (for example, kingdom) categories arranged in a descending series of ranks. In botany, the listing is usually divided into 12 ranks (listed in order): kingdom (some classifications also list a subkingdom), division, class, order, family, tribe, genus, section, series, species, variety, and form. In zoology, seven ranks are used: kingdom, phylum, class, order, family, genus, and species, with additional categories introduced using the prefixes "sub-," "infra-," or "super-."

The following lists how humans, tigers, and sweet peas are categorized according to taxonomy:

Human Taxonomy

Name	Taxonomic Level	Example Feature
Animalia	kingdom	animal
Chordata	phylum	spinal cord
Vertebrata	subphylum	segmented backbone
Tetrapod	superclass	four limbs
Mammalia	class	suckle young
Theria	subclass	live births
Eutheria	infraclass	placenta
Primates	order	most highly developed
Hominoidea	superfamily	humanlike
Hominidea	family	two-legged
Homo	genus	human
sapiens	species	modern human

Tiger Taxonomy			**Sweet Pea Taxonomy**		
Name	*Taxonomic Level*	*Example Feature*	*Name*	*Taxonomic Level*	*Example Feature*
Animalia	kingdom	animal	Plantae	kingdom	plant
Chordata	phylum	spinal cord	Anthophyta	class	angiosperms
Vertebrata	subphylum	segmented backbone	Dicotyledonae	subclass	dicotyledons
Mammalia	class	suckle young	Rosales	order	rosales
Carnivora	order	carnivores	Leguminosae	family	leguminous plants
Felidae	family	cats	*Lathyrus*	genus	pea
Panthera	genus	big cats	*odoratus*	species	sweet pea
tigris	species	tiger			

8. Climate and Weather

Introduction

Many people use "climate" and "weather" interchangeably, but they have specific meanings. The climate is the long-term weather conditions prevailing over a certain place. It is usually most controlled by geographic characteristics, including altitude above sea level and changes in elevation or amount of moisture above the locality. Deserts have hot, dry climates; the Amazon Rainforest has a hot, humid climate; and the Antarctic continent has a cold, dry climate. Weather is the short-term changes in air pressure, temperature, rainfall, and sundry atmospheric factors. Both climate and weather have been, and will continue to be, important to the natural history of our planet, especially the connection between climate and how life survives in certain localities because—or even in spite—of the climate and weather.

Timeline

Date	Event
Modern Events and Discoveries	
43 A.D.	Roman geographer Pomponius Mela (c.10 B.C.– ?) divides the Earth into five climate zones—North Frigid, North Temperate, Torrid (equatorial), South Temperate, and South Frigid.
late 1400s	Italian scholar, artist, engineer, and inventor Leonardo da Vinci (1452–1519) keeps journals that contain numerous studies of weather phenomena and designs for meteorological instruments, including a hygrometer (to measure humidity).
~1592	Italian mathematician and astronomer Galileo Galilei (1564–1642) develops what he calls a thermoscope, the first thermometer; he leaves no records of any observations he may have made with the instrument.
1644	Italian mathematician and physicist (and the father of hydrodynamics) Evangelista Torricelli (1608–1647) constructs the first barometer.
1644–45	Reverend John Campanius Holm, a Lutheran minister living in the American colonies, makes the first recorded weather observations in the Western Hemisphere, near present-day Wilmington, Delaware.
1654	The Grand Duke Ferdinand II of Tuscany, Italy, sets up the first meteorological observation network.
late 1600s	English physicist Robert Hooke (1635–1703) uses mercury in a barometer and is believed to be the first to use the words, "very dry," "clear," "variable," "rain," and "stormy," on a barometer; he also invents the rain gauge.
1730s	A Scottish-born resident of Charleston, South Carolina, John Lining, takes the first thermometer and barometer measurements in the Western Hemisphere.
1740s	The first pressure-tube wind gauges are invented.
1752	U.S. author, inventor, scientist, and diplomat Benjamin Franklin (1706–1790) proves that lightning is an electric discharge through an experiment he conducts in France.
1753	Franklin further proves that lightning is a electric discharge in a famous experiment, in which he launches a kite with a key attached; that same year, he presents the lightning rod, which conducts lightning bolts from buildings and allows them to pass harmlessly to the ground.
1781	The Mannheim Society establishes a large network of weather observatories in Russia, Europe, Greenland, and North America.
1802	English scientist John Dalton (1766–1844) demonstrates that the amount of water vapor varies with temperature, eventually leading to the concepts of relative humidity and vapor pressure.
1828	German meteorologist Henrich Dove (1803–1879) develops the science of climatology.
1830s	The development of the telegraph by Samuel Morse (1791–1872) makes the rapid collection and analysis of weather observations more practical.
1849	Secretary of the Smithsonian Institution, Joseph Henry (1797–1878), links the first meteorological network by telegraph in the United States.

Date	Event
1853	The first International Meteorological Conference is held in Brussels, Belgium.
1857	Christoph Hendrik Diederik Buys Ballot (1817–1890), a Dutch meteorologist, discovers a new rule for determining areas of low pressure based on wind direction, which is now called the Buys Ballot law.
1863	English anthropologist and explorer Sir Francis Galton (1822–1911) publishes *Meteorographica*, which introduces modern weather mapping techniques.
late 1800s	U.S. oceanographer Matthew Maury (1806–1873), as head of the Navy's Depot of Charts and Instruments, prepares charts showing the weather patterns over the oceans, as well as favorable sea routes; his efforts reduce time spent at sea by ships.
1870	The U.S. Weather Bureau is established, mostly due to the efforts of meteorologist Cleveland Abbe (1838–1916).
1873	The International Meteorological Organization is founded, partly due to the influence of Maury; it becomes the World Meteorological Organization in 1950.
1896	Swedish scientist Svante Arrhenius (1859–1927) determines that carbon dioxide is important to the Earth's heat balance and that more of the gas would increase the average global temperature.
1899	U.S. climatologist Thomas Chamberlin (1843–1928) confirms Arrhenius' theory about how carbon dioxide increases affect the Earth's global temperatures.
1904	The first study that takes a scientific approach to weather forecasting, *Weather Forecasting as a Problem in Mechanics and Physics*, is published by Norwegian meteorologist Vilhelm Friman Koren Bjerknes (1862–1951).
1919–1936	U.S. scientist Andrew Douglass (1867–1962) publishes his *Climatic Cycles and Tree Growth*, connecting tree rings to climate changes and showing how they can be used for age dating.
1920s	Bjerknes and his son, American-Norwegian meteorologist Jacob Aall Bonnevie Bjerknes (1897–1975), prove that the atmosphere is made up of air masses of different temperatures, with the sharp boundary between them called fronts.
1932	Swedish-born Carl-Gustav Arvid Rossby (1898–1957) announces the Rossby diagram for understanding air masses; he also discovers the Rossby waves, the wave patterns of westerly air flow in the upper troposphere.
1940s	Jacob Bjerknes discovers the jet stream.
1955	Computer weather forecasts are being made on a regular basis.
1957	Researchers in 70 countries cooperate in a coordinated study of the Earth and its atmosphere during the International Geophysical Year.
1960	The first polar-orbiting weather-observing satellite, TIROS 1 (Television Infrared Observation Satellite), is launched.
1966	The first geostationary satellite is launched in the United States.
1976	The first Geostationary Operational Environmental Satellite (GOES) is launched to study the Earth's atmospheric interactions.
1984	NOAA-9 becomes the first satellite to measure a minor atmospheric constituent when it is launched; the mission of this satellite was to monitor ozone.
1990s	Some of the first satellites using advanced microwave sounding units to track storms are launched.
1993	Studies of ice at the bottom of Greenland's ice sheet reveal that the Earth's climate has a habit of changing fast—even within decades—and is therefore less stable than previously thought.

History of Climate and Weather

Early Climate and Weather Studies

From the days of early humans, climate and weather have been important—but in different ways than today. Weather was no doubt watched with superstition. Without the knowledge of the mechanisms behind the weather machine, phenomena such as lightning were feared, and storms were thought to be precursors to trouble or bad times, or a punishment from a higher authority.

Some early meteorologists did try to define climate and weather based on the science of the day. In his book *De situ orbis* (*A Description of the World*), Roman geographer Pomponius Mela divided the Earth into five climate zones: the North Frigid, North Temperate, Torrid (equatorial), South Temperate, and South Frigid. He believed the torrid region was so hot that people could not cross the zone to get to the south—and thus peoples on either side would never know about one another.

By the end of the Middle Ages, the interest in and studies of Earth's climate and weather grew. Many new instruments were invented, some of which are still used to this day. (At least the basic concepts have not changed.) In 1644, Italian mathematician and physicist (and the father of hydrodynamics) Evangelista Torricelli, along with his colleague Vincenzo Viviani, showed that atmospheric pressures cause a fluid to rise or fall in a tube when it is inverted over a saucer of the same liquid—the first barometer.

The eighteenth century also brought on a new interest in weather observation, mainly by people settling in the New World. Some of the first recorded weather observations were made by the Reverend John Campanius, near present-day Wilmington, Delaware; Benjamin Franklin was a keen weather watcher and the inventor of the lightning rod. Even the presidents were involved in the study of weather: Thomas Jefferson and James Madison made the first simultaneous weather observations in North America; and George Washington kept a regular weather diary. (In fact, his notes about the weather for December 13, 1799, are thought to be the last words he ever wrote.)

Because of the importance of weather, especially to farmers, several meteorological networks were set up in Europe under patronage societies, such as the Royal Society in Britain. In 1723, observations were being received from England, continental Europe, North America, and India. The most famous and significant network—especially for its role in establishing procedures for the emerging weather forecasting field—was founded by the Mannheim Society. In 1781, the society had 14 stations; it gradually grew to 39 observatories in Russia, Europe, Greenland, and North America, but it was closed in 1799.

Daily weather studies—in particular forecasting the weather by rapidly collecting and analyzing data—increased dramatically in the 1830s. One major reason for this increased awareness in weather forecasting can be attributed to the use of the telegraph in the United States. This invention allowed the exchange of weather information on a wide—not local—scale. Although predictions were still not accurate, the exchange of data helped build records of weather events and sequences all over the country—and as more telegraph cables were laid all around the world.

Progress was made in weather studies elsewhere, too. For example, in 1802, English scientist John Dalton demonstrated that the amount of **water vapor** required to saturate the air varies with temperature, a discovery that lead to the concepts of relative humidity and vapor pressure. In 1857, Dutch meteorologist Christoph Hendrik Diederik Buys Ballot discovered a new rule for determining low pressure using wind direction. He formulated the Buys Ballot law, which states that in the Northern Hemisphere, if people stand with their back to the wind, low pressure lies to their left and high pressure to their right. (The reverse is true in the Southern Hemisphere.) In 1863, English anthropologist and explorer Sir Francis Galton published *Meteorographica*, which introduces modern weather mapping techniques. Galton was also the first to propose the concept of anticyclones, the high pressure systems in the atmosphere. But much of the data collected by observers was just stored, with the majority of people relying on local folklore for weather forecasting.

Modern Climate and Weather Studies

By the turn of the twentieth century, meteorologists had worked out many major intricacies in weather and weather prediction. In the 1920s, Vilhelm Bjerknes and his son, Jacob Aall Bonnevie Bjerknes, proved that large groupings of the atmosphere with different temperatures form **air masses,** which have sharp boundaries between them called **fronts.** By 1922, Lewis Fry Richardson proposed the idea of numerical weather forecasting; his first calculations to forecast the weather 24 hours in advance would take months to complete. In the 1930s, one of the most influential meteorologists of the twentieth century, Carl Gustav Rossby, carried out pioneering work on the general circulation of the atmosphere and air mass properties.

Other important discoveries followed, including the discovery of worldwide **jet streams**—high atmospheric **winds**. The existence of these winds was postulated in the early 1940s—but the actual discovery was made serendipitously: At the end of 1944, when high-flying U.S. B-29 aircraft flew over Japan during World War II, pilots found these west-to-east moving rivers of fast air, as more than 100 planes were sent north from Saipan to bomb industrial sites near Tokyo. During the approach to the city, at a height of about 6 miles (10 kilometers), the airplanes' normal speeds were altered. In fact, the as-yet-unknown jet stream pushed the planes at speeds of up to 150 miles (241 kilometers) per hour—which would have carried nearly all the bombs out to sea if they had been dropped.

From there, computers and satellites became an important part of weather forecasting and continue to be to this day. By 1955, computer forecasts in the United States were being conducted on a regular basis. Even more discoveries about the climate and weather were found as Earth-orbiting satellites were launched into space. The first polar-orbiting satellite TIROS-1 (Television Infrared Observation Satellite) was launched in 1960 and took 23,000 images of Earth and its clouds. The value of satellites was soon known: In 1961, satellite images of the movement of cyclone Clara led to the evacuation of more than 350,000 people along the Gulf of Mexico. By 1963, the first images from satellites were obtained; and by 1966 the first geostationary satellites—in which the craft's speed matches the Earth's rotation, so it is always over the same point on the surface—were launched. Today, we are especially in debt to the Geostationary Operational Environmental Satellites (GOES), which watch the United States and produce the images we see on daily weather maps. They are especially useful for tracking the Pacific and Atlantic Ocean **tropical cyclones** that affect North America.

Weather prediction seemingly improves—and continues to improve—with each advance in technology. One new technique is radar—and in particular Doppler radar, which helps in the detection of **lightning**, **tornadic-type winds**, and sundry other violent storms and **precipitation** events. Modern weather prediction is not perfect, but it is estimated that meteorologists are correct in their interpretations about 85 percent of the time. The Internet has also become a prime spot for locating information on slow- and quick-moving storms—available to anyone with a modem and powerful enough computer.

Although weather forecasting has improved in the past few decades—even in terms of our daily forecasts—we still have a long way to go when it comes to completely understanding our weather and the climate. Today, studies of the weather and climate involve the general circulation of the atmosphere, a major task that can only be carried out using computers. The use of parallel processing and supercomputers are necessary to do the "number crunching" of so many weather variables—sometimes performing more than a billion calculations per second. These GCMs (general circulation models) are used to produce a set of forecasts quickly. Supercomputers also are being used to determine the future of our climate—especially to understand the effects of humans and their by-products on the global atmosphere—by analyzing data collected from Earth-orbiting satellites.

And although not every mechanism is known, we do know a great deal more about why weather and climate changes occur. For example, no one actually knows how **droughts** occur, but computer models of shifting climate patterns have allowed scientists to determine where, when—and often why—certain droughts will occur. *See also* ATMOSPHERE, EARTH, and OCEANS.

Topic Terms

air masses—Large groupings of air in the atmosphere that have common characteristics. These air masses, or systems, can contain cold or warm air, and dry or moist air. An air mass with cold, dry air is known as a high pressure system; low pressure systems have warm, moist air. The boundary between these air masses is known as a front, and the movement of air from high pressure to low pressure air masses causes winds. The properties of an air mass are derived partially from the region over which it passes, as seen in the following chart:

Major Air Masses—United States		
Air Mass	*Origin*	*Effects*
arctic	winter	very cold temperatures, very dry air
continental polar	winter	cold temperatures, dry air
	summer	cool temperatures, variable humidity
maritime polar	winter	mild temperatures, moist air
	summer	cool temperatures, moist air
continental tropical	summer	hot temperatures, dry air
maritime tropical	winter	warm temperatures, moist air
	summer	very warm temperatures, moist air
equatorial	summer	very warm temperatures, moist air

drought—A prolonged period of below-average rainfall over a wide area. This term is relative, based on the normal amount of rain expected for a specific region. These are naturally (normally) occurring events that happen irregularly, although some scientists believe they can be linked to other periodic natural events. For example, droughts over the Americas and Australia have been linked to sea-surface temperature patterns across the Pacific and along the west coast of South America, including one of the most well-known patterns, El Niño. Such links may eventually pave the way for long-range forecasting of droughts.

The Palmer Drought Severity Index compares the balance between incoming water, such as rainfall, and stored moisture in the soil with outgoing water in the form of evaporation, absorption, and transpiration of soil moisture by plants. This information is then compared against climatic data. The area's water level is expressed as a negative or positive number; the larger the negative number, the more severe the drought.

Severe droughts are more than just a scientific curiosity. Such drought conditions may have been responsible for the decimation of thousands of species over Earth's natural history. Scientists believe that changes in the positions of the continental plates may have led to drought conditions in certain regions. If this was the case, die-outs of certain species unable to adapt or move on to other territories were probably common. Today, the consequences of a prolonged drought is catastrophic, including decimated crops and livestock, ruined economies, and widespread famine.

fronts—Fronts are boundaries, or leading edges, between air masses. The four major types of fronts are as follows:

cold front—When a cold, dry air mass (a high pressure system) moves into an area of warm, moist air, a cold front is said to be approaching. The heavier, drier air forces the lighter, moister air to rise rapidly, forming clouds and potentially thunderstorms along the boundary. As the rest of the high pressure air mass moves in to an area, the weather clears and fair skies dominate.

warm front—An approaching warm front (low pressure system) gradually rises over the cold, dry air in a region, leading to cloud formation and rain over a wide area. The first indication of an approaching warm front is usually high cirrus clouds.

stationary front—A stationary front is just that—one that does not move very quickly.

occluded front—An occluded front occurs when cold, warm, and cool air run into each other, leading to the formation of warm or cold occlusions. Rain usually occurs where the cool and cold air meet.

hurricanes—See tropical cyclone

jet streams—The jet streams are currents of fast-moving air located high in the atmosphere that separate the warm flow of air (usually from the midlatitudes) from the cold air (usually from the poles). These currents steer air masses and change weather patterns. They are usually found at about 30,000 to 35,000 feet above the Earth's surface (in the upper and middle troposphere), and may reach speeds of 180 miles (300 kilometers) per hour. They are the result of significant temperature and pressure differences between the hot and cold air and are normally in the shape of flattened tubes. Although they travel in a general west-to-east direction in both hemispheres, they can temporarily swoop northward and southward.

lightning—Lightning is the explosive release of electrical energy in a thunderstorm cloud. During a thunderstorm, a cloud can build up an extremely large electrical charge due to the action of air currents. This charge is transferred through the cloud by raindrops, hailstones, and ice pellets. As a result, a large negative electrical charge accumulates in the lower part of the cloud, while a large positive charge accumulates in the upper part.

Opposing electrical charges are strongly attracted to each other; lightning is the reaction that neutralizes these charges and can take on many forms. For example, when the insulating qualities of the air between the negatively charged cloud bottom and the positively charged ground are no longer sufficient to keep these charges apart, a cloud-to-ground lightning discharge occurs. The negative charges move toward the ground, tree, or lightning rod in an invisible, jagged pattern called a stepped leader. When connection is made, a massive flow of positive current flows back to the cloud, becoming visible as the familiar lightning bolt. Other forms of discharges are cloud-to-cloud and in-cloud lightning. The bolts of lightning travel quickly—about 60,000 miles (96,000 kilometers) per second—and can heat the surrounding air to around 54,000° F (30,000° C). It is this sudden heating and expansion of the air that creates the thunder that accompanies lightning.

precipitation—Precipitation is defined as water that is heavy enough to fall from clouds to the ground. Water vapor condenses into clouds; then, depending on the processes that occur within the clouds, and the air temperature between the clouds and ground, the water may reach the ground as drizzle, rain, freezing rain, dry or wet snow, or a combination of these. The two main processes within clouds are the coalescence of small water droplets into larger ones that fall to Earth as rain, and the ice-crystal process that forms snow.

The importance of precipitation to the Earth's natural history is obvious: The rains carry much needed water to plants and animals on Earth, with runoff from the rains carrying nutrients into rivers and land, helping plants to grow. Precipitation is also instrumental in the deposition of sediment and the erosion of bedrock and other sediment over time. Lack of precipitation also creates its own conditions, such as the stark deserts around the world.

A wall of rain falls from this huge cumulonimbus cloud, as it travels over a lake in Texas.

tornadic-type winds—There are many tornadic-type winds—strong swirling winds that travel across the surface.

dust devil—Very localized tornadic-type winds are dust devils, a dust-filled, upward-spiraling funnel of air normally occurring in desert or semi-arid regions. Although dust devils resemble tornadoes, they are much smaller, less intense, and rarely cause damage. Winds create a rotating air mass in the atmosphere that combines with strong updrafts produced by the heated ground to create a rising funnel of air; this funnel draws up dust particles, which makes it visible.

waterspouts—A rapidly rotating funnel of air over water. Most waterspouts are nontornadic in nature, caused by a rotation of an air mass over water combined with an updraft. The water in the vortex is due to condensation, not a drawing up from the lake or ocean below. True tornadic waterspouts are much rarer and are essentially tornadoes over water.

tornado—A tornado is a spinning funnel of extremely fast-moving air that extends from the base of a cumulonimbus (thunderstorm) cloud to the ground. Tornadoes are an extremely violent phenomena, with the potential for large amounts of property damage, injuries, and deaths. They are often called twisters.

Unpredictable in nature, the reasons for tornado formation are still being studied. Tornadoes can occur as isolated incidents or in great numbers along a storm front. In extreme cases, they are capable of generating winds of more than 300 miles (483 kilometers) per hour and may travel over 200 miles; the average tornado is much weaker, lasting 5 to 10 minutes on the ground and traveling 2 to 5 miles. The swatch they normally cut on the ground is 0.125 to 1 mile (0.2 to 1.6 kilometers) wide, although the largest ones have much wider paths. The strength of tornadoes is measured by the Fujita and Pearson Tornado Scale, which is reproduced below.

The United States is the most tornado-prone country in the world, having approximately 750 yearly, with most occurring in the area of the Great Plains known as Tornado Alley. Australia also sees regular occurrences of tornadoes; and occasionally they are found in other parts of the world, such as the United Kingdom and China.

The Fujita and Pearson Tornado Scale		
Scale Number (ranking)	Wind Speed (miles per hour [kilometers per hour])	Damage
F-0	up to 72 [116]	Light
F-1	73–112 [117–180]	Moderate
F-2	113–157 [181–253]	Considerable
F-3	158–206 [254–331]	Severe
F-4	207–260 [332–418]	Devastating
F-5	261–318 [419–512]	Incredible
F-6	319–380 [513–611]	Theoretical, never observed (although such speeds were clocked in a 1999 tornado; the winds were higher up, not at the surface)

tropical cyclone—A tropical cyclone or hurricane (also known by other names outside of North America) is the pinnacle of a progression of weather events. The following list details the stages of a hurricane formation:

tropical depressions—Tropical depressions are areas of low pressure that form over tropical ocean regions, in which the Sun has heated the water surface to at least 80° F (27° C). Huge masses of moist air start to ascend and, if this happens at least 5 degrees from the equator, rotate. This rotation of the air mass is due to the Earth's rotation and is called the Coriolis effect. The rising air, as it ascends, condenses and releases energy that strengthens the system. This cycle of rising and condensing fuels the growth of the storm system, increasing the amount of wind and clouds and creating an area of low pressure called an eye. The sustained winds in a tropical depression, near the surface, are less than 39 miles (63 kilometers) per hour.

tropical storm—As a tropical depression increases in intensity and the winds grow stronger than 39 miles (63 kilometers) per hour, it is classified as a tropical storm. If the atmospheric conditions are favorable (warm, moist rising air), a tropical storm may lead to a hurricane, which has sustained winds of over 74 miles (119 kilometers) per hour.

hurricane—A hurricane is an intense, long-lived storm system that occurs in the Atlantic Ocean and the eastern Pacific Ocean. Identical storms are called cyclones when they occur in the Indian Ocean, typhoons in Southeast Asia, and willy-willys off the coast of Australia. Hurricane wind speeds exceed 74 miles (119 kilometers) per hour in any part of its weather system. These major storms have the potential for sustained winds of over 150 miles (250 kilometers) per hour and gusts up to 190 miles (118 kilometers) per hour. These winds, combined with the rain and ocean surges, give hurricanes the potential to cause large amounts of property destruction and numerous deaths. Unlike most severe weather, which tends to be short-term in nature, hurricanes can last for weeks and move over large distances. As with the original tropical depression that it formed from, the hurricane's characteristic shape, rotation, and movement are all a consequence of Earth's rotation, also known as the Coriolis effect. The storms have a clear, calm area at the center known as the eye; hundreds of bands of thunderstorms spiral around the eye and may extend up to 300 miles (483 kilometers) or more from the center.

Over past history, hurricanes have no doubt caused major changes along the natural coastlines in the oceans in which they form. Modern hurricanes not only erode the coastline, but often destroy structures and cause loss of life. This damage is mainly caused by the violent hurricane winds, which create huge waves and storm surges along the coasts, along with torrential rains that can flood coastlines. In addition, if a hurricane strikes a coastline at high tide, even more destruction from waves can be experienced.

water vapor—Water vapor, a gas present in the atmosphere, is made up of water molecules consisting of one oxygen atom and two hydrogen atoms. The majority of the atmosphere's water vapor comes from the evaporation of water from the oceans.

Saffir-Simpson Damage Potential Scale for Hurricanes		
Scale Number	Winds	
	(miles per hour)	[kilometers per hour])
1 (minimal)	74–95	[119–153]
2 (moderate)	96–110	[154–177]
3 (extensive)	111–130	[178–209]
4 (extreme)	131–155	[210–249]
5 (catastrophic)	>155	[>249]

The atmosphere can hold varying amounts of water vapor, depending on the temperature: The higher the temperature, the more water vapor that can be held. There is a specific limit to the absolute amount of water the atmosphere can hold at each temperature, above which no more can be held. This limit is known as the saturation point or dew point. Any addition of water vapor above this amount will result in the formation of liquid water, a process called condensation.

Condensation will also occur if the temperature of the air mass falls below the dew point. Since the temperature is lower, the atmosphere can no longer hold the same amount of water vapor that it could at high temperatures; this excess water vapor will condense, and depending on conditions, form a variety of familiar phenomena such as clouds, rain, fog, snow, hail, or frost.

wind—Wind is the movement of the air in the atmosphere from high pressure to low pressure areas. Cool, dry air, being heavier, sinks, leading to higher pressure while warm, moist air, being lighter, rises, leading to low pressure. The atmosphere is constantly working to restore equilibrium, so the air in the higher pressure areas flows to the lower pressure areas; this flow is known as the wind. The wind always flows in this high to low pressure direction, never the reverse. The greater the difference in pressure, the greater the flow and the stronger the wind. This difference in pressure is known as the pressure gradient force.

Localized winds are important to a specific region. For example, breezes around oceans and larger lakes and rivers form because of differences in pressure set up by temperatures differences. On a warm day, a sea breeze occurs as the winds flow from the water to the land, in response to the land heating up faster than the water (onshore breeze). A land breeze occurs at night, as the winds flow from the land to the water, in response to the land cooling and the water retaining the heat from the day (offshore breeze). Even more localized winds can occur. For example, the Bora winds occur in the winter months, as a cold mountain wind flows into the warmer regions along the Mediterranean coast.

Wind, similar to water (or precipitation) has always been an important process in Earth's natural history. The winds blow particles of sediment and sand, carving the landscape in the process of erosion and depositing sediment elsewhere. Winds carry seeds of certain plant species, ensuring the continuation of the species.

9. Cytology

Introduction

Cytology is the study of cells, which are the structural units of all organisms. They are usually divided into animal and plant cells, and encompass a range of sizes and shapes, depending on the species. The importance of the cell in the natural history of the Earth is obvious: Without the presence of cells, no life—plant, animal, or in between—could exist on our planet. Thus, cytology is vital to understanding Earth's long natural history of organisms.

Timeline

(note: bya=billion years ago)

Date *Event*

Prehistoric Events

~4 bya — Something sparks the first inorganic chemicals to develop into primitive "cells." Scientists are still unsure of what that "something" was.

~3.5 bya — Biomolecules begin to gather and somehow begin to self-replicate.

~2 bya — The first cells, prokaryotic, begin to form and develop into simple unicellular microorganisms.

~1.5 bya — The first eukaryotic cells form; they have a nucleus and complex internal structures; they are the precursors to protozoa, algae, and all multicellular life.

Modern Events and Discoveries

1590 — Dutch instrument maker Zacharias Janssen (1580–c.1638) invents the compound microscope, a tube with lenses at both ends (see 1609); it is an important turning point in the study of organisms' cells.

1609 — Dutch spectacle maker Hans Lippershey (c.1570–c.1619), independently from Janssen (see 1590), invents the compound microscope.

~1660 — Dutch microscopist and zoologist Anton van Leeuwenhoek (1632–1723) invents the single-lens microscope; he discovers the one-celled animals called protozoa, as well as sperm cells.

1665 — English biologist Robert Hooke (1635–1703) publishes the first drawings of cells; he is also the first to use the word "cell" to describe the living structures he sees under a compound microscope.

1824 — French biologist and physiologist René Joachim Henri Dutrochet (1776–1847) further advances the cell principle.

1831 — Scottish botanist Robert Brown (1773–1858) finds the cell nucleus in plants.

1835 — Czech physiologist Johannes Evangelista Purkinje (1787–1869) shows all animal tissues have cells.

1838 — German botanist Mathhais Jakob Schleiden (1804–1881) publishes a cell theory as it applies to plants, although he does not use the term in his thesis.

1839 — Drawings of German physiologist Theodor Schwann (1810–1882) help to prove cells make up all organisms; he also introduces the term "cell theory."

1846 — Hugo von Mohl (1805–1872), a German biologist, discovers protoplasm—now called cytoplasm—the main living substance in a cell.

1847 — Schleiden and Schwann publish, *Beiträge zur Phytogenesis*, the basis for modern cell theory; they believe cells are the fundamental organic units common to all life.

1858 — German pathologist Rudolph (Rudolf) Carl Virchow (1821–1902) states that all cells arise from other cells; he also believes that disease occurs when cells do not cooperate with each other (true of some diseases like cancer, but not of infectious diseases).

Date	Event
1866	German biologist Ernst Heinrich Haeckel (Häckel, 1834–1919) hypothesizes that the nucleus of a cell transmits its hereditary information.
1875	Polish-born German botanist Eduard Adolf Strasburger (1844–1912) and German cytologist Walther Flemming (1843–1905) discover the mechanism of cellular division and associated movement of the chromosomes.
1879	German biochemist Albrecht Kossel (1853–1927) studies nuclein, a substance in cells that leads him to discover nucleic acids.
1882	Flemming names the cell substance, nucleus, and cell division; he also discovers the process undertaken by chromosomes known as mitosis.
1882	Strasburger names cytoplasm and nucleoplasm in the cell.
1887	Edouard-Joseph-Louis-Marie van Beneden (1846–1910), a Belgian biologist, finds that each species has a fixed number of chromosomes; he also discovers haploid cells (which contain half of the chromosome pairs).
1888	German zoologist Theodor Heinrich Boveri (1862–1915) discovers the centrosome in the cell, which appears to control the process of cell division.
1901	Dutch plant physiologist and geneticist Hugo de Vries (1848–1935) discovers mutations, or how changes in a species occur in jumps.
1902	U.S. geneticist Walter Stanborough Sutton (1877–1916) discovers that chromosomes are paired and are the carriers of heredity.
1902–03	Sutton and German zoologist Theodor Heinrich Boveri (1862–1915) point out the parallelism between chromosome behavior and Mendelism, closing the gap between cytology and heredity.
1903	Sutton proves that sperm and egg cells have one of each pair of chromosomes.
1904	Santiago Ramon y Cajal establishes the theory that the nervous system is composed only of nerve cells.
1911	U.S. geneticist Thomas Hunt Morgan (1866–1945) maps the location of genes on the chromosomes of fruit flies.
1936	DNA (deoxyribonucleic acid) in its pure state is isolated for the first time.
1941	U.S. geneticist George Wells Beadle (1903–1989) and U.S. biochemist Edward Lawrie Tatum (1909–1975) discover that genes control the chemical reactions within the cells.
1956	Romanian-born U.S. cell biologist George Emil Palade (1912–) discovers ribosomes within the cells; eventually it is determined that proteins are manufactured in the ribosomes.
1967	U.S. biochemist Arthur Kornberg (1918–) and his colleagues synthesize biologically active DNA.
1974	The cytoskeleton (skeleton structure of a cell) is uncovered for the first time.
1990	The first genetically engineered cells are infused into a human for therapeutic purposes—genetically engineered white blood cells are injected into a four-year-old girl with an inherited immune disorder.

History of Cytology

Early Cytology

No one really knows what caused the first cells to form on early Earth more than 4 billion years ago. These early cells are thought to have come from a watery soup of organic compounds, often referred to as the "primordial soup." But more recently, some scientists have begun to think that the evolution of cells was not that easy; many researchers believe that the first cells may have developed deep in the oceans, at hot, volcanic vents, where life was recently discovered. Still others cite the possibility that the first cells were brought to Earth by comets or asteroids, the huge space bodies dropping organic compounds throughout the solar system. However the first cells began, the conditions on Earth were right—with the "correct" amount of water, ice, and temperatures—and life eventually flourished.

The first cells on early Earth were no doubt extremely primitive and similar to today's bacteria. The cells had no nucleus and the DNA was carried loosely throughout the cell. For more complex life to develop, some mechanism had to allow the cells to stick together—to eventually evolve into multicelled organisms and replicate. Numerous theories abound as to how this process could happen, most of them having to do with the rocks or other sites that would attract the cells and allow the small organic molecules to begin to **self-replicate**.

The next step was the accumulation of cells—especially important for multicelled organisms. About 1.5 billion years ago, eukaryotic cells (with a nucleus and complex internal structure) evolved. From these cells, single-celled protozoa and algae evolved—and eventually all multicellular life.

Modern Cytology

The study of cells emerged in the seventeenth century. Before that time, scientists knew that organisms were made of tissues and organs, but believed that they were composed of nonliving materials. Independently, and around the same time (the turn of the seventeenth century), Hans Lippershey and Zacharias Janssen invented the compound microscope—the beginnings of cellular discovery. By 1665, biologist Robert Hooke focused a microscope on cork. He saw something that he thought looked like the cells in a monastery, and he named them cells. Although finding these cells was a big discovery, Hooke did not interpret their importance to the lifecycle of organisms. In fact, all Hooke was viewing were the thick cell walls of the cork.

Even though the compound microscope had been around since the turn of the seventeenth century, scientists did not truly begin to interpret details about cells until the nineteenth century—a sort of revival of microscope studies, especially in Germany. But even as advances were made, the actual nature of cells was still in debate, and several theories abounded. One of the major ideas was the "globule theory," popular in the 1820s, in which animal tissues were thought to be made of protein globules. This misconception was probably due to the distortion of the cells' appearance under the microscope, as many of the more powerful instruments did not correct for spherical aberration—a lens problem that can distort the object and make it look much bigger than it really is.

One of the major advances in modern cellular studies came in 1824, when Henri Dutrochet further advanced the idea of cells as the major building blocks of life. He stated, "All organic tissues are actually globular cells of exceeding smallness, which appear to be united only by simple adhesive forces; thus all tissues, all animal (and plant) organs, are actually only a cellular tissue variously modified. This uniformity of finer structure proves that organs actually differ among themselves merely in the nature of the substances contained in the vesicular cells of which they are composed."[1]

Other advances soon followed. Botanist Robert Brown noted the nucleus ("little net") in plant cells in 1831, and Johannes Evangelista Purkinje showed that all animal tissues had cells. But none of these discoveries attracted much attention in the scientific world. It wasn't until 1838, when Matthias Jakob Schleiden introduced the idea that all plant tissues are made of cells, that the true study of cytology took hold—about 150 years after Hooke discovered cells. Theodor Schwann was next to make a breakthrough, noting that tissues of organisms were composed of cells; he also discovered that eggs were actually cells, and he believed that the life of all organisms started as a single cell. Thus was born the cell theory, based on the findings of these two men, Schwann and Hooke. Their cell theory was defined by three rules: (1) cells are the fundamental units of life; (2) cells are the smallest entities that can be called living; and (3) all organisms are made up of one or more cells.

By 1875, botanist Eduard Adolf Strasburger and cytologist Walther Flemming discovered the mecha-

---

[1] Dutrochet, R.J.H., 1824, *Recheres Anatomiques et Physiologiques sur la Structure Intime des Animaux et des Vogotaux et sur leur Motilit, Paris.* The quote is from a translated book by Michael Morange, *A History of Molecular Biology,* Cambridge, MA: Harvard Univ. Press, 1998.

nism of **cellular division** and associated movement of the chromosomes. From there, other scientists began to discover other parts of the cells and how cell division creates new cells. New techniques also helped to advance cellular studies. For example, in 1881, after German bacteriologist Paul Ehrlich discovered that staining of bacteria with methylene blue would emphasize internal structures, cytologists were able to discern cellular structures more clearly.

Other cellular discoveries continued through the twentieth century. For example, in 1901, the role of mutations was discovered, as the fruit fly provided a look at how its four chromosomes reacted to sudden changes in the transfer of inherited characteristics. (Chromosomes were discovered in 1882 by Walther Flemming, allowing the connections between the chromosomes and genes to be better understood.)

Many of these subsequent studies were also inevitably tied to advances in technology. In particular, the development of the electron microscope in the 1930s improved the ability to see smaller structures with greater magnification. (Although it must be noted that "older" technology still produced results—including the X-ray studies of DNA that eventually allowed James Watson and Francis Crick to determine DNA's overall structure.) In the early 1940s, biologists discovered that DNA was the substance that transmitted genetic information; by 1945, electron microscopy revealed such small details as those of the mitochondria, a major structure in the cell. By the 1950s, the further development of the scanning electron microscope led to the discovery of not only details about cellular structures, but also more information on the inner workings and mechanisms that allow cells to reproduce and survive.

Most cellular studies conducted from the middle of the century on have concentrated on the search for deciphering genetic codes within cells; the role of other organic substances—such as **proteins** and their subunits, the **amino acids**—in the life of specific cells; what structures set up certain conditions within a cell; the balance of enzymes, genetic material, and other chemical conditions within a cell; and cells in relation to genetic engineering. *See also* ANIMALS, ARTHROPODS, BIRDS, EVOLUTION, EXTINCTION, FUNGI, GENETICS, HUMANS, LIFE, PLANTS, PROTISTA, and VOLCANOES.

Topic Terms

amino acids–See GENETICS, Topic Terms

cells—All living organisms are organized as a collection of cells—from single-celled creatures such as bacteria, to human beings, who are estimated to have 50 to 100 trillion cells each. Although cells are more complex today than their primitive ancestors, ancient cells were similar in some ways, particularly in their ability to reproduce and adapt to the many changes occurring in their environment. Cells perform biochemical processes in all species, generate and process energy for use by the organism, and store the genetic codes necessary to grow a healthy organism, passing down genetic material to the next generations.

The largest single cell is the ostrich egg, which measures about 20 inches (51 centimeters) in diameter, or about 1,500 times the size of a human egg cell; the human egg is 14 times the size of a human red blood cell. The smallest cells are single-celled microorganisms that measure a few atoms across—the human red blood cell is about 35 times the size of these smallest creatures. The longest cells are certain nerve cells (neurons), which can reach over 3 feet (1 meter) in length.

The major activity of all cells is to transform energy. Cells gather this energy from the breakdown of food at the molecular level. They use the energy to grow, to eliminate waste material, to reproduce by mitosis, to move about a body, and to perform certain specialized tasks within the body.

The action of the cell on the cellular level is a staggeringly complex sequence of internal events that enable it to reproduce—and to survive—without any conscious help from the main organism. To complete any cellular action, the series of events must take place in the correct order—and within a certain time frame—to coordinate with the other cells in the body. The resulting products must be appropriate for a certain bodily function or they might be useless or even harmful. For example, if a person cuts a finger, the body's system immediately sends the necessary cells to repair the damaged area. The action is totally independent of any conscious thought—but the body's cells know what to do. The damaged area will not be covered with fur or feathers, but with new skin. The cells of each creature have this ability. Even in the case of growing young from a single cell, the cells know what to create. In other words, the bear will give birth to a cub, not a rabbit—showing that cells have certain inherited sets of rules.

Animal cells have certain structures or features. The cytoplasm includes everything within the plasma membrane except the nucleus. The cell membrane surrounds the cytoplasm at the cell's surface; the nucleoplasm is everything within the nuclear membrane. The nucleus contains long, thin structures composed of deoxyribonucleic acid (DNA) and protein (the chromosomes);

these structures contain genes, which are the individual units of information that tell the cell what to do, how to do it, and when to do it—all in terms of life processes. A dark area in the nucleus is called the nucleolus, which contains a high concentration of ribonucleic acid (RNA) and protein. Around the nucleus is the nuclear membrane; a series of double-layered membranes found throughout the cytoplasm is the endoplasmic reticulum. (See illustration, composite animal cell.)

Plant cells also have many of the same organelles that animal cells contain—but many structures are different and include additional features. For example, plant cells have a cell wall, large vacuoles, plasmodesmata, and chloroplasts (in association with chlorophyll to produce food for the plant). (See illustration, generalized plant cell.)

The following list compares animal, plant, and prokaryotes (bacteria) and their cell parts:

Cellular Comparisons of Modern Animals, Plants, and Prokaryotes

Cell Part	Animal	Plant	Prokaryote
cell membrane	yes	yes	no
cell wall	no	yes	yes
centrioles	yes	no	no
cilia or flagella	often	no	no
chloroplasts	no	yes, photosynthetic cells only	no
chromosomes	multiple	multiple	single
golgi bodies	yes	yes	no
lysosomes	often	similar structures	no
mitochondria	yes	yes	no
nucleus	yes	yes	no
ribosomes	yes	yes	yes
vacuoles	small or none	one large (mature)	no

cellular division—Cellular division has to do with cellular reproduction. Overall, when an organism grows or replaces old or damaged tissues, it often forms new cells by cellular division. In a process called mitosis, the chromosomes within the cells are duplicated and another complete set of genes is passed on unchanged to the "daughter cells." Thus, before a cell divides, the genetic information in the nucleus (chromosomes) must be duplicated.

A typical animal cell may contain between 10 and 50 chromosomes; these chromosomes are duplicated in the cell nucleus by a set of special replicating enzymes, coiling up so as not to become tangled. From there, each chromosome contains two identical units joined by special areas known as centromeres. Simply put, the nuclear membrane breaks down, allowing the two identical units to separate, each drawn toward one end of the cell. Each end of the parent cell then has a full set of chromosomes; the parent cell then "pinches itself off" into the two daughter cells, and a new cell membrane regrows around each new cell and its "new" chromosomes. The normal cells formed by this process of mitosis are called diploids.

Cells are also an especially important part of an organism's reproduction. Some single-celled organisms, plants, and simple animals reproduce asexually by simple cell division. Thus, for these organisms, the offspring are genetically identical to each other and to the single parent.

But most multicellular plants and animals reproduce sexually—and have special sex cells (gametes) that are essential: ova (eggs) from the female animal parent, and sperm from the male animal parent. In plants, these cells are called pollen and ovules. These sex cells are brought together to form a zygote, the initial single cell from which an embryo develops.

The sex cells, formed by a special type of cell division called meiosis, are called haploids (meaning that each cell contains only one copy of each chromosome, as opposed to the cell formed by mitosis, which contains two copies of each chromosome). Thus, since each parent contributes only one copy (or half) of their chromosomes, meiosis keeps the same number of chromosomes in each generation.

Meiosis also carries out another major function: They allow two sets of parents' genes to be "reshuffled." This reshuffling is not done in a haphazard way, but it is "controlled" by the cells, resulting in genetically different offspring from the parents. Thus, meiosis—along with random mutations—plays a major role in a species evolution, allowing for genetic variation.

proteins—Proteins are the building-block molecules that form the structures and mechanics of the body. Proteins are long chains of amino acids that are linked by enzymes. Their assembly takes place at the ribosomes within the cell.

self-replication—The early development of life has always puzzled scientists. Some early theories touted the idea that life arose spontaneously around the planet. More modern theories include mechanisms for starting life—or self-replicating catalysts (or places that would encourage the cells to duplicate) for the first cellular life.

Although the origin of the organics (cells) is debated, scientists do know that for the biomolecules to self-replicate—and eventually to form higher forms of life— some mechanism had to encourage the early

Animal and Plant Cell Structures and Functions
(CELL – Smallest living unit of which all organisms are composed)

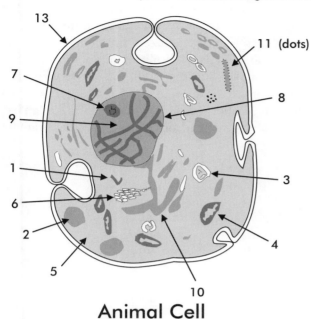
Animal Cell

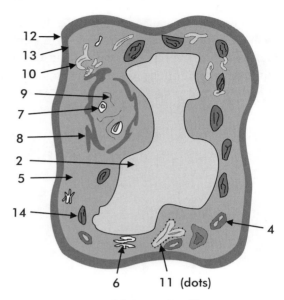
Plant Cell

STRUCTURE	MAIN FUNCTIONS	PRESENT IN ANIMAL OR PLANT CELL OR BOTH
centriole (1)	aids in cell reproduction	animal cells
vacuole (2)	storage	both
lysomes (3)	contain enzymes	animal cells
mitochondria (4)	cellular respiration {makes energy (ATP)}	both
cytoplasm (5)	site of most metabolism, jelly like substance	both
golgi complex (6)	packaging; secretion of protein	both
nucleolus (7)	production of ribosome, contains RNA	both
nucleus (8)	control; heredity: contains DNA in chromosomes (9)	both
ribosome (10)	protein synthesis	both
endoplasmic reticulum (11)	transport of proteins throughout cell	both
cell wall (12)	support; protection; made of cellulose	plant cells
cell membrane (13)	boundary "gate keeper" regulates entry and exit of materials: semipermeable	both
chloroplasts (14)	site of photosynthesis	plant cells

biomolecules to "stick" together. The latest theories point mostly to mineral and rock surfaces as the mechanism to encourage self-replication. For example, pyrite (fool's gold or iron sulfate) could have started self-replication, the surface of the mineral allowing the biomolecules to adhere and begin an energy-produc-

ing reaction. Another mineral with similar characteristics is zeolite, a silica-rich, hydrophilic (water-loving) mineral that tends to absorb water from its surroundings. Recently discovered, a naturally occurring zeolite-type mineral called mutinaite prefers to absorb organic materials out of water—and could have feasibly been a place for organics to accumulate and self-replicate.

As for rocks, some scientists point to the surfaces of clay (a type of sedimentary rock) that may have allowed the biomolecules to self-replicate. The main reason for this belief is a characteristic of modern clays: Clays can store energy and release it in the form of chemical energy. If this property was present on early Earth, the release of energy could have been harnessed by the early biomolecules to self-replicate.

Other suggestions and theories do not point to rocks or minerals, including one that may seem bizarre: Some scientists suggest biomolecules self-replicated through bubbles, similar to those you see in sea foam at the edge of the ocean waves washing on the shore. The bubbles could have "gathered" the biomolecules—acting as catalysts to self-replication. And of course, one of the simplest theories, but the hardest to verify since no humans lived during Earth's formation, is a spontaneous reaction—that the biomolecules had no catalyst, but just spontaneously started self-replicating.

10. Dinosaurs

Introduction

Dinosaurs were the most successful creatures on Earth to date, dominant for a period lasting approximately 150 million years; to compare, humanoids have only been around for about 4 million years. Dinosaurs were large land-dwelling reptiles that lived during the Mesozoic era, which lasted from about 225 to 65 million years ago; this era encompassed the Triassic, Jurassic, and Cretaceous periods. The name *dinosauria* was invented by British anatomist Richard Owen in 1841 and was used to describe fossil remains of his time that corresponded to no known living creatures. *Dinosauria* comes from the Greek words *deinos* and *sauros* and means "terrible reptiles or lizards."

Timeline

(note: mya = million years ago)

Date	Event
Prehistoric Events	
~300 mya	The reptiles, having evolved from the amphibians, rapidly spread, developing into thousands of species; of the four main lines, the diapsid line evolves into the archosaurs, or "ancient reptiles."
~250 mya	A great extinction event occurs, killing off nearly 96 percent of all living species; the resulting open ecological niches are eventually filled by the evolution of dinosaurs.
~230 mya	The first early dinosaurs emerge from the thecodont reptiles.
~205–145 mya	The Jurassic period—the dinosaurs become the dominant animals on land.
~145–65 mya	The Cretaceous period—dinosaurs continue to be the dominant animal on land; the animals diversify greatly during this time, evolving into a large variety of species.
~65 mya	A mass extinction occurs, leading to the disappearance of the dinosaurs from the known fossil record; this extinction marks the end of Mesozoic era, and beginning of the Cenozoic era.
Modern Events and Discoveries	
1787	A huge fossil bone found in New Jersey goes unreported and unverified, but it may have been the first modern dinosaur bone ever scientifically collected.
1818	The fossil bones of the dinosaur *Anchisaurus* are found in the Connecticut Valley, but the find is not interpreted as a dinosaur.
1824	English cleric and geologist William Buckland (1784–1856) becomes the first person to publish a paper describing and naming a prehistoric reptile that subsequently came to be known as a dinosaur; it is the *Megalosaurus*, or "big reptile."
1825	English physician and amateur paleontologist Gideon A. Mantell (1790–1852), extrapolating from some fossil teeth found by his wife in 1822 in Tilgate Forest, England, describes an "extinct reptile," the *Iguanodon* (iguana-tooth).
1841	English paleontologist Sir Richard Owen (1804–1892) concludes that these giant prehistoric reptiles were part of a new sub-order of animals that he called *Dinosauria*, or "terrible reptile."
1854	The first life-sized model of a dinosaur is exhibited at the Crystal Palace in Sydenham, England; it is an *Iguanodon*, a large, bipedal herbivore that lived during the Cretaceous period.
1877	Longtime rivals Edward Drinker Cope (1840–1897), a U.S. paleontologist and herpetologist, and Othniel Charles Marsh (1831–1899), a U.S. professor of paleontology at Yale University, make the first of many separate collections of Jurassic dinosaurs from the Morrison formation (Colorado and Wyoming); their finds also lead to the "fossil rush" in the western part of North America.
~1882	Louis Dollo (1857–1931), a Belgian paleontologist, becomes the founder of ethnological paleontology, the study of the relationships of ancient animals to their environment.

Date	Event
1887	Harry Seeley determines that there were two suborders of dinosaurs, based on pelvic structure: The saurischian, or reptile-hipped, and the ornithischian, or bird-hipped, dinosaurs.
1901	Peabody Museum of Natural History (Yale University) mounts an *Edmontosaurus*, the first skeleton of real dinosaur bone to be mounted in the Western Hemisphere.
1902	U.S. paleontologist Barnum Brown (1873–1963) discovers the first *Tyrannosaurus rex* skeleton in the Hell Creek area of Montana.
1929	The first dinosaur footprint is excavated in Shanxi Province, China.
1941	Barnum Brown retires from the American Museum (New York), but continues to work; he is considered to be the greatest collector of dinosaurs of all time.
1956	M.W. De Laubenfels publishes an article in the *Journal of Paleontology* that proposes dinosaurs became extinct as a result of an asteroid impact.
1969	U.S. paleontologist John H. Ostrom publishes a description of the *Deinonychus*, a Cretaceous period carnivorous dinosaur; based on his study of the creature, he theorizes that dinosaurs may have been endothermic (warm-blooded).
1975	U.S. paleontologist Robert T. Bakker publishes an article in *Scientific American* creating what he called a "dinosaur renaissance"; he proposes dinosaur endothermy, and by bringing the idea to the scientific forefront, begins a new era of dinosaur paleontology.
1978	U.S. paleontologists John R. Horner and Bob Makela discover the first known fossilized nest of baby dinosaurs, a new species named the *Maiasaura*.
1980	U.S. physicist Luis Alvarez (1911–1988), his son, Walter (1940–), and several colleagues propose that a large asteroid or comet hit Earth about 65 million years ago, causing a mass extinction, including that of the dinosaurs; their evidence included a previously unknown layer of iridium in Cretaceous period rock layers, an indication of an impact on Earth.

History of Dinosaurs

Dinosaurs and Their Origins

Dinosaurs did not exist during the early years of life on Earth—but the first evolutionary steps toward the creatures took place as the diversity of life increased on the planet. During the middle Carboniferous period of the Paleozoic era, about 300 million years ago, reptiles evolved from the dominant animals, the amphibians. The new animals rapidly spread, developing into thousands of species. The reptiles had distinctive skeletal characteristics, were completely land-dwelling, and took advantage of the plants, insects, and animals found in the surrounding environment.

For about 100 million years, from the middle Carboniferous to the end of the Permian period, early reptiles evolved along four main lines: anapsids, synapsids, euryasids, and diapsids. These classifications were based on the pattern of openings in the side of the animals' skulls. The diapsids, for example, had two temporal openings on the sides of their skulls, which served as attachment sites for jaw muscles. It was the diapsid line that evolved into the lepidosaurs and the **archosaurs**, or "ancient reptiles"—the animals that eventually led to the **dinosaurs**.

Approximately 250 million years ago, at the end of the Permian period (and end of the Paleozoic era and beginning of the Mesozoic era), a mass extinction of about 96 percent of all species occurred, almost ending all life on the planet. But for the reptilian survivors, this extinction set the stage for the rise of the dinosaurs in the Mesozoic era, or "age of reptiles." The reptiles rapidly moved into the now-vacant ecological niches, evolving into numerous forms. The archosaur reptiles evolved into the thecodonts, or "socket-toothed" reptiles. By the middle of the Triassic period, approximately 230 million years ago, the thecodonts rapidly evolved into groups that would eventually produce the pterosaurs, or now-extinct flying reptiles; the crocodilians, precursors to modern crocodiles; and the dinosaurs.

The earliest, primitive dinosaurs—thecodonts—were quite small compared with the giants that were to come. The early dinosaurs from the Triassic period were both herbivores and carnivores, evolving between 230 and 225 million years ago. One of the earliest was the *Eoraptor*, a small, bipedal carnivorous dinosaur with powerful hind legs and a long tail; an early herbivore was the *Thecodontosaurus*, one of the first Triassic dinosaur fossils ever found.

The thecodonts gave rise to two orders of dinosaurs, the saurischians and ornithischians. These dinosaurs eventually evolved and grew in diversity, dominating the Jurassic and Cretaceous periods, from around 205 to 65 million years ago. The saurischian dinosaurs had the usual reptilian **hip structure**, with the pubis and ischium bones pointing forward and backward, respectively. Dinosaurs that had this structure included the theropods, or meat-eating dinosaurs, and the sauropods, or large, long-necked plant-eaters.

The ornithischian dinosaurs, or bird-hipped dinosaurs, had a modified hip structure, with the pubis and ischium bones both pointing backwards and parallel to each other. These dinosaurs were all herbivores and were divided into ornithopod (two-legged), ceratopsian (horn-faced), stegosaurid (plate-spined), ankylosaurid (armored), and pachycephalosaurid (thick-headed) dinosaurs.

About 65 million years ago, a mass extinction occurred. Some scientists believe that an asteroid or comet (or several space bodies) collided with Earth, sending dust into the upper atmosphere, which spread around the planet, blocking out a large amount of sunlight. Others feel that large amounts of volcanic activity, such as that found in the Deccan Traps of India, were the cause of the **dinosaur extinction.** According to this theory, a large amount of dust, ash, and volcanic gases spewed into the upper atmosphere, cutting off sunlight. In addition, ash and gas could have smothered many of the dinosaurs. Still other scientists think the dinosaurs were dying out long before this event, and a catastrophe just pushed the animals over the edge. In any case, the fossil record shows that the dinosaurs became extinct in this relatively short time frame.

Modern Dinosaur Studies

Although dinosaur fossils had been present in Earth for millions of years, it was not until the nineteenth century that systematic digging and scientific theory sought to explain the fossils' origins. Initially, these fossils were thought to be animals of legend. For example, in China, the **dinosaur bones** were thought to be from dragons; in England, big bones were thought to be the remains of elephants brought to the islands by the Romans. Adding to the confusion, prevailing Western religious convictions (known as plenum) held that all species were created only a few thousand years ago, and none of these animals or plants had gone extinct. However, by the early twentieth century, absolute dating of certain fossil-bearing rocks found them to be millions of years older than the proposed creation date. Thus, the fossil remains of creatures contained in the rock layers were also much older.

Dinosaur fossil bones were a big mystery for many years, as they were totally unlike any living creatures on modern Earth. The first person to propose the existence of a giant, extinct reptile was Gideon Algernon Mantell, an English doctor and amateur paleontologist. In 1822, he began studying some large and unusual fossil teeth that had been found by his wife, Mary Ann Mantell, in some gravel near Lewes, England (although it has been suggested that Gideon Mantell was the actual discoverer). After tracing the source of the gravel to a local quarry, he was able to extract more fossils from the rock layers. Although leading experts of his day did not see any significance in these teeth, Mantell knew they were from the Mesozoic era and suspected the fossils were reptilian. He finally found that the fossil teeth were similar in appearance—though much larger—to those of the iguana, a living South American lizard. In 1825, he published a description of this giant, extinct relative of the iguana, naming it *Iguanodon* (iguana-tooth).

In this same period, other fossil remains were being discovered in England. William Buckland, an English professor of geology at the University of Oxford and dean of Christ Church, studied (with other experts) the bones of a creature discovered in England around 1818 that could not be identified. He eventually classified the animal as a giant reptile, naming the creature *Megalosaurus* ("big reptile") and estimating it had been about 40 feet (12 meters) long.

As more and more fossils of reptiles from the Mesozoic era were uncovered and named in England, Sir Richard Owen, an eminent English scientist and expert in animal anatomy, examined the remains of nine discovered reptiles. He concluded they corresponded to no known living reptile. He also decided that at least three of them, the *Megalosaurus*, *Iguanodon*, and *Hylaeosaurus*, were giant prehistoric reptiles representing a new suborder of animals. He called the suborder *Dinosauria*, from the Greek *deinos* ("terrible") and *sauros* ("reptile"). He announced this new

dinosaur classification in 1841 to the British Association for the Advancement of Science at their annual meeting.

Interest in dinosaurs remained high during the middle and later 1800s—so much so that, in 1854, under the supervision of Owen, a public exhibit of the new dinosaurs was held at the Crystal Palace in London. Although the dinosaurs were portrayed as giant four-legged lizards—now known to be incorrect—the exhibit attracted huge crowds. The imagination of natural scientists and the public ran high over these fascinating creatures, and Mantell, Buckland, and Owen were generally acknowledged as having "discovered" the existence of dinosaurs.

In 1878, a large collection of *Iguanodon* fossils was found in a coal mine beneath the village of Bernissart, Belgium. Paleontologists from the Royal Museum of Natural History in Brussels carefully excavated 39 almost complete skeletons of *Iguanodon*, as well as numerous other fossils of plants, fishes, and insects. Louis Dollo was a Belgian paleontologist responsible for the organization, preparation, and description of the *Iguanodon* fossils, which became his life's work. Dollo was one of the first scientists to use a holistic approach to describe dinosaurs; rather than simply naming them, he considered the creatures' biology and how they lived, reconstructing the complete skeletons in natural poses. By expanding the study of dinosaurs in this way, Dollo became the founder of ethnological paleontology, the study of the relationships of ancient animals to their environment. His work laid the foundation for the study of dinosaurs as living creatures, not just pieces of skeletons.

Europe wasn't the only place where **dinosaur fossil locations** were coming to light—the western part of the United States was quickly becoming a hot spot for dinosaur fossils during the late nineteenth century. This dinosaur fervor was best exemplified by the famous rivalry between U.S. paleontologists Edward Drinker Cope and Othniel Charles Marsh—two men who dominated the study of dinosaurs in the United States during this time. They were both brilliant, both ambitious in finding and describing dinosaurs, and although onetime associates, both hated each other.

Throughout the late 1800s, these men engaged in an intense rivalry to find new dinosaur fossils and to name new species. Their desire to outdo each other led to numerous expeditions out west, and the resulting "fossil rush" showed the incredible diversity of dinosaurs that existed in the Mesozoic era in western North America. But not all the rivalry was negative; the men also developed several methods of hunting and preserving fragile dinosaur bones. For example, Marsh developed the "toilet-paper and plaster method" of protecting the fossil bones, and a similar process is still in use today. In addition, Marsh and Cope were responsible for the collection, description, and naming of almost 130 new species of dinosaurs.

In the latter part of the nineteenth century, more dinosaur discoveries and conclusions were made. In 1887, Harry Seeley determined that dinosaurs should be classified into two suborders, based on the animals' pelvic structures: the saurischians, or reptile-hipped, and the ornithischians, or bird-hipped dinosaurs. Discoveries continued, including those in 1884, by Canadian paleontologist Joseph Burr Tyrrell, who discovered the skull of an *Albertosaurus*, a previously unknown Cretaceous carnivore. In 1908, paleontologists Charles H. Sternberg and his sons, Charles M., George, and Levi, discovered the first known fossilized dinosaur skin, an impression on a rock from a duck-billed dinosaur; in 1922, in Mongolia's Gobi Desert, Roy Chapman Andrews found the first known clutch of dinosaur eggs in a sedimentary layer known as the Nemget Formation.

More recently, modern paleontologists have focused on the social associations, feeding habits, and physical abilities of dinosaurs based on the new fossil finds, giving us a more complete picture of these fascinating reptiles. In 1969, paleontologist John H. Ostrom publishes a description of the *Deinonychus*, a Cretaceous period carnivorous dinosaur; based on his study of the creature, he theorized that dinosaurs may have been endothermic (warm-blooded).

After this announcement, many paleontologists began to explore this idea, using new technologies to analyze dinosaur bones. The idea that emerged was that dinosaurs were endothermic (warm-blooded) and homeothermic (constant body temperature), contrary to the widely held view that dinosaurs, being reptiles, had to be ectothermic (cold-blooded) and poikilothermic (variable body temperature)—or slow and sluggish. By 1975, paleontologist Robert T. Bakker who, along with others, had also been working on the idea, published an article in *Scientific American*, creating what he called a "dinosaur renaissance." He proposed dinosaur endothermy, and by bringing the idea to the scientific forefront, began a new era of dinosaur paleontology that has continued ever since. The many mammalian characteristics that dinosaur fossils exhibited—such as complex bone structures, an upright posture, and a predator-prey ratio closer to mammals than reptiles—was the basis for Bakker's theory.

Other dinosaur discoveries helped to turn around behavioral theories. For example, in the 1970s, David Norman, a British paleontologist, proposed, with Peter Galton, that the upright stance Louis Dollo attributed to the *Iguanodon* was incorrect. They theorized these dinosaurs must have held their backbones in a horizontal stance, to retain balance while in motion. More recently, dinosaur bones were found in previous polar regions (such as Australia), indicating that some dinosaurs could survive the cold and thus, may truly have been endothermic. In 1978, U.S. paleontologists John R. Horner and Bob Makela discovered the first known nest of baby dinosaurs, a new species named the *Maiasaura*, or "good mother lizard." By 1979, Horner would also determine that the maiasaurs exhibited colonial nesting and herding behavior—one of the first insights into the social behavior of dinosaurs.

One of the major theories in recent years concerning the dinosaurs was stated in 1980, when Luis Alvarez proposed that a large asteroid or comet hit Earth about 65 million years ago, causing a massive extinction, including the demise of the dinosaurs. His son, geologist Walter Alverez, discovered a high concentration of iridium at the Cretaceous-Tertiary (nicknamed "K/T") boundary in Italy, an element associated with extraterrestrial impacts; thus, Luis and Walter, along with colleagues Frank Asaro and Helen Michel, proposed that the extinctions at the K/T boundary were caused by the impact of a large space object. Since that time, the iridium anomaly has also been found in over 50 K/T boundary sites around the world. Although many scientists do not believe this scenario, the scientific world seems to have embraced this theory as the explanation for the dinosaur ruin—at least until another plausible explanation (or explanations) is discovered.

By the 1980s, as the number of known dinosaur species increased, debate intensified as to the nature of a common ancestor. Were all dinosaurs actually in the order *Dinosauria*, or were there multiple ancestors? The evidence soon was overwhelming—all of the species did come from a common ancestor and the existence of the saurischians and ornithischians was confirmed. One recent modification, though, is that saurischians are not classified based on hip structure (which is shared by other reptiles), but on other features of the skull, backbone, and hands.

Another debate also increased in the 1990s—were birds truly dinosaurs or just another offshoot of a reptile group?

Dinosaur fossil expeditions continue throughout the world today. Each discovery sheds more light on the history of these reptiles—and it seems, with each new dinosaur fossil found, the debate as to the actual character, features, behaviors, and descendants also increases. It is the nature of science to formulate fresh ideas and theories based on new findings—and the study of dinosaurs is no exception. *See also* AMPHIBIANS, ANIMALS, BIRDS, CLASSIFICATION OF ORGANISMS, EVOLUTION, EXTINCTION, FOSSILS, GEOLOGIC TIME SCALE, PLANTS, REPTILES, and SOLAR SYSTEM.

Topic Terms

archosaur—The archosaurs, or "ruling reptiles," were one of the major groups of reptiles abundant in the Mesozoic era. This group included dinosaurs and pterosaurs, all of which became extinct at the end of the Cretaceous period. The only living descendant of this group is the modern crocodile.

dinosaur—Not all of reptiles from the Mesozoic era (and there were many) were dinosaurs—only those that lived on land, had certain characteristics, and were incapable of flight were given this classification. Dinosaurs were unique among the reptiles in that their legs were tucked in beneath their bodies, allowing them to walk and run very efficiently. Their distinct traits and adaptability made them the perfect candidates to dominate the planet during the Mesozoic era. For the most part, the climate was hotter and drier; but the climate was also changing, due to the splitting apart of the large land masses into the continents we know today. And for close to 150 million years, the dinosaurs adapted well to these conditions.

The dinosaurs disappeared from the known fossil record about 65 million years ago at the end of the Cretaceous and beginning of the Tertiary periods (K/T boundary), as did many other species. Today, some scientists feel that no living dinosaurs exist today. Other scientists disagree, though, based on recent and past fossil discoveries: they believe that birds were small dinosaurs that arose in the middle Mesozoic, survived the mass extinctions at the K/T boundary, and evolved into the forms we see today. This debate will continue for a long time, until fossil evidence—or even genetic evidence—can prove or disprove it otherwise.

Currently, approximately 700 species of dinosaurs have been named. But only about half of these specimens are complete skeletons—thus, the other fossils are incomplete, often making it difficult to say the bones are from a unique or separate species. About

300 dinosaur genera (*Tyrannosaurus*, *Stegosaurus*, etc.) have been verified, although about 540 have been named; many scientists believe close to 700 to 900 more have yet to be found.

The following lists some of the major dinosaurs that lived during the Mesozoic era:

Mesozoic Era Dinosaurs

Triassic Period (245–208 million years ago)

Name	Approximate Age (mya=million years ago)	Locality	Length (in feet/meters)
Anchisaurus	200–190 mya	USA	up to 6.5/2
Coelophysis	225–220 mya	USA	up to 10/3
Eoraptor	225 mya	Argentina	up to 3/1
Herrerasaurus	230–225 mya	Argentina	up to 10/3
Plateosaurus	~210 mya	France, Germany & Switzerland	up to 23/7

Jurassic Period (208 to 146 million years ago)

Name	Approximate Age	Locality	Length
Apatosaurus	154–145 mya	USA	up to 70/21
Allosaurus	150–135 mya	USA	up to 50/15
Archaeopteryx (This reptile's origin/nature has been debated since the first fossil was found in 1861. See Birds)	147 mya	Germany	up to 1.5/0.5
Barosaurus	155–145 mya	USA	up to 80/24
Brachiosaurus	155–140 mya	USA/Tanzania	up to 75/23
Camarasaurus	155–145 mya	USA	up to 65/20
Camptosaurus	155–145 mya	USA	up to 16/5
Coelurus	155–145 mya	USA	up to 8/2.4
Compsognathus	147 mya	Germany	~2/0.6
Dacentrurus	157–152 mya	France, England & Portugal	~20/6
Diplodocus	155–145 mya	USA	up to 90/27
Dryosaurus	155–140 mya	USA, Tanzania	up to 13/4
Kentrosaurus	140 mya	Tanzania	up to 10/3
Mamenchisaurus	155–145 mya	China	up to 72/22
Massospondylus	208–204 mya	South Africa	up to 13/4
Megalosaurus	170–155 mya	England, Tanzania	up to 30/9
Ornitholestes	155–145 mya	USA	up to 6.5/2
Pelorosaurus	150 mya	England	unknown
Scelidosaurus	203–194 mya	England	up to 13/4
Stegosaurus	155–145 mya	USA	up to 30/9
Tuojiangosaurus	157–154 mya	China	up to 21/6.4

Cretaceous Period (146 to 65 million years ago)

Name	Approximate Age	Locality	Length
Albertosaurus	76–74 mya	Canada	up to 30/9
Avimimus	~75 mya	Mongolia	up to 5/1.5
Baryonyx	~124 mya	England	up to 34/10
Centrosaurus	76–74 mya	Canada	up to 16/5
Chasmosaurus	76–74 mya	Canada	up to 16/5
Corythosaurus	76–74 mya	Canada, USA	up to 33/10
Craspedodon	86–83 mya	Belgium	unknown
Deinocheirus	70–65 mya	Mongolia	unknown, arms found are 10/3
Deinonychus	110 mya	USA	up to 11/3
Dromaeosaurus	76–74 mya	Canada	up to 6/1.8
Dryptosaurus	74–65 mya	USA	~16/5
Edmontonia	76–74 mya	Canada	~13/4
Edmontosaurus	76–65 mya	Canada	up to 43/13
Euoplocephalus	about 71 mya	Canada	up to 20/6
Gallimimus	74–70 mya	Mongolia	up to 18/5.5
Gilmoreosaurus	80–70 mya	China	~20/6
Hadrosaurus	83–74 mya	USA	up to 26/8
Hylaeosaurus	150–135 mya	England	up to 13/4
Hypsilophodon	~125 mya	England	up to 7.5/2
Iguanodon	130–115 mya	USA, England, Belgium, Spain, Germany	up to 33/10

Cretaceous Period (146 to 65 million years ago) (cont'd.)			
Name	Approximate Age (mya=million years ago)	Locality	Length (in feet/meters)
Kritosaurus	80–75 mya	USA	up to 26/8
Lambeosaurus	76–74 mya	Canada	up to 30/9
Maiasaura	80–75 mya	USA	up to 30/9
Ornithopsis	~125 mya	England	unknown, perhaps 65/20 ?
Orodromeus	~74 mya	USA	up to 6.5/2
Ouranosaurus	~115 mya	Niger	up to 23/7
Oviraptor	85–75 mya	Mongolia	up to 6/2
Pachycephalosaurus	~67 mya	USA	up to 26/8
Pachyrhinosaurus	76–74 mya	Canada, USA	up to 20/6
Parasaurolophus	76–74 mya	Canada, USA	up to 33/10
Parksosaurus	76–74 mya	Canada	up to 10/3
Protoceratops	85–80 mya	Mongolia	up to 6/2
Psittacosaurus	124–97 mya	China, Mongolia, Russia	up to 6/2
Rhabdodon	83–70 mya	Austria, France, Spain, Romania	up to 10/3
Saurolophus	74–70 mya	Canada, Mongolia	up to 40/12
Saurornithoides	80–74 mya	Canada	up to 6.5/2
Scartopus	~95 mya	Australia	unknown
Segnosaurus	97–88 mya	Mongolia	up to 13/4
Struthiosaurus	83–75 mya	Austria, Romania	up to 6.5/2
Styracosaurus	85–80 mya	Canada, USA	up to 18/5.5
Tenontosaurus	110 mya	USA	up to 21/6.4
Triceratops	67–65 mya	USA	up to 30/9
Troodon	75–70 mya	Canada, USA	up to 8/2.4
Tyrannosaurus	67–65 mya	USA	up to 40/12
Velociraptor	84–80 mya	China, Mongolia	up to 6/2

dinosaur bones—When some dinosaurs died, their remains were buried under sediment, which became rock over the intervening millions of years. Their skeletons were chemically altered and preserved, and in many cases are now an integral part of the surrounding rock. Paleontologists, after traveling to a promising area and finding a potential location for dinosaur bones, begin carefully digging out the fossilized remains. Many times the fossils are just clumps of bones from many dinosaurs, but in some instances almost complete skeletons of one dinosaur have been found.

The paleontologist use tools ranging from a drill to the finest brush to expose the sought-after fossil bones. Depending on conditions, this process may take weeks or months. During the dig, the location of each bone is carefully recorded for future reference; then the fossil-containing rocks are packed in foam or plaster-of-Paris for shipment back to the laboratory. There, the protective material is removed and the hard work of exposing the fossilized bones begins. The surrounding rock, or matrix, must be removed slowly and carefully, using mechanical and/or chemical means. The process can take weeks.

Dinosaur bones are then often pieced together to re-create the morphology of these great reptiles. After

Chipping away at dinosaur bones is a delicate operation, as seen here with this paleontologist working in Utah. *(Photo courtesy of U.S. Department of the Interior, National Park Service)*

the bones have been prepared, the paleontologist then must reconstruct the skeleton, similar to putting together a three-dimensional jigsaw puzzle. Missing or damaged pieces can sometimes be modeled and built. While reconstruction can be slow for a known species of dinosaur, a completely new type will take even longer, because there is no guide. Finally, the puzzle is complete, and the size, shape, even the stance are now visible. Dinosaur skeletons are on display in numerous natural history museums and are the basis for our speculations concerning their behavior, locomotion, and social structure.

dinosaur classification—For as long as dinosaur fossils have been studied, their classification has been debated. Scientists know that dinosaurs were reptiles; but as to what kind, and where to put them in the myriad other species, is still debated. In general, the higher order classification is as follows: Dinosaurs are in the Kingdom Animalia, the phylum Chordata, the class Reptilia, and subclass Archosauria. They were a monophyletic group, having originated from a common ancestor—and thus all are classified under the order Dinosauria.

dinosaur extinction—Extinction is the dying out of a species. In the case of the dinosaurs, an extremely successful group of species that had adapted to changing conditions for millions of years, extinction was relatively, in geologic time, sudden and complete. The cause, or causes, for the extinction of the dinosaurs is an ongoing subject for research and speculation, but it is generally agreed that a sudden, catastrophic event or events affected Earth's climate so quickly that the dinosaurs did not have enough time to adapt. Evidence of a large asteroid impact on Earth around the time of the dinosaur's disappearance has been discovered, as well as evidence of a large amount of volcanic activity. Either or both of these phenomena could have drastically changed Earth's climate, leading to the extinction of the "terrible reptiles." These events define the K/T boundary between the Cretaceous period of the Mesozoic era and the Tertiary period of the Cenozoic era.

dinosaur fossil locations—Though fossils have been found on every continent, the best locations for dinosaurs are found in specific areas around the world, most notably the western United States, Mongolia, western Europe, and eastern Australia. These areas have a combination of factors that make them the best places to discover these fossils; they have sedimentary rocks originating from lakes or rivers, the rocks are from the Mesozoic era, and they are exposed due to erosion. Many of these areas are remote and desolate, necessitating large and well-organized expeditions to discover, remove, and transport the bones.

hip structure—One of the early criteria used to classify dinosaur fossils into two groups, saurischians (reptile-hips) and ornithischians (bird-hips), was hip structure. This characteristic was first noted by Harry Seeley in 1887. Modern classification makes use of other shared characteristics such as hands and feet to distinguish between groups.

trackways—Trackways of dinosaurs (and other creatures) are the fossil imprints of the animal's footprints—a type of trace fossil. These trackways sometimes give scientists an idea about a dinosaur's stride, if the dinosaur was walking or running—and even if the animals were stampeding. For example, in Australia, the Lark Quarry Environmental Park, south of Wilton, has hundreds of dinosaurs prints. Scientists interpret the many tracks as a group of large carnivores hunting plant-eating animals along a muddy shoreline. In one set of tracks, it appears that a carnivore is chasing a herbivore, causing the other plant-eating animals to run away in a stampede.

The classic layers of rock in which dinosaur bones are found in the Morrison formation. This dinosaur-collecting spot is located in western Colorado.

11. Earth

Introduction

Earth is the name we humans have given to the planet on which we live. The Earth's formation is thought to have occurred about 4.55 billion years ago along with the formation of the other planets in our solar system. Geology, the study of the Earth's composition, structure, processes, and history, directly relates to the natural history of the planet. The movement of the continents across the face of the planet and the rising and falling of landmasses are especially significant. This movement not only changes the habitats of marine and terrestrial organisms, but also affects ocean currents and climates around the world—but over hundreds of thousands of years. It is not known if the continents will ever stop moving as the Earth continues to cool down from its billions-year-old formation, due to radiation of its natural heat. But in our time, and probably for hundreds of millions more years, the continents will move across the face of the planet, continuing to change the morphology, and natural history, of our Earth.

Timeline

(note: bya=billion years ago; mya=million years ago)

Date	Event

Prehistoric Events

Date	Event
~4.55 bya	Matter collides within a coalescing nebula; this forms the rocky bodies known as the terrestrial planets, including Earth.
~4.35 bya	The Earth's atmosphere and oceans are established; the Earth is also being almost continually bombarded by objects from space.
~3.8 bya	The heavy bombardment of the planet dies down, allowing the planet to continue to cool down; life probably begins to form around this time.
~700 mya	The supercontinent Rodinia, a collection of small landmasses that hovered around the equator, is established; this date is still debated.
~500 mya	The large landmass of Gondwanaland (or Gondwana) forms to the south, and Laurasia (or Laurentia) forms around the equator.
~250 mya	Another supercontinent forms called Pangea (or Pangaea), a large landmass composed of Gondwanaland and Laurasia huddled around the equator.
~200 mya	Pangea breaks apart again into Laurasia and Gondwanaland; these landmasses begin to drift apart even more and eventually form the familiar continents we see today.
~50 mya	The familiar continents we see today are relatively in place.

Modern Events and Discoveries

Date	Event
~240 B.C.	Greek librarian and astronomer Eratosthenes of Cyrene (c. 276 B.C.–c. 194 B.C.) is the first to correctly measure the circumference of the Earth; he accomplishes this by measuring noontime shadows at Alexandria and Syene, two localities in Egypt of differing latitudes.
late 1400s	Italian scientist Leonardo da Vinci (1452–1519) recognizes that fossils of sea creatures found at the top of the Alps were the result of material carried by rivers to the sea slowly being compacted into sedimentary rock; and that the rock eventually was uplifted to form mountain chains.
1600	English physician William Gilbert (1544–1603) suggests, based on his studies in magnetism, that the Earth is a giant magnet.
~1650	Irish archbishop James Ussher (1581–1656) sets the date of Earth's creation and the creation of humans at 4004 B.C., which is later put into various editions of the Kings James version of the Bible.
~1654	Englishman John Lightfoot takes the calculations of James Ussher farther, proposing that Earth was created at 9:00 A.M. on October 16, 4004 B.C.
1665	German geologist Athanasius Kircher (1602–1680) writes *The Subterranean World*, a standard geological text of its time, although many of his ideas were incorrect.
1669	Danish geologist and anatomist Nicolaus Steno (Niels Stensen) (1638–1686) identifies some of the common fossils, and postulates the three basic geological principles that are still held to be true today.

Date	Event
1719	Swedish scientist Emanuel Swedenborg (1688–1772) deduces that Scandinavia was once covered by an ocean, sighting evidence such as gravel ridges, sedimentary deposits, and the fact that the land is rising along the Baltic coast.
1719	English stratigrapher John Strachey (1671–1743) identifies strata in the Earth's crust, especially in describing layers in the coal mines of southwest England.
1759	Italian geologist Giovanni Arduino (1714–1795) divides the Earth's crust into four divisions.
1774	English scientist Nevil Maskelyne (1732–1811) determines the mass of the Earth; he estimates the average density as being 4.5 times the density of water (the current estimate is actually about 5.5 times the density of water).
1774	German scientist Abraham Werner (1749–1817) publishes his first edition on mineral classification, based on external characteristics such as form, color, hardness, cleavage, and luster.
1785	Scottish geologist James Hutton (1726–1797) introduces the theory of uniformitarianism into geology.
1795	Hutton publishes *Theory of the Earth*, in which he postulates that the origin of granite is volcanic; certain mountains have a glacial history; valleys are created by rivers; and volcanoes are old and the result of heat being released from within the Earth.
1798	English scientist Henry Cavendish (1731–1810) determines a new number for the mass of the Earth, using a method suggested by English clergyman John Michell (1724–1793); his value is 5.5 times the density of water, an improvement over the measurements obtained by Nevil Maskelyne; this is close to the accepted modern figure.
1798	Scottish geologist James Hall (1761–1832) demonstrates how lava forms different rocks as is cools.
1799	English geologist William Smith (1769–1839) lists and describes the strata in a stratigraphic section of England, including fossils found in certain layers; his description helps correlate other similar strata elsewhere in England; he is among the founders of the study of stratigraphy.
~1800	Called the "heroic age" of geology, as the "Plutonists" such as James Hutton clash with the views of the "Neptunists" such as Abraham Werner.
1802	Scottish geologist John Playfair (1748–1819) argues that valleys are carved by streams; this theory goes against the idea of many of his contemporary scientists who believe valleys formed during cataclysmic upheavals, and later had rivers flowing through the area.
1823	English cleric and geologist William Buckland (1784–1856) determines that the flood mentioned in the Bible occurred 5,000 to 6,000 years earlier, based on fossils, large boulders, and irregular deposits of gravel, sand, and clay in Great Britain.
1830	Scottish geologist Charles Lyell (1779–1875) publishes *The Principles of Geology*, expanding on James Hutton's theories; he is one of the first scientists to describe geology as the study of the Earth's composition, structure, processes, and history, and to theorize that geological features are eroded, shaped, and reformed at a constant rate over time.
1837	Swiss-American geologist Louis Agassiz (1807–1873) gives his famous speech on the widespread ice age conditions that existed long ago.

Date	Event
1839	U.S. geologist Timothy Conrad (1803–1877) discovers evidence of erratic boulders, striations, and polished rocks in western New York, thus supporting the theory of Louis Agassiz that glaciation was extensive all over the world during the Ice Ages.
1842	French scientist Joseph Adhémar (1797–1862) suggests that the Earth's ice ages are the result of the 22,000-year precession of the equinox (movement of the Earth that switches seasons over thousands of years); it is the first attempt to explain the Ice Ages using an astronomical connection.
1851	French scientist Jean Bernard Léon Foucault (1819–1868) uses the Foucault pendulum to demonstrate experimentally that the Earth rotates.
1857	U.S. geologist Richard Owen (1810–1897) suggests that the continents drifted apart from one another over time; his essay, *Key to the Geology of the Globe: An Essay,* is largely forgotten for decades.
1885	Austrian scientist Eduard Suess (1831–1914) postulates the existence of a supercontinent called Gondwanaland from which the current landmasses in the Southern Hemisphere have all separated.
1891	U.S. astronomer Seth Carlo Chandler (1846–1913) determines that the Earth's axis—which is tipped about 26.5 degrees from the plane of the planet's orbit and points to an area around the North Star—wobbles as the planet spins.
1912	Alfred Lothar Wegener (1880–1930), a German geologist and meteorologist, first publishes his ideas about the displacement of continents in the *Marburg University Science Journal.*
1958	James Van Allen (1914–) discovers belts of the planet's magnetosphere around the Earth, now called the Van Allen Belts.
1960	U.S. geologist Harry Hammond Hess (1906–1969) discovers that the sea floor is spreading, which eventually leads to the theory of plate tectonics.
1962	English geologists Fred Vine and Drummond Matthews propose that the sea floor around mid–ocean ridges holds a record of past magnetic reversals of the Earth.
1965	Canadian geophysicist John Tuzo Wilson, (1908–), a pioneer of plate tectonics, describes the origin of the San Andreas fault, which is responsible for the great number of Earthquakes that occur in California.
1980s	Satellites, lasers, and the positions of distant galaxies allow scientists to accurately measure the movement of the continents—more than 1 inch (2 centimeters) per year.
1989	U.S. scientists complete an ocean-mapping project; it reveals that the Mid-Atlantic Ridge is actually a string of 16 spreading centers.
1990	The oldest part of the Pacific plate is found.

History of Earth

Early Earth

It is thought that Earth's formation coincided with the formation of the other planets throughout our solar system. About 4.55 billion years ago, matter condensed and then collided in a huge gaseous nebula, eventually creating the rocky bodies we now know orbiting our Sun. The innermost bodies became the terrestrial planets—of which Earth is a member. Initially, Earth is thought to have been a homogeneous, hot ball, with an ocean of magma on its surface formed from the hot, original nebula.

As the Earth started to cool about 4.55 billion years ago, it began to form its internal layers—an iron-rich core, a molten mantle, and a very thin crust. The initial crust was fragile, breaking open as the interior of the planet released gases—especially water vapor that would eventually form not only an atmosphere, but the oceans. By about 4.35 billion years ago, the atmosphere and oceans were probably established. Throughout this time, Earth was being almost continually bombarded by objects from space (similar to today's comets and asteroids, but larger); volcanic eruptions on the planet were frequent, not only from the impacting bodies, but also from the active, forming crust and interior. By about 3.8 billion years ago, the bombardment from space died down, allowing the planet to continue to cool down. In addition, life probably began to form around this time.

Because of the reworking of the Earth's crust over billions of years—from erosion, deposition, and the movement of the continental plates—the number and shape of the **continents** before 700 million years ago are not known. And our ideas about what the continents were like millions of years ago are often theoretical, because the information is based on our interpretations of a few outcrops of rock.

In general, it is thought that continents have moved across the surface of the planet for billions of years. Most scientists begin the discussion of continents with the supercontinent Rodinia, a collection of landmasses that hovered around the equator about 700 million years ago. About 500 million years ago, two large continents formed—Gondwanaland (or Gondwana) to the south, and Laurasia (or Laurentia) on the equator. Eventually about 250 million years ago, another supercontinent formed called Pangea (or Pangaea), with even larger landmasses huddled around the equator. And by about 200 million years ago, Pangea began to separate once again into the northern and southern continents (Gondwanaland and Laurasia). By about 50 million years ago, the familiar continents we see today were relatively in place.

Studies of the Earth and Its Origins

Because we have found very few written records, it is thought that most early cultures paid little attention to the Earth's characteristics, except in regard to how the Earth could supply them with shelter, food, and protection. The first records we have about humans determining more scientific features of the planet came during the latter centuries B.C. For example, Eratosthenes of Cyrene was the first to correctly measure the circumference of Earth; he measured noontime shadows at Alexandria and Syene, two localities in Egypt of differing latitudes. His results were amazingly accurate: Measured in stadia (the measurement of length in his day), he calculated that the Earth's circumference was about 24,055 miles (40,320 kilometers). In actuality, the equatorial circumference of the Earth is 24,900 miles (40,064 kilometers); while the polar circumference of the Earth is 24,857 miles (39,995 kilometers).

In the fifteenth century, Leonardo da Vinci studied many topics and was one of the major earth scientists of his time. Da Vinci was a true representative of the Renaissance, and was, in a way, a science generalist. His interests covered biology, astronomy, engineering, and geology. In earth science, he recognized that fossils found at the top of the Alps were the result of material being carried by rivers to the sea and then slowly compacted into sedimentary rock; the rock eventually uplifted to form mountain chains.

By the seventeenth century, earth science studies began to proliferate. For example, besides identifying some of the common fossils, Nicolaus Steno stated the three basic geological principles that are still held to be generally true today: (1) layers of rock are younger as you go upward, now called the law of superposition; (2) layers of rock are initially formed horizontally, now called the law of original horizontality; and (3) every outcrop of rock in which only the edges of the layers are exposed can be explained by some process (for example, mountain building, erosion, Earthquakes, etc.), now called the law of concealed stratification. Each of the laws is very general and is usually verifiable under ideal circumstances.

Some very illogical conclusions about the Earth were also put forth at this time, however. For example, in 1696, William Whiston declared, in his book *New Theory of the Earth*, that the planet was stationary until the Fall of Man; it then began to rotate on its axis. Whiston said that Noah's Flood began on November 18, 2349 B.C., because a comet flew over the equator, causing the water from the planet's interior—and torrential rains—to inundate the land, and he claimed that the strata of rock found on the planet were merely sediments from the Flood.

Other more logical theories based on the scientific methods versus nonscientific methods, also developed about the Earth. For example, in 1759, Giovanni Arduino determined that the Earth's crust was divided into four parts: mountains of mostly rock with metallic ores; mountains of marble and limestone; low mountains and hills of gravel, sand, and clay; and the alluvial materials washed down these mountains by streams and rivers.

In the late eighteenth century, James Hutton, one of the outstanding geologists of his time, was the first to describe the principle of uniformitarianism, which states that the past history of the Earth can be explained by the present; he is also considered the founder of modern geology and geomorphology. He determined the principle of cross-cutting relationships. For example, Hutton stated that an igneous intrusion, a vein of hot magma cutting across a rock layer, must be younger than the rock it intrudes.

The end of the eighteenth century was also known as the "heroic age" of geology, when about three major camps clashed with each other concerning how surface features developed on Earth. The first was called Neptunism, championed by such geologists as Abraham Werner, in which Earth was thought to have formed after being totally submerged under a global ocean. All the rocks were deposited by chemical precipitation; later rock emerged and was deposited by chemical means. A lesser, and more controversial, camp believed in catastrophism, in which all Earth's features were produced by catastrophic events over time. James Hutton rejected catastrophism and proposed the third idea called Plutonism (which also included Vulcanists), recognizing that rock could be created from both aqueous and subterranean processes, such as river deposition and volcanic activity. Thus, for example, heat from molten rock within the Earth forced mountains to uplift. His book, *Theory of the Earth*, proposed this idea—becoming a classic, with many ideas that are still in favor today.

The heroic age was followed by a number of important discoveries in the nineteenth century. In a famous speech to the Swiss Society of Natural Sciences ("Discourse at Neuchâtel") in 1837, Louis Agassiz, an influential geologist, announced that he believed there was an ice age (a term Agassiz adapted from Karl Schimper, who coined the word the year before), with ice sheets that ran from the North Pole to the Mediterranean and Caspian Seas. Although others had suggested this possibility, Agassiz's theory was much grander, with ice caps and huge ice sheets covering Europe—and he especially noted the widespread annihilation of animal life during this period.

Other, more subtle discoveries about the Earth were made in the late nineteenth century. For example, in 1891, Seth Carlo Chandler determined that the Earth's axis—which is tipped about 26.5 degrees from the plane of the planet's orbit and points to an area around the North Star—wobbles as the planet spins. This wobble was and still is thought to be caused by the pull of the Moon as it swings around Earth. The 14-month period of the wobble was also discovered by Chandler and is called the Chandler wobble or period.

By the early twentieth century, one of the first true steps toward the discovery that the Earth's plates moved around the planet was taken by Alfred Wegener. Although his ideas were greatly debated at the time, Wegener first published his theories about the displacement of continents in the *Marburg University Science Journal* in 1912. In 1915, his book, *Die Entstehung Der Kontinente und Ozeane (The Origins of Continents and Oceans)*, expanded on this idea. Between 1903 and 1910, Wegener had investigated the idea that the continents had moved over the face of the Earth over time. He noted that the coastlines of some continents, most notably South America and Africa, appeared to fit together, like pieces in a jigsaw puzzle.

Wegener studied data and papers on geological formations around the world collected by Austrian geologist Edward Suess and others. His resulting theory held that the continents move and continue to move slowly across the surface of the planet. In fact, one anecdote, which may or may not be true, has Wegener realizing that the continents could move by watching ice floes drifting on the sea. At one time, all of the continental landmasses were joined in one large supercontinent, which he named Pangaea (also Pangea, or "all land"). They drifted apart and formed the continents we see today. Collisions between the landmasses produced uplifting, or mountains; the gaps filled with water to form oceans. This theory was controversial, refuting, as it did, most of the leading knowledge of the time—especially that the continents were always in the same place. Also, the properties of rock were thought to preclude this type of slow movement and no obvious driving force for the movement of the continents could be found.

But Wegener never did discover why the continents had moved, and his theory would remain controversial until the 1960s. By then, scientific measurements showed that the continents were indeed moving on giant plates of the Earth's crust. Eventually, scientists discovered one of the mechanisms for their movement—the spreading of the crust along mid-ocean ridges. Thus, Wegener's theory of continental displacement became **plate tectonics**, the basis for modern geology.

One of the major pioneers of plate tectonics was J. Tuzo Wilson, who, in 1965, described the origin of the San Andreas fault—a feature responsible for the great number of Earthquakes that occur in California. He theorized that this fault, where the Pacific plate moves northwest approximately 1 to 1.5 inches (2.5 to 4 centimeters) per year relative to the North American plate, is a transform, or strike-slip, fault between two spreading oceanic ridges within the Pacific plate. In this case, neither plate subducts under the other; rather, they are butted up against one another and moving in opposite directions.

Another pioneer of plate tectonics was Harry Hess, an academic at Princeton University whose theories on the spreading of the crust along the mid-ocean ridges provided a mechanism for Wegener's continental displacement theories. A former naval officer who served in the Pacific during World War II, Hess explored the ocean floor with the new echo-sounder, which was used to detect objects underwater. He believed that the composition of the ocean floor rocks must be denser, because they were lower than the continents. By the late 1950s, he proposed an explanation for what he called a flat-topped guyot or a seamount, which he had seen in deep waters. He believed that molten rock deep in the Earth's mantle pushes its way to the ocean floor, then moves sideways, away from the crest in a continual "conveyor belt" manner. It slowly cools, contracting and sinking all the while. Thus, mountains are carried down below sea level, and are eroded away to eventually become the flat-topped seamount.

The conveyor belt theory soon became known as "sea floor spreading," which not only described the mid-ocean ridges, such as the Mid-Atlantic Ridge, but also explained why the ocean floor is young in geologic terms. Adding to this knowledge was Fred Vine, who determined that the magnetic fields around the mid-ocean ridges reversed in polarity over time. As volcanic rock cools, it becomes magnetized parallel to the magnetic field of the Earth—thus, not only did it prove sea floors do spread, but it also contained records of the Earth's many magnetic field reversals over time. Based on the latest data, scientists believe that none of the rocks on the deep ocean floor are more than 200 million years old. In other words, the continents on the average are 10 times older than the present ocean basins. They also believe that the Earth's crust around the mid-ocean ridges increases by about 1.35 square miles (3.5 square kilometers) of material every year.

The exploration of the Earth in the latter part of the twentieth century has centered on discovering the minute details of the planet. The **layers of the Earth** are being explored by tracking the seismic waves from earthquakes through the planet; more detailed "sonar" type instruments that see deep into the crust are also being used to determine the extent of the crust and locate the major faults that can lead to earthquakes. Many depositional and erosional features are being examined to determine how these geologic features form; in particular, findings from these studies are being used to describe similar features on the planets, moons, and other satellites of the solar system. Even seemingly old geologic features, such as **caves**, are being looked at in a new light; new discoveries about the depth and size of some terrestrial and oceanic caves have recently been made.

Probably just as important as physical aspects of the planet are the **habitats** on the Earth's crust—from the deserts to the oceans. All these habitats developed as a result of geologic changes over time. These important and diverse habitats are continually being studied with the survival of organisms in mind—especially our own human survival. *See also* ATMOSPHERE, CLIMATE AND WEATHER, EARTHQUAKES, EVOLUTION, FOSSILS, GEOLOGIC TIME SCALE, LIFE, MOON, MOUNTAINS, OCEANS, PLANTS, SOLAR SYSTEM, and VOLCANOES.

Topic Terms

cave—A cave is a natural, lightless, underground opening in the Earth's surface. It can be formed in several ways, including the following: (1) the wearing away of soft rock underground by the action of groundwater. (Karst topography—areas of soft limestone rock, usually form extensive cave systems, such as those found at Carlsbad Caverns in New Mexico, and Luray Caverns in Virginia); (2) volcanic tubes that did not collapse, such as those found around Sunset Crater, Arizona; and (3) cliffs worn away by ocean waves, such as those found along the Baja Coast of North America.

Once thought to be mostly devoid of life, caves actually have a very complex food web, and though little vegetation grows in them, caves are teeming with life. Fungi, bacteria, and animals inhabit many caves. The animals within the cave either venture out of the cave to feed or are permanent residents (troglodytes) of the cave. Many of the animals are so specialized in their shelter and food needs that they cannot even move from one cave to another.

The food webs in caves are very fragile, as animals have few food resources. In fact, many of the animals switch from secondary to primary consumers when they enter a cave environment. For example, outside of cave environments, an owl is considered a secondary consumer in the food web, because it is usually eaten by another, higher predator; however, in a cave environment, owls become primary consumers in the food web, as they are usually the primary predator of creatures that come out of the cave at night in search of food.

In general, in a North American or European cave, the food web may include owls preying on mice from the cave, droppings from the mice supporting a dung beetle and millipede population, and cave crickets feeding on carrion of the mouse. Guano (dung) from brown bats is eaten by the blind cave beetles and millipedes, and beetles and millipedes feed the predatory spiders. Even more complex are the relationships among creatures in cave pools, including interactions between specialized bacteria, flatworms, crustaceans, and shrimp that live in the cave.

continents—The continents are the major landmasses on the Earth. Today's seven familiar continents are as follows:

The Continents	
Continent	Approximate Square Miles [Square Kilometers]
Africa	11,670,000 [30,230,000]
Antarctica	5,500,000 [14,240,000]
Asia	17,130,000 [44,370,000]
Australia	2,965,000 [7,680,000]
Europe	4,000,000 [10,350,000]
North America	9,360,000 [24,240,000]
South America	6,880,000 [17,810,000]

habitats—Habitats are defined as having a certain physical characteristic present over a large area. They are separated not only by their diverse physical and geological characteristics, but also by the diversity of organisms—animals and vegetation—that live in the habitat. Each habitat is governed by a certain set of circumstances that arose during the Earth's natural history and each will continue to change slowly over time.

Human activities over the past few centuries have caused and continue to cause changes in natural environment, including changes in the geography of landscapes. Structural developments—from homes to industry—along coastlines often stymie the natural shifting of the land in these areas. Rivers are dammed, creating changes in a region's water supply and in wetland regimes as the water becomes depleted. Mountain regions are strip-mined, leading to changes in land and vegetation that cause top layers of the soil to wash away.

One of our greatest concerns about the environment is how artificial changes affect habitats, such as grasslands, rain forests, and natural forests, especially in South America, the Far East, and Africa. There, land is clear-cut of forests, burned to accommodate growing populations, or used for cattle grazing. But in areas in which there is no clear planning to compensate for the changes, such activities can be detrimental. For example, in northern Africa, overgrazing, deforestation, and burning has caused the topsoil to blow away in several regions. And now, the desert is taking over, a process called desertification. Elsewhere on the planet, the elimination of tropical rain forests, is having a direct negative effect on the species that live in the rich, temperate forests. The final Earth frontiers that have yet to be drastically changed by human activity are the polar regions, especially the continent of Antarctica at the South Pole.

Scientists have divided the modern world into seven major habitats: (1) mountains, (2) grasslands, (3) temperate forests, (4) tropical forests, (5) deserts, (6) polar regions, and (7) oceans. The following definitions describe these major habitats, with the exception of mountains and oceans, which are covered in separate chapters in this volume:

forests—The Earth's land has a diverse selection of forests, based mostly on which vegetation and animals have adapted to a certain geographical region. With the exception of mixed forests (having characteristics of more than one kind of forest) and transitional forests (such as a deciduous forest in the process of overtaking an evergreen forest), forests of the world can be divided into the following general categories:

boreal forests—Boreal forests are the great conifer woods of the north, made up especially of pine, fir, and spruce trees. The climate there includes long, harsh winters and brief summers. The range of this type of forest includes the northern regions of Siberia, Eastern Europe, Scandinavia, Alaska, and Canada—a wide swath of land that stretches more than 7,500 miles (12,000 kilometers) around the northern parts of the planet and that includes the least disturbed habitats in the world. These forests are less diverse than the temperate or tropical forests, as the boreal forests have only grown since the end of the last Ice Age, about 10,000 years ago.

Organisms in the boreal forests vary greatly. Plants other than conifer trees are sparse, because the conifers' needles are highly acidic (making the soil less condusive to plant growth), and the tree limbs shield most of the forest floor from sunlight. This creates a poor environment for plants other than mosses, lichens, and specialized grasses. Most of the animals are well prepared to survive the hard winters and short summers. Larger, major species include bears that hibernate to survive the winter; elk and bison that graze off sparse grasses; and carnivores such as wolverine, lynx, and great gray owl, which have thick fur or feathers to protect them from the cold winters.

temperate forests—The temperate forests are the deciduous woodlands of the temperate regions, including oak, hickory, ash, beech, maple, and numerous other trees. In most temperate forests, even though many types of trees may be present, two or three types of deciduous trees dominate. Temperate evergreen forests are also prevalent in coastal areas with mild winter and heavy rainfall, such as the Gulf Coast and Pacific Northwest of the United States. In general, the temperate forest climate experiences four distinct seasons, with cold winters and warm summers. Rainfall is seasonal, and does not dominate like it does in the rain forests.

The largest temperate forests are found in eastern North America, western Europe, eastern Asia, with some smaller temperate regions that dot the Southern Hemisphere. Another type of deciduous forest includes the tropical deciduous forests found in India, southern Africa, and parts of Central and South America. They are similar to rain forests in appearance and animal life, but experience less rainfall.

Organisms in the temperate forests vary greatly. Plants other than trees include flowering plants and grasses. And because the ground often warms and dries from the Sun, a natural recycling takes place, making the temperate forests home to fungi and bacteria that aid in the decomposition of plants and animals. Larger, major species in the temperate forests include bears, moose, and deer; flying animals such as bats and birds; and climbing animals, such as squirrels and raccoons.

tropical forests—The tropical forests, or tropical rain forests, include the richest forest canopies in the world. Tropical rain forests are found in Central and South America, central Africa, and the Far East, mostly near the equator. The largest continuous, and most well-known, rain forest is the Amazon Rainforest in northern South America. The tropical forests' climate is warm and wet all year long.

The trees within the rain forest vary widely. A small area of a tropical forest may yield close to a thousand different tree species; the same size area in a temperate forest may hold only two dozen different tree species. These forests have distinct layers: The tallest trees are the emergents that break through to the top sunny layer and photosynthesize the most. Most of the wildlife, flowers, and fruits live in this layer of the forest canopy. Beneath the tallest trees, smaller plants, such as younger trees, shrubs, vines, ferns, and mosses grow, forming the understory of the canopy. On the dark, humid forest floor, few plants have gained a foothold because of the limited sunlight. Natural recycling takes place here, as debris rapidly decays in the warm climate through the action of bacteria and fungi.

Organisms in the tropical rain forests are the most varied on Earth. It is estimated that 50 percent of all plant and animal species on the planet live in such regions, because food is plentiful. Plants other than trees include a wide diversity of lush flowering and fruiting plants. Listing the varieties of organisms is beyond the scope of this book. In general, major species in the tropical forests include large animals, such as jaguars, leopards, giant boars, deer, and anteaters; reptiles, such as iguanas; amphibians, such as tree frogs; flying animals, such as birds and bats; and climbing animals, such as monkeys.

deserts—Deserts are geographical regions defined by low precipitation, usually less than 10 inches (25 centimeters) per year; high evaporation rates, exceeding the average yearly rainfall; or certain regional landforms (such as dunes) or vegetation (such as cacti).

About 20 percent of the Earth is covered by deserts. They are found on all the continents. The overall size of a desert can change each season. For example, during the summer, the Sahara Desert of Africa expands by 40 percent; in the winter, more frequent rain causes the desert to shrink its borders. The largest true desert is the Sahara, measuring an average of 3.5 million square miles (9 million square kilometers).

Not all deserts are filled with great shifting sands similar to the Sahara Desert. Most of the world's deserts are rocky and barren, with sparse vegetation. One of the largest "deserts" is not covered with sand or rock, but ice; Antarctica is considered a barren, dry desert measuring an estimated 5 million square miles (13 million square kilometers).

Animals that live in the deserts have to be specially adapted to withstand the arid conditions, daytime heat, and scarcity of food and water. Most desert animals remain inactive during the day to avoid the heat and conserve water. They get their water from eating insects, other animals, or vegetation. Desert vegetation such as cacti and other succulents, exist under arid conditions by special adaptations that allow them to store water. Other desert plants, such as those called "resurrection plants," shrivel and turn brown as they reduce their water content as much as possible. During the occasional desert rainstorm, they rapidly absorb water and the leaves rehydrate, so that they are ready to carry on photosynthesis again.

The desert habitat holds a variety of plants and animals—an entire ecosystem based on limited precipitation and high temperatures.

Deserts are categorized into five types based on the reason for their dryness:

subtropical—These deserts lie between 15 and 30 degrees north and south of the equator and are caused by the circulation patterns in the Earth's atmosphere. As the air is heated and rises along the equator, it carries a great deal of rain; but as it cools toward the tropic of Cancer and Capricorn (both at 15 degrees north and south latitude, respectively), the air cools and dries, then descends and warms. This descending air mass does not allow precipitation to develop, creating desert conditions. Subtropical deserts include the Sahara in North Africa, the deserts of Australia, and southern Africa's Kalahari Desert.

coastal—Air moving over cold ocean currents produces a layer of fog that is carried onto the land. Here, the humidity is high, but the conditions do not cause precipitation to develop, creating an arid land condition. The fogs that reach the shoreline of California carry enough moisture to allow the ground to be damp, thus some succulent-type plants—and even some redwood trees—may grow. Such fogs are also prevalent in one of the driest places on Earth, the Atacama Desert of Chile, South America, in which parts only receive about 0.02 inches (0.05 centimeters) of precipitation annually.

rain shadow—As storms run up the windward sides of mountain ranges, they drop almost all their rain as the air cools from the increase in altitude; thus, the air that moves over the crest and down the leeward side of mountain ranges, is much drier. Several areas along the eastern side of the Rocky Mountains in the United States are rain shadow deserts, as is the desert in Patagonia in southern Argentina, South America.

interior—Asia has the greatest number of interior deserts, caused by the lack of moisture-laden winds reaching deep into the interior of the continent, mostly because the mountains "squeeze" out all the moisture. The Gobi Desert is the largest interior desert.

polar—Polar deserts are found at the south and north polar regions, on the continent of Antarctica and above the Arctic Circle, respectively. Despite precipitation being less than 10 inches (25 centimeters) annually, the polar deserts contain great quantities of water, but in a frozen state. Because of this lack of water and the cold, few animals and plants can survive there. Those that do usually live close to the edge of the ice, near the ocean.

polar regions—Glaciers (or ice sheets) are huge masses of thick ice that cover and slowly move across the land. Glaciers are found in high mountain areas and at the north and south poles (also called the polar ice caps). They form as layers of ice and snow accumulate in the cold regions; the ice and snow eventually compress, crystallizing to form the ice sheet. Continental ice sheets cover extremely large areas.

Mile-plus thick ice sheets once covered the northern regions of the Earth, during a time called the Ice Ages. Peaking in size about 18,000 years ago, the ice sheets covered about one-third of the planet. The sea levels were an estimated 330 feet (100 meters) lower than today because water was tied up in the ice sheets. About 10,000 years ago, the last of the Ice Ages ice sheets retreated. Today, the continental glaciers cover only Greenland and Antarctica, and sea-ice in the Arctic is often considered part of the planetary ice sheet left over from the last Ice Age. (There has been more than one ice age.)

layers of the Earth—Geologists divide the Earth into several major layers, all determined by the way seismic waves from earthquakes penetrate them and by their magnetic, thermal, and gravitational characteristics.

Seeing the Earth from space has always given us a feeling of "the big picture"—or that everything on the planet is tied together.
(Photo courtesy of NASA)

core—The cores, both the inner and outer, make up about 15 percent of the Earth's volume and about 32 percent of its mass. The entire core is extremely dense and is probably made of iron.

inner core—The inner core is thought to be solid, with pressure reaching about 3 million atmospheres (1 atmosphere is the pressure at sea level), and temperatures almost as hot as the surface of the Sun. It is approximatey 800 miles (1,287 km) thick.

outer core—The outer core is thought to have the characteristics of liquid. It extends from the outside of the inner core to the Gutenberg discontinuity, or the border between the mantle and the outer core It is approximately 1,400 miles (2,253 km) across.

mantle—The mantle makes up about 84 percent of the Earth by volume and about 67 percent of the mass. It is thought to be rich in silica, and also contain iron, magnesium, and other metal-rich minerals. The mantle is about 1,802 miles (2,900 km) thick.

crust—The crust makes up about 0.6 percent of the Earth by volume and about 0.4 percent of its mass. It is separated from the mantle by the Mohorovičić discontinuity, and its thickness varies. The ocean crust, mainly basalt, is between 3 and 6.8 miles (5 and 11 kilometers) thick; the continental crust, mainly light rocks such as granite, is between 12 and 40 miles (19 and 64 kilometers) thick.

Other divisions of the Earth's interior include the following:

lithosphere—The lithosphere is the upper layer of the Earth, including the ocean and continental crusts and part of the cooler, more solid mantle.

asthenosphere—The asthenosphere includes part of the upper mantle and lower crust, and separates the crust-mantle lithosphere from the mesosphere. It is thought to be the area where the crustal (continental) plates are slowly "carried" across the planet.

mesosphere—The mesosphere includes part of the upper and all of the lower mantle.

plate tectonics—See MOUNTAINS, Topic Terms

12. Earthquakes

Introduction

The Earth, because of its immensity, seems solid and stable to us. But in reality, the Earth is constantly moving and changing; the pace is so slow compared to our own lifetimes that we are fooled into thinking that everything has been and will remain the same. But when the solid, stable ground beneath us sways and cracks, creating destruction on a massive scale, we are made instantly aware of the powerful forces constantly at work beneath our feet. Earthquake is the overall name that defines these quick, localized movements of the Earth's crust.

Timeline

Date	Event
Modern Events and Discoveries	
~600 B.C.	Greek philosopher, geometer, and astronomer Thales (c.625–c.547 B.C.) suggests that earthquakes are caused by the action of powerful waves hitting shorelines.
~350 B.C.	Greek philosopher and scientist Aristotle (384–322 B.C.) develops his erroneous theory about the origin of earthquakes, a theory that would continue to be accepted, like most of his theories, for centuries.
~132 A.D.	The earliest known seismometer, invented by Zhang (Chang) Heng (78–139 A.D.), records an earthquake that happened 400 miles away, in the area northwest of Loyang, China.
1705	Some of English scientist Robert Hooke's (1635–1702) writings are published posthumously; they advance the theory that earthquakes have major consequences in raising and lowering parts of the Earth's surface.
1760	John Michell (1724–1793), English clergyman at Cambridge University, proposes the theory that earthquakes are waves generated by the movement of rock deep below the Earth's surface.
1783	The first earthquake commission is formed in response to several earthquakes that devastated Calabria, Italy; it leads to the first intensity scale to measure Earthquakes.
1795	Italian naturalist and clockmaker Ascanio Filomarino (1749–1799) builds a seismoscope using a pendulum, bells, and a clock.
1856	Italian seismologist, Luigi Palmieri (1807–1896), the director of an observatory near Mt. Vesuvius, invents the first electric seismograph that produces a trace, or seismogram.
late 1800s	English mathematician and physicist Augustus Edward Hough Love (1863–1940) discovers the earthquake waves that would be named after him, the Love (surface) waves.
1884	English physicist Lord John William Strutt Rayleigh (1842–1919) discovers another surface wave, now called Rayleigh waves.
1857	Irish engineer Robert Mallet (1810–1881) generates isoseismal maps showing the contours of damage and intensity of an Italian earthquake; he also publishes a world map, based on over 20 years of historical earthquake data.
1890s	English geologist John Milne (1850–1913) develops a practical recording seismograph; he is often thought of as the first modern seismologist.
1902	Italian seismologist Giuseppe Mercalli (1850–1914) invents an intensity scale to quantify earthquakes, now known as the Modified Mercalli Intensity (MMI) scale.
1906	British seismologist and geologist Richard Dixon Oldham (1858–1936) uses earthquake waves to determine the Earth's layers.
~1907	U.S. seismologist and glaciologist Henri Victor Reid (1859–1944) develops a mechanism, called "elastic rebound," to describe the origin of the large surface rupture associated with the 1906 San Francisco earthquake.

Date	Event
1912	Alfred Lothar Wegener (1880–1930), a German geologist and meteorologist, first publishes his ideas about the movement of the Earth's continents in the *Marburg University Science Journal*; his idea is ridiculed by scientists until about the 1960s, when many scientists show proof for it through earthquake and other studies.
1925	The Mid-Atlantic Ridge is discovered by a German expedition using sonar.
1935	U.S. seismologist Charles F. Richter (1900–1985) proposes a mathematical formula to scientifically measure the amount of energy released in an earthquake from seismographic readings, and a scale to compare them to each other.
1954	Hugo Benioff publishes a study of earthquakes found beneath oceanic trenches; he believes the zones beneath oceanic trenches are areas of giant thrust faults, where movement of the Earth on both sides of the fault causes earthquakes.
1960	U.S. geologist Harry Hammond Hess (1906–1969) discovers sea floor spreading, which eventually leads to the theory of plate tectonics.
1965	Canadian geophysicist John Tuzo Wilson (1908–1993), a pioneer of plate tectonics, describes the origin of the San Andreas fault, which is responsible for the great number of earthquakes that occur in California.
1975	The first successful prediction of an earthquake is accomplished by Chinese scientists at the State Seismological Bureau in Beijing, China.
1980s	Changes in the natural electric currents in the ground caused by impending earthquakes are studied by Greek scientists; they use the changes to successfully predict the magnitude and location of numerous earthquakes in Greece in 1988 and 1989.
1998	Scientists detect the largest oceanic intraplate earthquake ever and the largest seismic event worldwide since 1994; the 8.2 magnitude quake occurred near the Balleny Islands between Australia and Antarctica.

History of Earthquakes

Earthquakes in Human History

In our recorded history, earthquakes have brought about the fall of civilizations, destroyed large cities, changed the existing geography, and killed and injured countless numbers of humans. No one knows when the first **earthquake** occurred, but it must have happened sometime after the Earth's crust cooled enough to become solid. Over the approximately 4.55 billion years of our planet's history, the tectonic crustal plates that float on the Earth's surface have been slowly moving. Where the plates meet and interact, enormous pressure and stress builds up; it is the sudden movement of these plates relative to each other, with the accompanying release of this built-up energy, that causes vibrations in the Earth, commonly known as earthquakes.

Of the 4.55 billion years of Earth's history, humans in their earliest forms only appeared in the last 2.5 million years. And anatomically modern humans only appeared approximately 100,000 years ago. Cave paintings have been dated at some 22,000 years old, and the earliest form of writing is approximately 5,500 years old. Thus, recorded history, at least by humans, only spans a very short time span, and therefore preserves limited accounts of earthquakes. In China alone, records have only been kept for 3,000 years, during which time more than 13 million people have died in earthquakes.

Early humans had various explanations for why the Earth shook at times. Some thought that the Earth was carried on the back of a giant tortoise, frog, or some other enormous creature, whose movement caused the earthquakes. Other early theories included winds coursing through caves in the center of the Earth, burrowing giant moles, and dancing gods. Each culture, it seems, based their own explanations on their own set of unique circumstances and religious beliefs.

With the age of logic around 600 B.C., as exemplified by the Greek philosophers, the first scientific theories were proposed as to the origin of earthquakes. The Mediterranean area was, and continues to be, an extremely active seismic area, so they had many examples to analyze. From their understanding of the world, they developed numerous theories as to the origin of earthquakes. Thales suggested that earthquakes were caused by the action of powerful waves hitting shorelines, but this theory could not account for the numerous quakes that occurred a long distance inland. Anaximenes proposed that earthquakes were vibrations that occurred as a result of large masses of rock falling from the roofs of great, underground caverns. Archelaus felt that the release of highly compressed gases from deep in the Earth was the cause of earthquakes.

In fourth century B.C., Aristotle, the most influential natural scientist of his time, proposed that warm, still air moved down into large underground caverns in the Earth, where fires superheated it. When the pressure had grown sufficiently, the gases erupted out of the Earth, causing earthquakes and volcanic eruptions. This theory, like most of Aristotle's theories, was taken as the last word in natural science for many centuries, at least in the West.

In China, around, A.D. 132, Zhang (Chang) Heng invented the world's first seismometer to detect earthquakes; it consisted of a suspended pendulum, with linkages to the outside of a pot. The vibrations from an earthquake would shake the pot; the pendulum would remain stationary, precipitating the release of a ball from one of the dragon's mouths that were located around the circumference of the pot. The ball would fall into the mouth of a frog statue underneath it. Which ball was released determined the direction of the center of the earthquake. This device allowed Zhang Heng to record the occurrence and direction of an earthquake that happened 400 miles away, in the area northwest of Loyang, China, before horse-bound messengers brought the news. But no further scientific study went into the causes of earthquakes at that time, and their origins were attributed to the various deities.

Until the mid 1750s, explanations for earthquakes were limited to supernatural, quasi-scientific, or religious reasons. In 1755, many in Europe felt the Lisbon, Portugal, earthquake, which killed approximately 70,000 people, was God's punishment for sin. The survivors of the quake were subsequently killed by the Inquisition for their supposed transgressions. But, a new age of scientific inquiry was starting to flower in the late 1700s, an age that over the next 250 years would completely change the outlook of all the sciences, including the study of earthquakes.

Modern Earthquake Studies

It's only in the last 200 or so years that our understanding of the Earth advanced enough for us to begin understanding the mechanisms and forces behind earthquakes. And, it has only been in the twentieth century that we have come to develop the theories and **earthquake measuring devices** that enable us to explain the occurrence of earthquakes. These theories and devices may someday allow us to predict when and where the next earthquake will strike.

The energy and motion associated with earthquakes are transmitted throughout the planet in waves. The waves are similar to those seen in the ocean except that they move through the ground. In the 1750s, the first theory of earthquakes as waves was put forward by John Michell, a clergyman at Cambridge University. He proposed that earthquakes were waves generated by the movement of rock deep below the Earth's surface, and that there were two types of waves, one preceding the other. He also postulated that the center and speed of an earthquake could be determined by measuring the arrival times of these waves at different locations, a method still in use today—but his theories were ignored for almost a century.

To measure and record these waves, instrumentation had to be invented. The first modern seismoscope was invented in 1795 by Ascanio Filomarino, using a pendulum, bells, and a clock. He used the same principle in his instrument that Zhang Heng used some 1,700 years earlier. A heavy, suspended pendulum would stay stationary during an earthquake, but the rest of the instrument would shake, ring bells, and start a clock. Unfortunately, Filomarino never got to improve on his design, for he was killed by an angry mob on Mt. Vesuvius, who did not like his experiments. They also burned his workshop and destroyed the seismoscope.

By the late eighteenth century, several people became more concerned with how earthquakes affected the local population. For example, in 1783, the first earthquake commission was created in the kingdom of Naples, in response to several quakes that devastated Calabria in southern Italy. The commission's final report led to the creation of the first intensity scale, based on loss of life and damage, that attempted to measure this phenomena.

In 1855, Luigi Palmieri was the director of an observatory near Mt. Vesuvius. He realized that having an instrument that could measure small ground tremors might help in the prediction of impending volcanic eruptions. His resulting **seismograph** has been recording earthquakes since 1856. This instrument has two parts, and makes use of the then new concept of electricity to function. The first part, which has springs, tubes of mercury, and pendulums, detects the earthquakes. The second part has a clock and a mechanism for recording the quakes on a paper strip. With this instrument, Palmieri discovered that small foreshocks were sometimes a precursor to larger earthquakes, and that volcanic eruptions were accompanied by tremors.

Such discoveries also led to other detailed earthquake studies. For example, in 1857, after another earthquake struck Naples, Italy, Irish engineer Robert Mallet examined the damage and generated isoseismal maps showing the contours of damage intensity. He also published a world map, based on more than 20 years of historical earthquake data, that revealed the clustering of earthquake incidences in specific, localized swaths around the planet—one of the first known maps of its kind.

With the advent of better instruments, scientists began to discover distinct types of **seismic waves**. In particular, Augustus Love discovered a surface wave now called the Love wave, which would eventually allow scientists to measure the thickness of the Earth's crust. By 1884, Lord Rayleigh discovered another surface wave, now called the Rayleigh wave.

In the 1890s, John Milne (1850–1913) invented his own practical recording seismograph. Milne, an Englishman, was a geologist who spent much of his career in Japan, studying earthquakes while acting as a professor at the Imperial College of Engineering. At first, he relied on written reports to gather data about quakes. Dissatisfied with this approach, he developed instrumentation that would detect and record the occurrence of earthquakes. His seismograph consisted of three pendulums supported in frames and oriented so that the vertical (Z) and both horizontal (X and Y) motions of an earthquake could be measured. The frames were attached to the ground and would move during an earthquake; the pendulums would stay relatively motionless, the same principle used by others before him. A stylus attached to a pendulum would make a trace on a revolving drum of smoked paper, which was moving with the frame. In this way a record of the motions induced by an earthquake in all three directions could be recorded for later analysis.

In 1893, Milne improved upon his instrument, enabling the trace to be recorded on revolving photographic film. Today, this same instrumentation has been modified to convert the motions into electrical signals, which are then recorded on paper or other media. Because of his continuing studies of earthquakes and his development of instrumentation to measure them, John Milne is considered to be the first modern seismologist.

When the vibrations associated with earthquakes could be accurately and reliably detected and recorded, the traces began to be studied and analyzed by seismologists. As a result, a wealth of information was obtained from the recordings, or seismograms, of an earthquake's waves. In addition to the magnitude and duration of an earthquake, the precise location of the epicenter and focus could be determined. Also, the direction, magnitude, orientation, and extent of the precipitating fault—and sometimes details of the Earth's structure and materials between the location of the seismograph and the earthquake—could be discerned.

By 1912, Alfred Wegener proposed that the continents had slowly moved during the course of Earth's history, but his peers scoffed at the idea. It was not until the work of many scientists came together that Wegener's theory would be proven correct. This work included Hugo Benioff's study of the nature of earthquakes beneath oceanic trenches; Harry Hess's theory of sea floor spreading; Tuzo Wilson's description of the mechanism of the San Andreas fault; and findings small and large by numerous scientists around the world. These studies and discoveries led to our present day understanding of how the continents have moved (now called the theory of **plate tectonics**)—and continue to move—and how that movement produces earthquakes.

More recent work deals with the long-term prediction of earthquakes—a field of earthquake study still in its formative stages. One success story in this field took place in 1975: The first successful prediction of an earthquake was accomplished by Chinese scientists at the State Seismological Bureau in Beijing, China. The Liaoning province in Manchuria began to experience numerous tremors, uplifting, and magnetic field changes during early 1974. Based on this data, scientists predicted a moderate to strong earthquake in the region within two years. After another series of tremors in late 1974, the forecast was refined; they predicted a magnitude 5.5 to 6 (Richter Scale) earthquake to occur in the first half of 1975 in the region around Yingkou, Manchuria. Forewarned by this prediction, and after a large group of tremors, a magnitude 4.8 quake, and then silence, the authorities ordered more than 3 million people in southern Liaoning province to evacuate their homes on the afternoon of February 4, 1975. At 7:36 P.M., a large earthquake struck, destroying most of the buildings in Yingkou and Haicheng, and damaging bridges and roads. It could have been worse—only approximately 300 people lost their lives thanks to the advance warning and evacuation.

Although triumphs, like the 1975 prediction, have occurred, there have been many more failures—quakes that did not occur as predicted or occurred without warning. It often depends on accurately measuring and interpreting the precursors to earthquakes, such as foreshocks, changes in ground strain, tilt, resistivity and elevation, changes in magnetic and gravitational fields, ground water level shifting, radon gas emission, sounds, flashes of light, and the behavior of animals. In China, the appearance of panicky rats is an official earthquake precursor.

More advances have been made on making structures safe during earthquakes. Most modern buildings in quake-prone areas are now constructed to be resistant to the waves that emanate out from the focus and epicenter. New construction techniques, materials, and designs have been developed to prevent collapse and structural damage from occurring.

As our modern, technology-driven civilization has expanded worldwide, the potential for damage and loss of life has increased. Earthquakes, in and of themselves, are not directly responsible for the many deaths associated with them; it is the effect they have on structures and the surrounding areas that normally leads to loss of life. For this reason, scientists have extensively studied the causes of past earthquakes, the forces at work in the depths of our planet, how the energy is distributed from the center of an earthquake, and how that energy interacts with and affects human-made and natural structures. Their hope is to find a way of predicting where and when the next quake might occur, to be able to predict the effects of a quake on the surrounding area, and to design and build structures that can withstand the forces associated with earthquakes. *See also* EARTH, FOSSILS, GEOLOGIC TIME SCALE, HUMANS, MOON, MOUNTAINS, SOLAR SYSTEM, and VOLCANOES.

Topic Terms

earthquake—An earthquake is simply a vibration at the Earth's surface and can be triggered by numerous events, from nuclear weapons tests to volcanoes. Most earthquakes, however, are a direct result of the slow, natural processes taking place within the Earth itself. Most, but not all, earthquakes occur along faults, or cracks, in the crust of our planet. Some happen deep within the crust at depths measured in miles, while others happen relatively close to the surface. Faults between sections of the ground are not normally completely vertical or horizontal, but can be at any angle. Movements of the ground along these faults can be in a vertical direction, called dip-slip; a horizontal direction, called strike-slip; or a complex combination of both.

When both sides of the ground become locked together at a fault, normally due to friction, the pressure and strain begin to build up. The forces at work in the Earth that slowly move the ground are much too great to be stopped by this blockage, and so eventually they overcome the friction. If the amount of friction is small, the ground will move in small increments, in a process known as creep. This produces small earthquakes, or tremors. But when the amount of friction is very high, and the ground becomes firmly locked together, the amount of energy stored in the deforming rocks as they are squeezed greatly increases, until the blockage is overcome in one large movement. This results in the great earthquakes.

Two major points are used when describing an earthquake. The first is the focus, or hypocenter—the precise spot where the initial movement of the ground occurs. This spot is usually found some distance, sometimes miles, below the surface of our planet. It is here where the built-up energy is converted into waves that move throughout the interior of the planet toward the surface. The second is the epicenter of an earthquake, a more familiar term—the spot on the surface of the Earth directly above the focus. Because this spot on the surface is closest to the focus, the interior waves arrive here with the most energy; and the interior waves become surface waves carrying the most energy and destructive potential.

Many large urban areas are located in earthquake-prone parts of the world. A sampling includes Athens, Beijing, Cairo, Hong Kong, Istanbul, Jakarta, Los Angeles, Manila, Mexico City, Rome, San Francisco, Singapore, Teheran, Tokyo, and many others. As their populations and structures grow, the potential for damage and loss of life increases dramatically. Japan, with its dense population centers, has nearly 1,000 felt quakes per year. It receives nearly 10 percent of the world's total annual seismic energy. To mitigate these circumstances, scientists and engineers are looking for ways to predict the occurrence of quakes, and to design and build earthquake-resistant structures, given our current understanding of the mechanisms and consequences of earthquakes.

Ancient Earthquakes

Date	Location	Approximate lives lost
115, 458, 526 A.D.	Antioch, Asia Minor	?
856	Corinth, Greece	45,000
893	Ardabil, Iran	100,000
1138	Aleppo, Syria	230,000
1290	Chihli, China	100,000
1456	Naples, Italy	60,000
1556	Shensi (Shansi), China	830,000
1693	Naples, Italy	93,000
1731	Peking (Beijing), China	100,000
1737	Calcutta, India	300,000
1755	Lisbon, Portugal	70,000
1868	Ibarra, Ecuador	70,000

Modern Earthquakes

Date	Location	Approximate lives lost
1908	Messina, Italy	160,000
1920	Kansu (Gansu), China	180,000
1923	Tokyo, Japan	143,000
1927	Tsinghai (Xining), China	100,000
1935	Quetta, Pakistan	60,000
1960	Agadir, Morocco	14,000
1970	Ancash, Peru	67,000
1976	Guatemala	23,000
1976	Tangshan, China	reports range from 255,000 to 650,000

earthquake measuring devices—Over the last two hundred years, attempts have been made to measure the amount of energy released during an earthquake and its effects on a local area. Two scales have achieved prominence, the Mercalli Scale and the Richter Scale. Both measure two different facets of an earthquake; although the Richter Scale is the one used most prominently today, the combination of the two gives the most information about an earthquake and its effects.

Intensity is a somewhat subjective measurement of the strength of an earthquake based on the effects on the local population and structures. It is classified primarily on direct observations and interviews. Intensity depends on many factors: the amount of energy released, or magnitude, of an earthquake; distance from the epicenter; type of ground; type and quality of construction; characteristics of the seismic waves; and other details. The Mercalli Scale is the scale most commonly used to measure these characteristics, categorizing the intensity on a scale of I to XII.

This was the result of an earthquake-induced rock and snow avalanche on Mt. Huascaran, Peru in 1970. The rock and snow buried the towns of Yungay and Ranrahirca, resulting in the deaths of more than 66,700 people. *(Photo courtesy of the National Geophysical Data Center, NOAA)*

Modified Mercalli Intensity Scale (1931)

Number	Comments
I	Not felt except by a very few under especially favorable circumstances.
II	Felt only by a few persons at rest, especially on upper floors of buildings. Delicately suspended objects may swing.
III	Felt quite noticeably indoors, especially on upper floors, but many people do not recognize it as an earthquake. Standing motor cars may rock slightly. Vibration like passing truck.
IV	During the day felt indoors by many, outdoors by few. At night some awakened. Dishes, windows, doors disturbed; walls make creaking sound. Sensation like heavy truck striking building. Standing motor cars noticeably rocked.
V	Felt by nearly everyone; many awakened. Some dishes, windows, etc., broken; a few instances of cracked plaster; unstable objects overturned. Disturbances of trees, poles, and other tall objects sometimes noticed. Pendulum clocks may stop.
VI	Felt by all; many frightened and run outdoors. Some heavy furniture moved; a few instances of fallen plaster or damaged chimneys. Damage slight.
VII	Everybody runs outdoors. Damage negligible in buildings of good design and construction; slight to moderate in well-built ordinary structures; considerable in poorly built or badly designed structures; some chimneys broken. Noticed by persons driving motor cars.
VIII	Damage slight in specially designed structures; considerable in ordinary substantial buildings, with partial collapse, great in poorly built structures. Panel walls thrown out of frame structures. Chimneys, factory stacks, columns, monuments, and walls may fall. Heavy furniture overturned. Sand and mud ejected in small amounts. Changes in well-water levels. Persons driving motor cars disturbed.

Modified Mercalli Intensity Scale (1931) (cont'd)	
Number	Comments
IX	Damage considerable even in specially designed structures; well-designed frame structures thrown out of plumb; great damage in substantial buildings, with partial collapse. Buildings shifted off foundations. Ground cracked conspicuously. Underground pipes broken.
X	Some well-built wooden structures destroyed; most masonry and frame structures along with foundations destroyed; ground badly cracked. Rails bent. Landslides considerable from river banks and steep slopes. Shifted sand and mud. Water splashed over banks.
XI	Few, if any, masonry structures remain standing. Bridges destroyed. Broad fissures in ground. Underground pipelines knocked completely out of service. Earth slumps and land slips in soft ground. Rails bent greatly.
XII	Damage total. Waves seen on ground surfaces. Lines of sight and level distorted. Objects thrown upward into the air.

** This scale, first developed in 1902 by geologist Giuseppe Mercalli, and later modified, attempts to quantify the intensity of earthquakes by direct observation of the types of damage inflicted on an area. It is not commonly used today.

Magnitude is an objective determination of the amount of energy released by an earthquake, as measured by a instrument called a seismograph. It does not depend on local observation like the Mercalli Scale, but rather the amplitude of the resulting seismic waves at a fixed distance from the epicenter, and therefore it can be measured and calculated from any measuring location in the world. The Richter Scale is the scale most commonly used for this purpose.

The Richter Scale was first proposed in 1935 by Charles F. Richter as an objective measurement of the magnitude, or energy released by an earthquake; it is a mathematical construct, not a physical scale like a ruler. The numbers associated with this scale are representative of the maximum amplitude of seismic waves that occur 62 miles (100 kilometers) from the epicenter of an earthquake. Since seismographs are never located at this exact distance from an earthquake, the measurements taken by remote instruments must take into account the difference in time between the arrival of the primary and secondary waves, as well as account for the attenuation of the waves as a function of distance.

Also, the scale is not linear, but logarithmic in nature; that is, each increase in whole number value represents 10 times the increase in power. An earthquake with a magnitude of 8 on the Richter Scale has waves with amplitude 10 times greater than one with magnitude of 7, with an energy release approximately 31 times greater.

Richter Scale of Earthquake Magnitude

Magnitude	Energy Equivalent (in weight of TNT)	Earthquake Example
1.0	170 grams	
1.5	900 grams	
2.0	5.9 kilograms	
2.5	28 kilograms	
3.0	179 kilograms	
3.5	450 kilograms	
4.0	5.5 metric tons	
4.5	29 metric tons	Denver, Colorado (1965)
5.0	181 metric tons	
5.3	455 metric tons	San Francisco, California (1957)
5.5	910 metric tons	
6.0	5.7×10^3 metric tons	
6.3	14.4×10^3 metric tons	Long Beach, California (1933)
6.5	28.7×10^3 metric tons	San Fernando, California (1971)
7.0	181×10^3 metric tons	
7.1	228×10^3 metric tons	El Centro, California (1940)
7.5	910×10^3 metric tons	
7.7	1811×10^3 metric tons	Kern County, California (1962)
8.0	5706×10^3 metric tons	
8.2	$11,421 \times 10^3$ metric tons	San Francisco, California (1906)
8.5	$28,711 \times 10^3$ metric tons	Anchorage, Alaska (1964)
9.0	$181,999 \times 10^3$ metric tons	

Well-Known Earthquake Examples

Richter Number	Year	Location
9.6 (not official)	1960	Chile
8.3–8.6 (estimate)	1899	Yakutat Bay, Alaska, USA
8.5	1964	Alaska, USA
8.0–8.3	1811–1812	New Madrid, Missouri, USA
8.3	1994	Bolivia, South America
8.2	1976	Tangshan, China
8.1	1985	Mexico City, Mexico
7.7–8.25 (estimate)	1906	San Francisco, California, USA
7.7	1990	Northwest Iran
7.1	1989	Loma Prieta, California, USA
6.9	1988	Armenia
6.8	1995	Kobe, Japan
6.8	1994	Northridge, California
6.5	1971	San Fernando, California, USA

elastic rebound—This mechanism, first proposed by H.F. Reid, is still the most widely used today to explain earthquakes and surface ruptures, though it is not without its faults. Reid observed that physical objects, such as roads, streams, and fences, were deformed in the areas surrounding San Francisco for years before the 1906 quake; after the quake, they had mostly returned to their normal state, but were offset by as much as 21 feet (6.4 meters). He theorized that at the fault itself, the ground had become locked together due to friction, but that the ground on each side of the fault had kept slowly moving in opposite directions, leading to elastic deformation. When the buildup in force became great enough, the friction at the fault was overcome, and the ground moved, snapping back to its origin state. This original state was, however, then offset due to the motion of the ground on each side of the fault over the course of many years. This quick motion, and its accompanying release of stored energy, was the origin of the great San Francisco earthquake of 1906. The fault Reid described is now known as the San Andreas fault, which runs through the southern half of California.

This old photo was taken in 1946, when a huge tsunami—gigantic waves thought to be produced by earthquakes on the ocean floor—struck Hawaii. The arrow points to a man who was never seen again. *(Photo courtesy of the National Geophysical Data Center, NOAA)*

plate tectonics—See MOUNTAINS, Topic Terms

seismic waves—The two types of seismic waves that are generated by an earthquake are called body waves and surface waves. Body waves are transmitted through the Earth from the focus, or hypocenter, of the quake, spreading out in all directions, similar to material ejected into space from a star that has exploded. There are two kinds of body waves, known as P waves and S waves. The P, or primary, waves travel the swiftest, moving away from the focus at speeds up to 4 miles per second, and are the first to show up on a seismograph. They behave similarly to sound waves in that they compress and expand the materials they travel through in the same direction that they travel. The effect, when they reach the surface, is to move everything in vertical directions, up and down. These are the waves that, if powerful enough, will compress the air and create a characteristic roar like a train.

The S, or secondary waves, travel slower through the Earth than the P waves, since they displace the materials that they travel through in a side to side, or shearing motion. They can be thought of as analogous to an ocean wave rippling through the Earth, from the focus to the surface. They are the next waves to be registered on the seismograph. On reaching the surface, they move everything vertically and horizontally, resulting in more damage than the vertical-only P waves.

When the body waves finally reach the surface, they give birth to movements along the ground, called surface waves. There are two kinds, called Love and Rayleigh waves. The Love waves move the ground in a side to side motion, perpendicular to the direction of the wave's travel. They are similar to the motion of a snake as it travels along and only register as horizontal movements on the seismograph. The Rayleigh waves, on the other hand, can be thought of a ocean waves that ripple along the ground, creating vertical, as well as horizontal, movement on the seismograph.

seismograph—Seismographs capture earthquake motion, also called seismic waves. In general, the different types of waves arrive at different times. They are detected by a delicately balanced seismograph, which is anchored to the ground. When the ground shakes, a device suspended within the seismograph makes lines (or if done electronically, a light beam makes lines) on a slowly scrolling paper or film; the lines make a record of the quake called a seismogram. The seismogram becomes a record of how much energy was released at the quake's focus. By comparing the arrival times of the different waves on various seismographs throughout the world, scientists can determine the exact location of the earthquake. Other information obtained from the seismogram includes the strength of the quake and how long it lasted.

A more recent method of earthquake analysis is called seismic tomography—a kind of X-ray of the Earth's interior. Scientists hope to determine more about the planet's structure by using seismic tomography along with seismography.

13. Evolution

Introduction

Evolution simply means changes in organisms and physical features of the Earth over time. That change happens every day is obvious to us; but evolution means change in the very long term. Indeed, the evolutionary time scale, measured in billions of years, is normally beyond human comprehension, since our own lifetimes are so relatively short. In addition, much of the evidence of evolution has been lost, eroded by the Earth's natural physical processes, such as erosion of fossil-filled rock. Because of this, more controversies and conflicting data are seen in evolution (of flora, fauna, and landscape development) than in most scientific fields. Current controversies include the continuing debates on the nature of evolution; the evolution of humans, other animals, and plants; and the evolution of physical structures on Earth and other bodies of the solar system. The various facets of evolution encompass some of humankind's most fundamental mysteries and most interesting challenges.

Timeline

(note: bya=billion years ago; mya=million years ago)

Date	Event

Prehistoric Events

Date	Event
~3.75 bya	The earliest form of life is thought to have evolved around this time; although recent unconfirmed findings show it may be closer to 3.85 billion years ago.
just before ~544 mya	The first primitive multicelled organisms evolve around this time.
~544 mya	The Cambrian Explosion occurs, in which the ancestors of most modern animal groups suddenly (geologically speaking) evolved in the oceans.

Modern Events and Discoveries

Date	Event
1550	Italian scientist Girolamo Cardano (1501–1576) writes *De subtilitate rerum* on natural history, implying a belief in the idea of evolutionary changes.
1686	English naturalist John Ray (1627–1705) defines the term "species" based on common descent.
1748	French natural philosopher (? –1738) Benoît de Maillet proposes a theory of evolution based on a retreating sea.
1749	French scientist Comte de Georges-Louis Leclerc Buffon (1707–1788) modifies John Ray's ideas on the term "species," paving the way for the modern definition; such a definition helps to further the theory of evolution.
1751	Although Swedish naturalist Carl von Linné (Carolus Linnaeus) (1707–1778) is the first to introduce a classification system to keep track of organisms on Earth, he rejects the idea of evolution.
1800–1812	French naturalist and anatomist Baron Georges Léopold Chrétien Frédéric Dagobert Cuvier (1769–1832) suggests, based on his studies of fossil bones of extinct animals, that there have been many "revolutions or catastrophes" in Earth's long history; Cuvier is one of the first proponents of catastrophism.
1802	In his book *Natural Theology*, English scientist William Paley (1743–1805) proves the existence of God using evidence of adaptation in nature.
1809	French philosopher Jean Baptiste de Lamarck (1744–1829) proposes that the environment causes species to evolve; he is considered one of the first evolutionists.
1844	English scientist Robert Chambers (1802–1871) publishes a theory on the development of species; it is published anonymously in *Vestages of Creation*, a book that is attacked by many famous scientists of the time.
1850s	English naturalist Alfred Russel Wallace (1823–1913), a contemporary of English naturalist Charles R. Darwin (1809–1882), independently develops the theory of evolution by natural selection at the same time as Darwin; his conclusions are based on his study of the flora and fauna of Brazil and Southeast Asia in the 1840s.
1856	Fossils of Neanderthal (or Neandertal) man are found in a cave in the Neander Valley of Germany.

Date	Event
1858	Wallace and Darwin make public the theory of natural selection in a joint paper.
1859	Darwin publishes his book, *On the origin of species by means of natural selection of the preservation of favoured races in the struggle for life* (shortened to *On the Origin of Species*).
~1860s	English biologist Thomas Huxley (1825–1895) and French anthropologist Pierre-Paul Broca (1824–1880) defend Charles Darwin's theory of evolution, pointing out that the recently discovered Neanderthal fossils are an extinct human form.
1861	English naturalist Henry Walter Bates (1825–1892) proposes that a non-toxic species can evolve to look like a toxic species to avoid being eating by a predator, called Batesian mimicry.
1864–1868	Wallace rejects the possibility that man's higher faculties have arisen, or "evolved," by chance.
1871	Naturalist Charles Darwin publishes *The Descent of Man*, applying his theories to humans.
1925	The Scopes "Monkey Trial" begins in Dayton, Tennessee.
1942	English biologist Julian Huxley (1887–1975) proposes the Modern Synthesis concept of evolution.
1967	U.S. geneticist Sewall Wright (1889–1988) is awarded the National Medal of Science for his mathematical theory of evolution, which argues that mathematical chance—and mutation and natural selection—affect evolutionary changes.
1972	Stephen Jay Gould and Niles Eldredge propose the hypothesis of punctuated equilibrium.

History of Evolution

Early Evolution

The **evolution** of organisms is thought to have begun more than 3.75 billion years ago. The first organisms were simple single-celled organisms. Billions of years later, multicelled organisms evolved. (See chapter on life.)

But no one truly knows when multicelled animals evolved. Some scientists point to the existing fossil record, noting that fossils of the first multicellular organisms appear in the record just a few million years before the Cambrian Explosion, a time of profuse animal growth. Other researchers who have looked at the DNA sequences of genes in certain animals say animals evolved much earlier. One group claims that the branching of several invertebrate phyla took place about 1.2 billion years ago; while invertebrates diverged from chordates (the phylum that contains vertebrates) about a billion years ago. As is usual in many ideas in paleontology, there is not enough data to support the theories. In particular, early fossils are hard to find because these animals only had soft parts and thus, would not have survived hundreds of millions of years of burial to form fossils.

The Theory of Evolution

The theory of evolution is a field in its infancy, with roots going back less than two centuries. And its attempt to describe a long history is fraught with much debate, interpretation, and discovery.

The idea of evolution started only after explorers and scientists began to seek out new plants and animals in the Western Hemisphere. For example, although Carl von Linné introduced a classification system to keep track of new organisms being constantly found on Earth, he still rejected the idea of evolution. He and many of his contemporaries believed in the religious concept of omnipotent creation—stating there were just as many **species** present today as there were in the beginning of creation. Another person in this field was William Paley, the eighteenth-century theologian and creationist who proposed one of the most famous arguments in defense of this theory: that of the watchmaker. Essentially, the theory holds that a watch must have been designed and built by a watchmaker, being too complex and functional to have come into existence by chance. Paley believed so too must life, as it was even more complex, have been purposely designed and built by some divine authority—or watchmaker.

And of course, some attempts at theories of evolution never took hold. For example, in 1748, Benoît de Maillet believed that evolution was the result of the retreating sea. He thought that the universe was filled with "seeds" that fell into the oceans. When the ocean gradually retreated, the seeds would grow and evolve into land animals.

The nineteenth century brought about an explosion in naturalist studies, which precipitated a turnabout in ideas on evolution. The rich plant and animal life were welcome—and expansive—additions to collections of specimens. For example, scientist-explorers such as German Baron Alexander von Humboldt, naturalist Henry Walter Bates, and naturalist Alfred Russell Wallace working in South America's tropical rain forests, and with others in Africa, made amazing discoveries in natural history. Naturalists searched for new species of organisms from the myriad shells, plants, birds, moths, butterflies, and mammals. These many searches brought on new theories—including thoughts on the evolution of organisms.

In 1809, Jean Baptiste de Lamarck—considered one of the first evolutionists—proposed the existence in organisms of a built-in drive toward perfection; the capacity of organisms to become adapted to "circumstances" ("environment" in our own modern terminology); the frequent occurrence of spontaneous generation; and the inheritance of acquired characters, or traits. Not everyone agreed with the last statement, including German biologist August Weissman (1834–1914), who believed it was impossible. Such disagreements about Lamarck's statements on the evolution of organisms helped shape Darwinism.

Although Charles R. Darwin was not the first to come up with the idea of evolution, he had the biggest influence on it as a theory. He changed the way scientists thought of evolution and **natural selection**. One of the most well-known stories of science concerns Darwin's reluctance to announce his theory on evolution—at least until Alfred Wallace mailed the same conclusions to Darwin. Darwin arranged to present the idea of evolution with Wallace, so that neither would have the credit over the other. Some researchers believe that it was Darwin who had the greatest effect on the theory's popularity—and that had Wallace proposed the idea on his own, it would not have made as much of an impact.

Darwin's book, *On the origin of species by means of natural selection of the preservation of favoured races in the struggle for life* (shortened to *On the Origin of Species*), published in 1859, helped to change the field of biology forever—and went against the beliefs of Victorian England by contradicting the biblical version of creation. The effect was almost immediate, with the first edition sold out on the day of publication (November 24, 1859)—and thus, the book was referred to as "the book that shook the world."

Darwin had formulated his theory as a result of his studies of animals and plants during the round-the-world voyage of the H.M.S. *Beagle* (1831 to 1835) that took him to remote corners of the globe. The most noteworthy place he visited was the Galapagos Islands. There, his study of life-forms that had been isolated from the rest of the world for a long period of time gave rise to his theory of evolution. It also sparked Darwin's idea of **adaptive radiation**—organisms adapt to their immediate surroundings (including adapting genetically to them). It took Darwin 20 years to conceive and refine the theory of evolution.

The other conceiver of the evolution theory was naturalist Alfred Russel Wallace, a Welsh-born scientist who devised the theory of evolution independently of Charles Darwin. Similar to Darwin, Wallace had read *Essay on Population*, by Thomas Malthus, which helped him develop an evolution theory. In about two days, during a bout of malaria in the jungles of Borneo in Southeast Asia, Wallace wrote his evolution theory.

Other evidence soon came out in support of evolution. For example, in 1825, English naturalist Henry Walter Bates developed a theory of insect mimicry, an idea that eventually contributed to the acceptance of Darwin and Wallace's theory of evolution. Bates continued his work in 1861, by suggesting that nontoxic species, or those that usually pose no toxic threat to other animals, can look toxic—especially in terms of special colors and color patterns. He believed that certain species did this to survive being eaten by a predator, in a process now known as **Batesian mimicry**. In other words, mimicry was the result of an evolution of a species, in which it adapted a favorable survival trait to meet a certain set of circumstances.

Even the 1856 discovery of the fossils of Neanderthal (or Neandertal) man in Germany seemed to support evolution. It is now accepted that these fossils represent an extinct species of humanity; but it took decades to accept that conclusion, with the biggest problem being the unknown age of the bones. Three years after the bones' discovery, Darwin's book was

> **Interpreting Evolution Evidence**
>
> The idea of evolution to most of us means hundreds of thousands or millions of years worth of time to "create" different species of higher orders of complexity. But in reality, evolution may be occurring right in front of our noses.
>
> One example of evolution often sighted by gradualists is that of a type of moth called *Biston betularia*, found in England. This moth was historically mottled with a brown and white coloration that acted as a camouflage similar to lichens found on the local trees. As the Industrial Revolution reached England, the trees and lichens became covered with gray soot from the coal-fired industries and homes. The mottling no longer allowed the moths to blend in and predators could easily see them. By the 1860s, these now dark gray-colored moths began to appear in increasing numbers; this coloration allowed them to blend in with the soot-colored trees and avoid the predators. As a result, by the early 1900s, the dark gray moths were predominant. As the use of coal has diminished in the last half of the twentieth century and the air has become less sooty, so have the trees and lichens. Now, the mottled-colored moths are increasing in number.
>
> Is this an example of Darwin's theory of evolution? Yes, say supporters of gradualism. Because of the changes in the environment, natural selection has weeded out those organisms with negative adaptive characteristics—for example highly visible coloration—and allowed those with the camouflage coloration to survive. No, say others. No evolution has occurred, since there has been no increase in complexity and the moths are still the same species. There has only been an emphasis or de-emphasis of traits, i.e., different colorations, that were always present in the populations. The changes were only in the percentages present in the populations, not the organisms themselves.

published and the controversy heated up. People like English biologist Thomas Huxley and French anthropologist Pierre-Paul Broca proposed that Neanderthal man represented an extinct form of humans. On the other end of the spectrum, people like German anatomist F. Mayer believed the bones were those of a Cossack cavalryman who deserted the Russian army as Napoleon retreated across the Rhine in 1814—complete with bowed legs from sitting on a horse for years.

For those who did believe in a form of evolution, there had to be an explanation of how fast or slow the actual process occurred. The theory of evolution that champions the slow, gradual build-up of changes was first put forward by Charles Darwin in 1859 with his publication *On The Origin of Species by Natural Selection* (another short way to write Darwin's wordy book title); this type of evolution is known as **gradualism**.

Other people who supported the idea of gradualism included Richard Dawkins. He did not propose a constant, steady rate of change, however; he felt there was an organized procession of changes that happened in small steps resulting in intermediate stages between large changes. His central thesis, like Darwin's, was that natural selection was unconscious, automatic, and random, with no purpose or design in mind. In fact, he compared evolution to a blind watchmaker, modifying William Paley's analogy of the watchmaker who designs complex life-forms.

Another theory of evolution was called **catastrophism**. Catastrophism opposed Darwin's theory and tried to explain the gaps in the fossil record. This theory was first proposed by Georges Cuvier, an early nineteenth-century French geologist and naturalist, and held that periodic devastations wiped out almost all of the previously existing species. New forms of flora and fauna replaced the extinct species. Though this theory was based on geological and fossil evidence—in the form of periodic catastrophes and species extinction—gaps in scientific knowledge at the time led to several problems with its interpretation. Because of this, and because many geologists and naturalists at that time had religious leanings, many flavored the theory with a creationist point of view. With the rise of Darwin's predominately scientific theory, catastrophism fell into disfavor.

Although gradualism emerged as the favored theory, increasing amounts of evidence brought this theory into question, such as gaps in the fossil record; it was not until 1980 that the theory of catastrophism was essentially reborn, based on solid, scientific results. In that year, the team of Walter and Luis Alvarez, Frank Asaro, and Helen Michel published the results of their study of the Cretaceous-Tertiary boundary in Italy. The thin clay layer that separates limestone formations was found to have a high concentration of iridium, a noble metal that is almost completely absent from the Earth's crust. These high levels of iridium, subsequently discovered at this boundary in areas all around the world, must have come from outer space, by means of large impacts, or from the Earth's core, by means of volcanic action. With the recent discovery of extremely large impact craters around the world, such as the Chicxulub crater in Mexico, the catastrophic theory has again gained prominence, although the inter-relationship between impacts and volcanic activity is still being sorted out.

Despite the development of these new theories, the belief in **creationism**—a theistic, or religious, theory explaining the appearance of life on Earth—remained strong, even among scientists. To help explain creationism in light of numerous fossil discoveries, still another theory developed, called **theistic evolution**, an amalgamation of religious and scientific attitudes toward evolution held by many modern theologians. This theory represented an adjustment to creationism, which was modified slightly over time in response to modern scientific findings; after Darwin's theory of evolution was published, many theologians accepted and adapted to these new findings. Theistic evolution holds that a divine force, or God, was behind creation, as prescribed by their religious beliefs. However, they believe this divine force works through the process of evolution, as revealed by modern science.

All of these theories are just that—theories that more or less postulate about the rise of the myriad lifeforms on Earth. None of them perfectly explains all of the current data available. Like life itself, the theories of evolution are continually adapting and changing over time.

In the twentieth century, evolution continues to raise questions and cause debate. Most of the debate arises from theological and cultural concerns. The rise of fundamentalist religion in the United States of America after World War I led to the passage of laws by some states that made the teaching of the theory of evolution unlawful. Tennessee was one such state; its Butler law made it unlawful for any teacher in a publicly supported school "to teach any theory that denies the story of the divine creation of man as taught in the Bible, and to teach instead that man descended from a lower order of animals." To challenge the legality of this law and to create publicity for the town of Dayton, a group of men that included the teacher John Scopes, decided to violate the law and thus create a

The Grand Canyon in Arizona shows how layer upon layer of sediment is deposited over time. The lower layers are billions of years old; the upper layers are younger, holding various species of fossils that show evolution of life during this period.

test case. The ensuing trial, which lasted from July 10 to 21 of 1925, attracted worldwide attention and was covered by scores of journalists and broadcast on the then new device called radio.

Arguing for the prosecution and fundamentalism was William Jennings Bryan, the most celebrated orator of his day. The defense was led by Clarence Darrow, at that time America's most famous criminal lawyer and a proponent of evolution. (The trial was the subject of the movie, *Inherit the Wind*.) From a legal standpoint, the trial centered around academic freedom, tolerance, science, evolution, and religion. From a publicity point of view, it centered around the emotional clash between the two great lawyers. In the end, Scopes was found guilty but only fined the minimum amount. On appeal, the Tennessee Supreme Court overturned the conviction on a technicality. The test case had fizzled out and the Butler law still stood.

Despite this sort of opposition to the theory of evolution, scientists continued to make new discoveries and adjustments to the theory. One of the most important steps in evolution theory occurred in the early 1940s, when people such as Julian Huxley proposed the Modern Synthesis concept of evolution. This idea assumed the pace of evolutionary change was slow and governed by natural selection that included small variations—with the variants being environmentally superior, allowing a species to better survive.

Expanding on these ideas and recent discoveries—especially in the fossil record—led to other important developments. One in particular, called **punctuated equilibrium,** proposed by Niles Eldredge and Stephen Jay Gould in 1972, explained the pattern of rest followed by abrupt changes in the evolutionary records. Thus, the gaps in the fossil record did not indicate incompleteness, merely sudden changes. In particular, this theory also stated that once a species is around, it remains unchanged for most of its history. But when a change occurs, it happens quickly. Not only does this theory try to explain gaps in the fossil record, but it also incorporates the latest findings on impacts from large objects from outer space. This theory is in contrast with gradualism, which states that changes result from the steady accumulation of small modifications to the species. To this day, these two diverse ideas have not been fully reconciled.

And other questions remain unresolved. For example, in evolutionary biological studies, it is difficult to separate function from adaptation, for example, how to identify the features of the nervous system in terms of adaptation versus function.

Besides these issues, other problems bog down the evolution discovery process. Although modern anthropologists continue to discover human, animal, and plant fossils to show how different species changed over time, there are still certain "missing links" for many animal species—even the ubiquitous "human from ape" missing link. Plus, time has not been on our side; many fossils did not survive burial or have been worn away by natural processes. It will take time to fill in the gaps in our overall knowledge—if the evidence exists at all.

But all the scientific discoveries do not mean that evolution is totally accepted around the world, much less in the United States. Many religious beliefs still run contrary to the idea of **human evolution**. Although a great deal of fossil evidence has been found to help confirm the idea of evolution, certain religions do not accept this point of view. *See also* AMPHIBIANS, ANIMALS, ARTHROPODS, BIRDS, CLASSIFICATION OF ORGANISMS, DINOSAURS, EXTINCTION, FISH, FOSSILS, FUNGI, GENETICS, GEOLOGIC TIME SCALE, HUMANS, LIFE, PLANTS, PROTISTA, REPTILES, and SOLAR SYSTEM.

Topic Terms

adaptive radiation—Adaptive radiation occurs when organisms split into diverse groups as a result of changes in environmental factors and genetics; this is also known as divergence. (Covergent evolution occurs when two unrelated groups of organisms experience similar evolutionary pressures and begin to resemble each other superficially.)

Batesian mimicry—Batesian mimicry was named by British naturalist Henry Walter Bates. It is the ability of a species to look threatening or toxic to predators—even if the species is not toxic. The classic example is the viceroy butterfly, a species that resembles the unpalatable monarch butterfly.

catastrophism—This theory of evolution emphasizes the effects of sudden, short-lived, catastrophic worldwide events and their effects on Earth's organisms. In essence, the modern catastrophic theory holds that these sudden, worldwide events caused mass extinction of most existing life-forms, resulting in low species diversity and a high number of ecological gaps. These gaps are filled by variants from the remaining life-forms, subsequently giving rise to further variations and adaptations. (Evolution is thus the process of extinction followed, after a pause, by the filling of empty ecological gaps by new species of the surviving life-forms. Natural selection is, in this theory, a stabilizing force rather than the engine of evolution.)

creationism—A view of the world through the lens of literal biblical, or religious, interpretation. In this world, a divine force, or God, is responsible for the creation of all things, not evolution. In contrast with theistic evolution, creationists hold that the large-scale change of one animal or plant order to another cannot be accomplished through mutation and natural selection. Although small changes do occur, they are not classified as evolution, merely the emphasis of some already-present traits over others. There is no common ancestor; each species was created whole by God.

This theory was predominate during the Middle Ages, but it was increasingly modified as more and more scientific evidence concerning extinct species and localized species became available during the eighteenth and nineteenth centuries. The theory of evolution through natural selection proposed by Charles Darwin in 1859 meant that God, or religion, was no longer necessary for the formation of species, including humans. Some theologians adapted to these new findings, while others resisted. As with any beliefs, there are many different camps in creationism. Some believe that the Earth and all flora and fauna were created by God only a few thousand years ago. Others feel that this single creation took place billions of years ago. Still others feel that the seven days of creation actually refer to seven ages that may have been widely separated in time. But the hallmark of all of these is the belief that one organism is incapable of evolving into another.

evolution—One definition of evolution is the theory that life developed gradually from one or several simple organisms into more complex organisms. This is also called organic evolution, or evolutionary biology.

This theory states that the complex life-forms of today, both plant and animal, all descended from the common, simple organisms that first originated on Earth. The sequence of events toward life and evolution seems to be as follows: The start of our universe with the big bang; the coalescence of the stars and galaxies; the formation of the Earth and other planets; the gradual changes to the Earth's surface and climate; the birth of life; the progression of life into more complex forms, both plant and animal; and finally, the adaptations, the fall and rise of species—all part of the invisible, but continual, evolution of organisms.

The evolution of complex life-forms on Earth came as a result of random genetic mutations during reproduction. The resulting organisms either died out or survived as a result of natural selection. This process was long and arduous, spanning billions of years; it was sometimes completely halted because of global catastrophes and climate changes, or progressed rapidly when conditions were favorable. It is still working today.

gradualism (evolutionary)—Also known as gradual evolution, this is the theory, first put forward by Charles Darwin, that small changes happen slowly to a species through random genetic mutation. Natural selection winnows out the undesirable mutations and the resulting changes gradually build up over immensely long periods of time.

human evolution—see HUMANS, Topic Terms

natural selection—The engine of evolution proposed by Charles Darwin in his *On The Origin of Species*. He postulated that certain factors in a species' environment, such as climate and food sources, cause those individuals with unfit random genetic mutations to die off and encourage the increase of those with favorable mutations. This process is based on the observation that there is, in nature, a continual struggle to survive and that only the fittest—or the best adapted—will survive to reproduce and pass on those characteristics that enabled them to survive. Those with adaptations that are unfit will die off.

Natural selection is all about how species change over time. Overall, the individuals of any given species are not exactly the same. Some organisms, whether plant or animal, are darker, larger, stronger, or even tolerate heat better than others; in other words, certain organisms have traits that make them better able to survive and reproduce in their surrounding environment than others of their own species. As animals evolve, the animals with the more desirable traits will produce offspring that share some of the more desirable traits of their parents. Thus, more of the young will survive, "changing" and adapting the species to the environment, allowing a better survival rate. This process that allows the traits of organisms to change over time is called natural selection.

punctuated equilibrium—A theory of evolution formulated by scientists Stephen Jay Gould and Niles Eldredge that attempts to explain gaps in the fossil record. These gaps show an absence of intermediate steps between forms of animals, and therefore are not compatible with a gradualist view of evolution. This theory proposes that species are stagnant until a sudden burst of geological time in which they undergo rapid evolutionary changes; this is followed by another period of stagnancy until the next burst. The periods of time in which this rapid growth takes place is on the order of hundreds of thousands of years; short in terms of geological time, long compared to human lifespans.

species—One of the categories used in the classification of organisms on Earth. The organisms in a specific species share some common attribute(s) and can interbreed or have the potential to interbreed. Fully 90 percent or more of all the species that ever lived on Earth are thought to be extinct. Currently, there are anywhere between 5 and 30 million species of flora and fauna on Earth; some types of organisms, such as mammals and birds, have only a few hundreds of thousands of species, while others, such as insects, have many more.

Over millions of years, new species formed (and continue to form) in a somewhat distinct way. In particular, scientists group animals and plants into distinct species when the organisms become so different that they cannot produce fertile offspring together. For example, say a certain group of birds live only on one island. One day, a hurricane rips past the island, the winds forcing many of the birds to be caught in the storm—then landing on another island. The birds on the original island, and the birds on the other island would gradually develop different traits as they continued to adapt to their surrounding environment. As time goes by, they may become very dissimilar to each other—and they would not be able to produce fertile offspring if the two bird types were to mate. Thus, the two bird groups would then be considered two distinct, separate species. The two groups may gradually develop different traits as they adapt to different environments. And there does not have to be a island separating the birds—the same scenario can occur on land as birds separate from others of their species to find new food sources. It is easy to see why so many plant and animal species have developed over the millions of years life has been on Earth.

theistic evolution—An amalgamation of religious and scientific attitudes toward evolution held by many modern theologians. The dominant view of creationism was modified slightly over time in response to modern scientific findings. After Darwin's theory of evolution was published, many theologians accepted and adapted to these new findings. Their beliefs, known as theistic evolution, hold that a divine force, or God, is behind creation, as proposed by their religious beliefs. However, this divine force works through the process of evolution, as revealed by modern science.

14. Extinction

Introduction

The theme of our planet and life is one of continual change. One process that contributes to this change is extinction, the sudden or gradual dying out of a flora or fauna species. Species disappear for a number of reasons—ranging from disease, human intervention, climate changes, volcanic eruptions, and objects from outer space that collide with our planet. Each one of these can cause the extinction of one or many species; sometimes a combination of reasons leads to the demise of an animal or plant. Although the process of extinction has occurred over the entire history of the Earth, our knowledge of its existence and causes is relatively new. We now realize that some species are threatened and endangered today. And we even know that, as a species, we are not immune to this process of change; we could become yet another extinct animal species in the Earth's fossil record that might be discovered millions of years hence.

Timeline

(note: mya=million years ago)

Date	Event

Prehistoric Events

Date	Event
~438 mya	A mass extinction occurs during the late Ordovician period of the Paleozoic era, killing off approximately 85 percent of the species on the planet.
~367 mya	A mass extinction occurs during the late Devonian period of the Paleozoic era, affecting about 82 percent of the Earth's species.
~245 mya	A mass extinction between the Permian and Triassic periods kills off approximately 96 percent of the living species; this marks, on the geological time scale, the dividing line between the Paleozoic and Mesozoic eras.
~208 mya	A mass extinction at the Triassic-Jurassic periods boundary leads to the demise of approximately 76 percent of the species on Earth.
~65 mya	A mass extinction at the Cretaceous-Tertiary boundary kills off approximately 76 percent of the species on Earth. The most well known victims are the dinosaurs, whose fossil remains are no longer found after this time.

Modern Event and Discoveries

Date	Event
16th & 17th centuries	Scientist uncover fossils of ancient plants and animals believed to represent living, though not currently discovered, species.
1796	French naturalist and anatomist Baron Georges Léopold Chrétien Frédéric Dagobert Cuvier (1769–1832) of the Museum d'Histoire Naturelle, Paris, proves that certain mammal fossils were the remains of extinct species, effectively making the idea of extinction a scientific reality.
1973	An agreement on endangered species at the Convention on International Trade in Endangered Species of Wild Fauna and Flora (CITES) is signed by over 80 nations.
1973	In response to CITES, the United States passes a broad-based Endangered Species Act, which applies to habitats as well as species.
late 1970s	The peregrine falcon is the first species to be listed on the endangered species list in the United States.
1980	Nobel prize–winning, U.S. physicist Luis Alvarez (1911–1988), his son, geologist Walter Alvarez (1940–), and colleagues propose that a large asteroid or comet hit the Earth about 65 million years ago, leading to the extinction of the dinosaurs and other species at that time.

History of Extinctions

Early Extinctions

By definition, a mass extinction affects a large and diverse range of species, over large geographic areas. The evidence we have for these major extinctions is found in the fossil record where, using dating techniques, scientists can determine when a species first arose and when it died out. Because of the millions of years that have passed since these events, it is very hard for researchers to pinpoint exact causes of mass extinctions, although more and more clues are being found that may help solve the mysteries.

The first known mass extinction occurred approximately 438 million years ago, during the late Ordovician period of the Paleozoic era. This extinction killed off approximately 85 percent of the species living on the planet. The next known mass extinction occurred approximately 70 million years later, or almost 367 million years ago, during the late Devonian period. This extinction was responsible for the loss of approximately 82 percent of the species.

An extremely large mass extinction, one that almost wiped out all life on our planet, occurred approximately 245 million years ago, and this extinction is used by scientists to delineate the division on the geological time scale between the Permian period of the Paleozoic era and the Triassic period of the Mesozoic era. Almost 96 percent of the living species were killed off in this extinction, the most extensive that we know of in the planet's history. The loss of these ancient life-forms opened up ecological niches for more modern life-forms.

Another mass extinction occurred between the Triassic and Jurassic periods, approximately 208 million years ago. In this event, approximately 76 percent of the species were wiped out. This also opened up ecological niches that the existing dinosaurs filled, leading to their dominance until the end of the Cretaceous period.

The last, and most well-known, mass extinction occurred approximately 65 million years ago and killed off 76 percent of the living species. The reason for the high visibility of this extinction is due to its most prominent victims, the dinosaurs. This event marks the boundary, on the geologic time scale, between the Cretaceous period of the Mesozoic era, and the Tertiary period of the Cenozoic era, also known as the K/T boundary. One of the reasons for the importance of this event is that the death of the dinosaur species eventually paved the way for the rise of mammals, including humans, to dominance on Earth.

Study of Extinctions

It was only about 200 years ago that extinction was finally accepted as fact—that some plants and animals that once inhabited the planet are no longer to be found in today's world. Even the modern dominance of mammals, including ourselves, was traced back to the mass extinction of the dinosaurs about 65 million years ago. The absence of the great beasts allowed the smaller mammals to survive and proliferate.

During the seventeenth and eighteenth centuries, scientists knew that fossils being uncovered were the ancient remains of plants and animals. However, the religious doctrines of the times did not include the possibility of any flora or fauna becoming extinct. Accordingly, most scientists felt that these unknown fossils represented living species—perhaps species that would shortly be discovered living in some remote, unexplored part of the globe.

This attitude began to change radically in the 1750s, when explorers in North America found the remains of what they thought were elephants. In reality, these animal remains were those of mastodons and mammoths, animals now known to have lived more than 10,000 years ago during the Ice Ages. As these and other fossils from the Western Hemisphere were examined, scientists realized that they were the remains of recently extinct species.

In 1796, Georges Léopold Chrétien Frédéric Dagobert Cuvier, the famous geologist and naturalist (and the first comparative anatomist) of the Museum d'Histoire Naturelle in Paris, published a series of papers in which he demonstrated that these "fossil elephants"—and giant mammal bones from other parts of the world—did indeed represent extinct species. Cuvier also introduced the concept of **catastrophism**, an idea that dominated European thought for several decades during the late eighteenth century. Cuvier proposed that six major catastrophes had occurred in the past, the last being Noah's flood; he also believed the Earth was relatively young, only about 6,000 years old. But it was his ideas on the extinct "elephants" that made extinction a reality, and combined with other new findings on the enormous age of the Earth, gave a more dynamic picture to the rise and fall of various types of life on our planet. Life was then seen to be in constant change, albeit on a time scale of millions of years.

But not everyone agreed with Culvier's explanation of extinctions. Naturalist Charles Darwin belonged to the school of thought that proposed **gradualism;** the gradual extinction of animals over time that meshed well with his newly proposed theory of evolution. (The terms catastrophism and gradualism are also used in the field of evolution to describe how certain species change over time.)

Currently scientists are still not sure whether or not animals and plants disappeared gradually or quickly. And even with new technologies, it has been impossible to come to a definite conclusion—or even decide how the extinctions occurred in the first place.

For example, with the advent of radiometric dating methods, such as carbon-dating, scientists have determined the precise times when species first appeared in the fossil record—and when they disappeared. Based on these findings, the average duration of a species is between 1 million and 10 million years, and there has been a constant "turn-over" in species over the millions of years of Earth's history. On the average, the number of species that go extinct is very low—approximately 1 to 2 species per species present, per year. This normal attrition rate is known as the background rate of extinction. Based on the number of named species, approximately 1.4 million species currently live on our planet (although some scientists feel the number may actually be between 5 and 50 million). The background rate of extinction, then, would be approximately two species per year, and could be caused by disease, climatic change, increased ecological competition, change in habitable area, or a combination of these factors. Mass extinctions occur during those times when the rate of extinction is much higher than this background rate, and they affect large numbers of diverse species over wide geographic regions—even globally. The duration of this extinction can cover anywhere from millions of years to an instant.

But studies of these periods of mass extinction have yielded a few clues as to the possible causes of widespread demise. **Extinction theories** involve sea level changes, extreme climate changes, extensive volcanic activity, environmental poisoning, and planetary collisions with outer space objects (such as comets, asteroids, and cosmic clouds). Again, combinations of causes may be the true explanation.

The most popular theory—albeit not completely agreed upon by everyone—came about in 1980, when physicist Luis Alvarez proposed that a large asteroid or comet hit Earth about 65 million years ago. His son, geologist Walter Alverez, led the team that discovered a high concentration of iridium, an element associated with extraterrestrial impacts, at the Cretaceous-Tertiary boundary (K/T) in Italy. Because of this find, and the realization that the dinosaurs and many other species died out at the end of the Cretaceous period, a new theory developed. Luis and Walter, along with colleagues Frank Asaro and Helen Michel, proposed that the extinctions at the K/T boundary were caused by the impact of a large space object, resulting in an impact crater, or large hole in the surface of the planet. The iridium anomaly has since been found in over 50 K/T boundary sites around the world, making it a truly global event.

The most compelling evidence so far in support of this theory is the buried Chicxulub crater in the Yucatan Peninsula, Mexico, an impact crater discovered by geologists in 1992 (although it was "imaged" many years earlier as an oil company searched for underground fuel). This 150-mile-wide (241-kilometer-wide) crater—although recent measurements of the underground crater reveal that it may be as large as 186 miles (300 kilometers) in diameter—is thought to be the result of the impact of a 6- to 12-mile (10- to 20-kilometer) diameter asteroid. The crater was created approximately 64.98 million years ago, in the right time frame for the extinction of the dinosaurs. It may be totally or partially responsible for the extinction of these and other fauna and flora.

An alternative theory for the presence of this iridium layer has been proposed, however. Some scientists feel that extensive volcanic activity was really the catastrophe that caused this mass extinction, pointing out the enormous lava fields found in India known as the Deccan Traps that formed around 66 million years ago, as the dinosaurs were dying out. A large amount of volcanic activity could have produced environmental effects similar to those of an asteroid or comet impact, with the tell-tale iridium being spewed out of the volcanoes from deep within the Earth. And yet another school of thought believes a large impact instigated the increased volcanism.

In the middle of the search for ancient extinctions came the knowledge these events—albeit not as alarmingly quick or all-encompassing as mass extinctions—are occurring during our own time. Because of this, many people began to speak out about our own time period's **endangered and threatened species**. The Endangered Species Act of 1973 was originally brought forth to save certain species from extinction. The act has not saved everything—and it is controversial because of its many different interpretations, the largest question being which species are "worthy or unworthy" of our protection. But many scientists believe, with humans increasing at an estimated rate of 30 mil-

Today's Extinction Problems

The current rate of species extinction is estimated to be between 4,000 and 27,000 species per year, much higher than the expected background rate of two species per year. What are the causes of this increased extinction? In a word, us. Humans have become dominant on the planet, spreading out to all the remote corners of the world and exerting increased pressure on species in numerous ways. The following chart summarizes some of the principal activities that have led, and are leading to, the recent high rate of extinction of plant and animal species around the world.

Habitat destruction (deforestation, transformation, desertification)—The main cause of species extinction is deprivation of habitat and significant loss of available areas.

Disrupting ecosystem by species introduction—Introducing a foreign species can drive a native species to extinction through predation or overcompetition. For example, the introduction of Nile perch into Lake Victoria caused the loss of 200 species of rare native fish through predation.

Pollution—The introduction of pollution into the ecosystem can kill off species. For example, the grayling disappeared in the 1920s from Michigan streams due to silt and sawdust being introduced from logging and lumber production.

Overharvesting—The taking of too many individuals of a species for food or sport either exterminates the population outright, or leaves few survivors whose demise is hastened by other factors. For example, the Great Lakes saw the extinction of three different species of whitefish during the first half of the twentieth century because of commercial overfishing.

Combination of activities—While many times one activity alone is not sufficient to drive a species to extinction, a combination may. For example, the abundant passenger pigeon of the Midwest was killed off by a combination of pollution, habitat destruction, and overhunting.

Which species are most vulnerable to these and other human activities and therefore have the highest potential for extinction? These species typically tend to be ones that are commercially valuable, have a small population or range, low reproductive potential, or are extremely specialized in terms of food, nesting requirements, or habitat. Some vulnerable species have a combination of these factors, such as China's panda, which eats mostly bamboo, has few offspring, and a low population. Most of the world's species live in the tropics and many of them have limited geographic areas. As humans continue to deforest the tropical rain forests at a rate of approximately 42 million acres per year, these species will be under increasing pressure to survive. In fact, most of the current extinctions are caused by tropical deforestation.

Species extinction has sometimes come about even though humans tried to employ actions that potentially could have saved the species. For example, three species could have been saved: the Maryland darter fish and two birds, the Kauai o'o and the dusky seaside sparrow. They were listed on the endangered list and probably had sufficient numbers to be saved, but there were problems. In the case of the dusky seaside sparrow, for example, even though there were 2,000 remaining birds, human errors, lax enforcement of the law, and other factors caused the destruction of the birds' Florida coastal salt marsh habitat—thus driving it to extinction.

As humans become more and more aware of the causes and mechanisms of species extinction, it becomes clear that we are not immune. We can drive our own species to extinction through some of the same factors that pressure other species, such as pollution and habitat destruction. Overpopulation seems to be a uniquely human problem, since we have no predators of note to control our numbers. The resulting contamination of water and air, along with a strain on the food supply, could eventually lead to the extinction of our species.

And as we have learned, factors beyond our control could cause our demise as a species. A random collision with an asteroid or comet, given the right parameters, could lead to the extinction of the human species. For example, if a large comet or asteroid more than 3 miles (5 kilometers) in diameter struck Earth's land surface or even a shallow water area, the impact would throw large amounts of dust into the atmosphere, blocking the rays of the Sun. The result would be darkness, a mini–ice age, disrupted weather and climate, and the cessation of plant growth for at least a year, if not longer. This would be a catastrophe without precedent in the history of the human race and would eliminate most flora and fauna from the planet, certainly destroying life as we know it. But, any remaining species would radiate out into the newly available ecological niches, just as they've always done. Humans might just not be there to witness it.

lion per year, it may be the only factor determining a species' existence in the future. The problem in keeping certain species in existence will be the cost—especially economically and socially—to humans. *See also* AMPHIBIANS, ANIMALS, ARTHROPODS, BIRDS, CLASSIFICATION OF ORGANISMS, DINOSAURS, EVOLUTION, FISH, FOSSILS, FUNGI, GENETICS, GEOLOGIC TIME SCALE, HUMANS, LIFE, PLANTS, PROTISTA, REPTILES, and SOLAR SYSTEM.

Topic Terms

catastrophism—See EVOLUTION, Topic Terms

endangered and threatened species—Endangered species are those thought to be on the brink of extinction. The conditions under which a species is considered "endangered" are difficult to determine—the process is complex and has no true set of fixed criteria that can be applied consistently to all the species around the world. Not even numbers of organisms within a species are a major factor. For example, a species with a million members living within a small area could be considered endangered, while on the other hand, a species smaller in number but spread over a wide area would not be considered endangered. Factors to be considered include the frequency of reproduction, the average number of offspring born, and the survival rate of the species—but even these constitute only a small portion of the information needed to determine whether a species is endangered.

Each country, and even states within the United States, deal directly with their own endangered species problems. For example, Tasmania's government promoted the protection of threatened species with their Endangered Species Protection Act of 1992 and their National Endangered Species Program that started in 1990.

In the United States, the Fish and Wildlife Service, under the Department of the Interior, determines which species are endangered. These lists are based on studies by experts in the field, including biologists and naturalists. According to the Endangered Species Act of 1973, a species can be listed if it is threatened by any of the following conditions:

(a) The presence of threatened destruction, modification, or curtailment of its habitat or range.

(b) Utilization for commercial, sporting, scientific, or educational purposes at levels that detrimentally affect the species (or numerous species).

(c) Disease or predation of the species.

(d) Absence of regulatory mechanisms adequate to prevent the decline of a species or degradation of its habitat.

(e) Other natural or artificial factors affecting the species' continued existence.

In general, if a species is threatened in any of the above ways, the director of the Fish and Wildlife Service can determine the "critical habitat" of the species, or areas containing the necessary physical or biological features necessary for the species' protection. These areas would then have possible protection to ensure the species' survival.

Since 1973, only about six species have recovered and been removed from the U.S. federal endangered and threatened species list. But during the same time period over seven species have also been removed from the list, because they have become extinct.

The following lists the most recent totals of endangered and threatened species presently in the United States compared to overall foreign counts:

Endangered Species *			
Group	Endangered U.S./Foreign	Threatened U.S./Foreign	Total Species
ANIMALS			
Mammals	59/251	8/16	334
Birds	75/178	15/6	274
Reptiles	14/65	21/14	114
Amphibians	9/8	7/1	25
Fish	68/11	39/0	118
Snails	15/1	7/0	23
Clams	61/2	8/0	71
Crustaceans	16/0	3/0	19
Insects	28/4	9/0	41
Arachnids	5/0	0/0	5
Subtotal (animal)	350/520	117/37	1,024
PLANTS			
Flowering	525/1	114/0	640
Conifers	2/0	0/2	4
Ferns & Others	26/0	2/0	28
Subtotal (plant)	553/1	116/2	672
GRAND TOTAL	903/521	233**/39	1,696

Total U.S. Endangered Species: 903 (350 animals, 553 plants)
Total U.S. Threatened Species: 233 (117 animals, 116 plants)
Total U.S. Endangered and Threatened Species: 1,136 (467 animals, 669 plants)
* From the U.S. Fish and Wildlife Service, June 30, 1998
** Eight animal species have dual status (seven within the U.S.); the olive ridley sea turtle is again considered to be a U.S. species.

extinction theories—Theories that explain mass extinctions over time abound. Some theories have been dismissed outright; others are highly debated. Some scientists believe that the impact theory is the only correct theory. But as more data are revealed, many are beginning to believe extinctions are caused by several reasons—not just one. The following lists the individual theories so far proposed by scientists to explain extinctions (note that for ease of understanding, we will use the dinosaurs for many of the examples):

climate theory—The idea that extinction resulted from changes in the climate has long been a major theory. For example, one of the greatest extinctions occurred 65 million years ago, when the dinosaurs became extinct (whether the extinction was slow or fast is still a matter of debate, too). At that time, a great deal of volcanism occurred on Earth. The release of dust, gases, and ash into the atmosphere could have cut down on the Sun's warming rays and cooled the climate. Also during this time, several regions were experiencing rapid mountain-building events, which could have uplifted certain regions enough to cause changes in the vegetation and climate. Many scientists believe that the "climate" theory of dinosaur extinction is more feasible than most other theories.

One of the theories of gradualism is that the movement of continents over millions of years brought about changes in the climate of the Earth. The climate changed due to such effects as the changing of oceanic currents; the spreading of deserts; the drying up of inland seas; shifts in the Earth's axis, orbit, or magnetic field; the spreading of polar ice caps; and the occurrence of volcanic eruptions. One or all of these led to slow climate changes that resulted in the gradual decline of the dinosaurs, who could not evolve quickly enough to compensate for the changes.

disease theory—One of the theories of gradualism is that species eventually died out because of disease. For example, in the case of dinosaurs, some scientists say that biological changes brought about by changes in their evolution made them less competitive with other organisms—including the mammals that had recently started appearing. Others say that major disease, from rickets to constipation, wiped out the dinosaurs, with some dinosaur bones showing signs of such diseases over time. Another idea is that overpopulation led to the spread of major diseases among certain species of dinosaurs—and eventually to them all.

human theory—Contrary to what many people believe, there is no human theory of extinction—unless the extinction occurred within the past million years. For example, some movies show humans and dinosaurs together, leading people to say that the dinosaurs could have become extinct because of humans. But in reality, humans did not exist at the same time as the dinosaurs—in fact, humans missed existing alongside the dinosaurs by more than 64 million years.

But the human theory of extinction has been suggested by many scientists as the cause of Ice Age animal extinctions. Although it is still a highly debated idea, it is thought that some animals were hunted to extinction as humans grew in number and needed the animals for food and clothing. Other scientists believe that the change in the climate, from colder to warmer, had already caused the animals to become extinct by the time humans were prolific enough to have an effect on their surroundings.

impact theory—The impact theory is one of the newest ideas in the catastrophic camp—especially to explain dinosaur extinction. This theory states that a large object, such as an asteroid or comet, collided with Earth, resulting in a large impact crater; giant waves in the oceans that smashed onto land at heights of 2 to 3 miles (3.2 to 4.8 kilometers); and radical, rapid changes in Earth's weather, temperature, amount of sunshine, and climate.

After a large impact, scientists theorize that several events would occur. Right after the impact, huge amounts of dust and debris from the impact would be thrown high into the atmosphere. The dust would be carried by the upper winds all around the world, filtering or blocking out the sunlight reaching the surface. Heat from the blast may also have created firestorms—huge forest fires that added smoke, ash, and particles to the already dust-filled atmosphere. If the dust-filled winds did not kill the animals, the lack of sunlight would kill off plants, creating a serious crisis: The animals that fed off of the vegetation would die, and in a domino-like effect, the rest of the other organisms in the food chain would die off—eventually reaching to the dinosaurs at the top of the food chain.

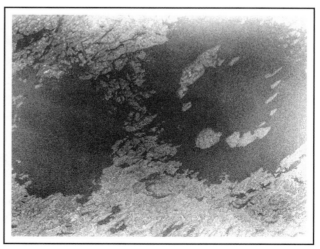

The Clearwater Lakes in Quebec, Canada, was formed by two impact craters, caused by a space body—probably asteroids, or a fragmented comet—striking the Earth's surface. *(Photo courtesy of NASA)*

To date, scientists have identified about 150 impact craters on our planet. The majority have been found on the surface; less than a dozen or so are buried on land or deep in the oceans. There were probably many more craters, but erosion—from wind, water, or the movement of the continental plates—has erased any evidence of their existence. In addition, there may be many more craters under the thick, vegetative growth of the jungles, in high mountains, or buried deep under sediment on land or in the oceans. The largest crater, the Vredefort crater in South Africa, is also one of the oldest, with an age of over 2 billion years. Another large impact crater, the Sudbury in Canada, is a major source of certain metals. Craters on the planet Mars dwarf Earth's craters. The largest impact crater (also called a basin) on Mars is Hellas Planitia, measuring 1,243 miles (2,000 kilometers) in diameter.

Major Impact Craters around the World

Name	Location	Diameter (miles/kilometers)
Vredefort	South Africa	186/300
Sudbury	Ontario, Canada	155/250
Chicxulub*	Yucatan, Mexico	150/241
Manicouagan	Quebec, Canada	62/100
Popigai	Russia	62/100
Acraman	South Australia, Australia	56/90
Chesapeake Bay	Virginia, USA	53/85
Puchezh-Katunki	Russia	50/80
Morokweng	South Africa	44/70
Kara	Russia	40/65
Beaverhead	Montana, USA	37/60

*Crater thought to be responsible, or at least partially responsible, for the extinction of the dinosaurs.

killer cloud theory—Recently, researchers have theorized another mechanism for mass extinction, specifically the one at the K/T boundary—the time of dinosaur demise. This is known as a killer cosmic cloud, found in large but localized areas of outer space. These clouds have much higher concentrations of hydrogen than normal. The typical region of space, such as our planet has been traveling through for the past 5 million years, is relatively empty, having a density of less than one particle (mostly hydrogen) per cubic inch. Killer clouds are found where new stars are being formed and have much higher densities, on the order of hundreds of particles of hydrogen per cubic inch. Supercomputer modeling has shown that encountering a cloud containing higher densities of hydrogen particles would lead to the slowing of the solar wind near Earth, with an accompanying collapse of the heliosphere, the electromagnetic shield that protects our planet and its life forms from cosmic rays. The increased levels of cosmic radiation striking the Earth would dramatically alter life, although we're not sure how much or in what ways. An encounter with a killer cosmic cloud could have precipitated a mass extinction.

mammal theory—Another theory of gradualism is that dinosaurs were slowly wiped out by mammals, animals that only appeared about the end of the Mesozoic era (or end of the Cretaceous period). There is a chance that the mammals ate too many of the dinosaurs' eggs for them to reproduce or that they took over territories from the dinosaurs. On a smaller scale, such events happen today, especially when an introduced species wipes out another species by eating the native organism's young, or by taking over the territory and eating the available food supply.

poison plant theory—The "poison plant" theory of dinosaur extinction involves the development of a new type of plant called angiosperms (flowering plants) that greatly flourished during the Cretaceous period. Some of the plants were no doubt poisonous to dinosaurs, as they probably had protective toxins (poisonous to animals) to avoid being eaten. This theory states that the more prolific plant-eating dinosaurs died out as the plants became more toxic to them. In turn, the carnivores had fewer plant-eating dinosaurs to eat, and so they too died out from starvation.

supernova theory—What would happen if a supernova explosion occurred close to Earth? It is thought that the high-energy particles would react with our atmosphere, especially breaking down atmospheric ozone. And since a blanket of ozone protects most living organisms on Earth—in the upper oceans and on land—massive extinctions of animals and plants could result. Could such a scenario have happened in Earth's past?

Scientists have found no conclusive evidence that Earth has been affected by a supernova explosion. The answer may lie in deep sea sediments, which contain geological records of millions of years. If true, scientists may be able to point to yet another method of past planetwide mass extinctions on Earth.

volcanic theory—Some scientists feel that an incredibly large volcanic eruption occurred around the time of the dinosaurs' extinction. This eruption, called the Deccan Traps, formed the highlands of India from its lava flows and could have produced enough ash to block out a good part of the Sun's rays, which would have led to changes in temperature and climate. Some catastrophists feel that there were two impacts in the same time frame that caused the extinction of the dinosaurs, one in the Yucatan, Mexico, and one in India, which triggered the volcanic eruption.

gradualism—See EVOLUTION, Topic Terms.

15. Fish

Introduction

Fish were probably one of the most important developments in life on Earth, as it was from the fish that amphibians evolved. The evolutionary connection is obvious, albeit more complex than stated here: The amphibians led to reptiles, then to mammals, and then to humans. Modern fish are also important to the overall food chain—both in freshwater and marine environments—and thus, to the natural history of the planet.

Timeline

(note: mya=million years ago)

Date	Event
Prehistoric Events	
~480–460 mya	The first jawless fish evolve.
~450 mya	The first jawed fish emerge.
~390 mya	The ancestors of the bony fish evolve.
~380 mya	The first shark-like fish evolve.
~360 mya	An early offshoot of the Osteichthyes (bony fish) eventually lead to the first amphibians.
~175 mya	The first true bony fish (teleosts) emerge.
~190–135 mya	The first modern sharks evolve.
Modern Events and Discoveries	
~350 B.C.	Greek philosopher and scientist Aristotle (384–322 B.C.) develops several ideas about the general structures of fish; he correctly distinguishes them from aquatic mammals and other sealife.
mid-1500s	French naturalist Pierre Belon (1517–1564) observes and records the fish species of the Mediterranean Sea.
mid-1600s	English naturalist Francis Willughby (1635–1672) and English naturalist and taxonomist John Ray (1627–1705) "classify" fish and birds, paving the way for the classification system of Carolus Linnaeus.
1656	Italian mathematician and physiologist Giovanni Alfonso Borelli (1608–1679) demonstrates that locomotion in fish is primarily by the motion of the tail, not the fins.
1709	Swiss geologist Johann Jakob Scheuchzer (1672–1733) publishes *Complaints and Claims of the Fishes*, in which fossil fish "argue" against the idea that they were never living, organic beings.
1738	The book *Petri Arted, seuci, medici, ichthyologia sive opera omnia de piscibus*, by Petrus Arted (1705–1735), edited by Linnaeus is published; it establishes Artedi as the father of ichthyology.
1749	Frenchman Comte de Georges-Louis Leclerc Buffon (1707–1788) begins the *Histoire Naturalle* (*General and Particular Natural History*), a 44-volume set of books explaining all that is known about the nature of animals and minerals at the time, including fish.
1788–1804	French scientist Bernard Lacépède (1756–1825) completes the 44-volume *Histoire Naturelle*, volumes initially started by George Buffon; eight of the volumes are about serpents and fish.
1833–1844	Swiss-American geologist Louis Agassiz (1807–1873) publishes a five-volume series that lists more than 1,000 fossil fish.
1910	Austrian zoologist, entomologist, and ethologist Karl von Frisch (1886–1982) determines that fish can see different colors.
1938	*Latimeria*, a living crossopterygian fish, is caught off the coast of South Africa; previously it was believed to have died out in the Cretaceous period.
1981	Chinese scientists successfully clone the first fish, a golden carp.

History of Fish

Early Fish

Fish were probably one of the most important stepping-stones to larger forms of animals on land. Their evolution is not truly known, but it is thought that, in general, the process followed a definite progression: About 480 to 460 million years ago, the jawless fish evolved; at about 450 million years ago, the first jawed fish emerged; about 380 million years ago, the shark-like fish evolved; and it was only about 175 million years ago, during the time of the dinosaurs, that the first true bony fish (teleosts) emerged.

The evolution of the fish themselves is not straightforward, and neither are their evolutionary offshoots—although scientists do have some ideas. Evolution continued after the initial appearance of fish. The first vertebrates to inhabit the land were fish that adapted by developing primitive lungs and fins for crawling.

Fish in the oceans continued to diversify and increase in numbers—but many did not change too much in overall form. For example, the class that contains the most fish species today, the bony fish, has bodies that are derived from ancestors that appeared some 390 million years ago. While others, such as the sharks, have relatively recent ancestors that evolved between around 190 and 135 million years ago (although other scientists believe sharks evolved as far back as around 350 million years ago).

Modern Fish Studies or Ichthyology

Although fish studies were probably conducted for centuries, one of the first known studies of fish is attributed to Aristotle, during a time that also coincided with his studies in the field of zoology. He had an accurate idea of the general structures of fish and correctly distinguished them from other aquatic animals, such as mammals. He also wrote about fish's methods of propagation, including the time of propagation and special characteristics of the animals—all of which were quite accurate. His collection of information on 115 species living in the Aegean Sea were accepted for many centuries as the classic way to write about such animals. There was one downfall in this methodology: Aristotle used the local names for fish—collected from the local fishermen—and the names changed from generation to generation.

The first true observational studies began around 1517 to 1575, with the work of Pierre Belon and several others; Belon observed and recorded the fish species of the Mediterranean Sea. By the mid-1600s, Francis Willughby and John Ray made the first real attempt to classify these animals, arranging them in the groupings based on the fish's structures.

The "father of ichthyology" was Swedish naturalist Petrus (Peter) Artedi, who made several investigative writings about fish and fish taxonomy in the mid-eighteenth century. His knowledge of the relationships of the various species and his concepts of fish groups have become classics in the field.

Characteristics of fish have been studied for centuries. However, some of the most important discoveries have been found in the past century, including how various fish **swim**, how they protect themselves, and how their **vision** functions. In addition, it has only been within the past two or so decades that scientists have been able to understand the mechanisms of how fish make sounds and how they detect sounds. And studies to discover and describe new fish territories are ongoing. For example, in 1978, a research group lowered baited traps and a camera into the water through the Ross Ice Shelf in Antarctica, to a depth of 1,960 feet (597 meters) below sea level. There, they found many new types of fish—adding to the already huge list of fish species.

But the overall importance of fish in the past century is their connection to the environment and humans. Fish's roles in the modern world include being important members of the marine and freshwater food chains. In particular, the tiny mobile organisms in the oceans in the upper levels of the sea—called plankton—are the foundation of a huge pyramidal food chain that supplies virtually all the dietary requirements of ocean dwellers. The primary producers supply food to the primary consumers—which include many fish; then the secondary producers, including fish, are the food for the higher consumers. In a progression from little to huge—with fish in the middle—the food chain keeps the ocean ecosystem in balance.

In addition, fish have been, and continue to be, one of the most important commercial products of many countries over the past few centuries. For many years, various species of fish (such as tuna, herring, and sardines) and certain "industrial fish" (monkfish, sea robin, squirrel hake, shark, and ray) have been used in livestock feed as sources of nutrition and oils. Others have been used as a major food source for humans, especially for people who live on islands or along coastlines—where it is usually more cost-effective to eat fish than other types of animals.

Today, more than 25,000 fish species inhabit the Earth (freshwater and saltwater). They represent about

36 orders and 400 families. *See also* AMPHIBIANS, ANIMALS, CLASSIFICATION OF ORGANISMS, EVOLUTION, EXTINCTION, FOSSILS, GEOLOGIC TIME SCALE, LIFE, OCEANS, and REPTILES.

Topic Terms

fish—There are certain general facts about fish: Most live in water, most are ectothermic (cold-blooded); they breathe by means of gills, and their hearts only have two chambers (as compared to the four of humans). The skin is kept "moist" by glandular secretions, and in many species, the skin is covered with scales; the fish's appendages are actually fins. Environmental and evolutionary changes have molded fish over time, creating diverse lifestyles and forms. For example, some fish are blind, others can crawl about on land, and some are electric; others build nests, change color in a few seconds, and even puff up to ward off predators.

Of the phylum Chordata and the subphylum Vertebrata (meaning segmented backbone), and often listed under the superclass Pisces, fish make up nearly half of the animals. Most classifications have two classes of fish: the Chondrichthyes (cartilage fish, such as sharks and rays) and the Osteichthyes (bony fish, often called true fish).

The majority of fish are the bony fish, or Osteichthyes, with skeletons made mostly or solely of bones. In general, on each side of a bony fish, an operculum (gill cover) protects the gills; many also have an internal organ called a swim bladder that helps the fish stay buoyant. These fish also generally spawn rather than mate, with the males fertilizing the eggs after the females lay them. They eat a wide variety of foods, including other animals (insects, larvae, and other smaller fish) and some plants. In addition, an early offshoot of the bony fish eventually led to the first amphibian.

Other fish have skeletons of cartilage, not bone (they are also called cartilaginous fish). These Chondrichthyes are the sharks, rays, skates, and chimaeras, all with paired fins, teeth, and small scales. These animals have multiple gill slits instead of a single covered opening; they have no swim bladder, and usu-

---

Fish From the Past

In the first part of the twentieth century, a small trawler brought up an amazing catch from the oceans: a coelacanth, once thought to have become extinct millions of years ago. Since this discovery and because of the uncharted depths of the ocean, scientists are beginning to believe that there may be all types of ancient-looking fish living deep in the ocean. Marine fish such as this one, whose continuing existence surprised scientists, are considered to be "living fossils."

As one of the most famous of all "living fossils," the coelacanth has a three-lobed tail and fins with arm-like bases. In fact, their fins move like the limbs of animals walking on land, and they bear live young from eggs about the size of grapefruits inside the female. The first such fish was found as a fossil in Devonian period rock. It was thought to have become extinct about 60 million years ago, at the beginning of the Cenozoic era.

But in 1938, fisherman Captain Hendrik Goosen caught the first known living coelacanth in the Indian Ocean off the coast of South Africa. The ichthyologist who identified the fish, Professor J.L.B. Smith of South Africa, offered a reward of £100 (about $50 in current U.S. currency) to anyone who found a second one. It took until 1952 for someone to catch another live specimen. Since that time, many have been photographed alive in water 200 to 1,310 feet (60 to 400 meters) deep off the coast of the Cormoro Islands near Madagascar.

Recently, in 1997 and 1998, two more coelacanth fish were found. (And in 1999, scientists believe they may have found another species of coelacanth.) One was discovered by a biologist working in Indonesia, who spotted a dead one on a handcart in a fish market. The other was found off the island of Manado Tua—of which the local lore talks of ancient-looking fish living in the island's reefs. These coelancanths were found more than 6,000 miles (10,000 kilometers) east of the first find, leading some scientists to believe that there may be more such animals living in the ocean depths.

This great white shark's tooth is from Aurora, North Carolina. It is almost 3 inches long and is from the Miocene period. Scientists estimate that this tooth came from a shark about 80 feet (24 meters) long.

ally have to keep moving to stay above the bottom of the ocean. To reproduce, these fish mate rather than spawn, with the females often producing live young, not eggs. Because of these and other characteristics, these fish live only in certain regions of the world—primarily in saltwater—and eat other animals for food.

It is interesting to note that other classification systems have three fish classes: the Agnatha (lampreys and hagfish), Chondrichthyes, and Osteichthyes. The Agnathans were originally filter-feeders that strained mud and water through their mouths and gills. The only living members have changed greatly over time; modern Agnathans have lost their bone and replaced it with cartilage. The lampreys and hagfish are jawless, and they lack scales and paired fins.

Fish range in size from the *Pandaka pygmaea* of the Philippines, which can reach about just under a half-inch (1 centimeter) at maturity, to the whale shark, which can measure up to 60 feet (18 meters) in length. (And amazingly, the whale shark only feeds on plankton, some of the smallest organisms.) The characteristics of fish range from the hagfish, a primitive, worm-like creature to a blue-fin tuna, that can regulate its own body temperature (similar to a mammal) and swim at speeds close to 50 miles (80 kilometers) per hour. Fish are found in anything from freshwater regions to deep ocean trenches; there are even some species that spend most of their time in the air.

swimming—How fish swim has fascinated scientists for centuries; but it has only been in the past century that most details have been discovered. Fish swim by pushing water aside; but there is no one definite way in which fish swim. The majority move by wiggling with the head, tail, and to some degree (depending on the species) by fins. Disk-shaped fish use pectoral fins to move; in other fish, the fins are mainly used for steering and guidance. The dorsal and pectoral fins are used to stabilize the fish, and the pectoral fins are used for balance or turning. Tail fins are multifunctional and are used for propulsion, steering, and stabilizing. In some more streamlined, fast fish, the dorsal and anal fins may even retract or fold into "grooves" that allow the fins to lay flush with the body. Other fish, like the rays, have huge fins on their sides that make them appear to "fly" through the water.

Speed of Fish	
Fish	*Top Recorded Speed (Per Hour)*
Barbel	11.2 miles (18 kilometers)
Carp	7.5 miles (12 kilometers)
Pike	15.5 miles (25 kilometers)
Shark	22.4 miles (36 kilometers)
Swordfish	60 miles (90 kilometers)
Trout	21.7 miles (35 kilometers)
Tuna	13.7 miles (22 kilometers)

vision—Fish vision varies, but in many of these animals the seeing apparatus appears to parallel human vision in many respects. This seems strange, especially since humans usually see in air, and fish see in water. Overall, the structure of fish eyes is dependent on the very limited ocean light and the absence of rapidly changing amounts of light. Thus, they have no eyelids; and they do not require an iris that changes the opening of the pupil to cope with variations in light (thus, the seemingly "bug-eyed" look of most fish). Fish eyes are spherical and almost rigid; the lens is filled with water. They have monocular vision (or the eye on one side is sensed by the opposite side of the brain). Because of where the eyes are located on each side of their head, they can see more than humans can in any one direction at a time.

As for fish seeing colors, there is much scientific evidence that they do and most sportsmen claim that fish can tell the difference in colors, noting that certain fish, such as trout, chase certain colored hooks. But there is still a question of which fish can see what (if any) colors. And although some fish have brilliant colors, some scientists believe that patterns on the fish may be more important to other fish—their own species or other species—mainly for mating or to keep predators away.

16. Fossils

Introduction

Fossils, the remains of ancient plants and animals, are our window into the past. Without these remains, carefully preserved in rock and other materials, the whole and wondrous history of life on our planet would be unknown to us. We would have no inkling that large, reptilian creatures called dinosaurs roamed the world millions of years ago. We would have few clues that would help us prove that the Earth's crust is continually moving, changing the shape and location of our landmasses. We would be in the dark about where we, the species *Homo sapiens*, came from. And, without a record of the way life has changed over the millions of years, we would have no concept of evolution, the fundamental principle that governs the growth and change of all living things. Without fossils, we would be adrift in the present, neither knowing where we have been, nor where we are going.

Timeline

Date	Event
Events and Discoveries	
600–500 B.C.	The Greek philosophers Xenophanes (570–475 B.C.) and Pythagoras (582–500 B.C.) are the first to record their ideas about fossils; they felt fossils were the remains of once-living animals or plants that were no longer present in the known world.
500–400 B.C.	Greek philosopher and historian Herodotus (485–425 B.C.) studies the tiny fossils embedded in the Pyramids' sandstone; he concluded that the ocean had once covered the present desert.
400–300 B.C.	Greek philosopher Aristotle (384–322 B.C.) states that fossils were the failures left behind when life spontaneously generated from mud.
early B.C. to A.D.	The Chinese attribute healing powers to fossils and crushed them to make medicines.
~400 A.D.	The church states that fossils were placed in the ground by the devil to tempt people into questioning the literal truth of the Bible.
1400s	Italian inventor, scientist, and artist Leonardo da Vinci (1452–1519) studies fossils and suspects their actual origin.
1517	Italian scientist Girolamo Fracastoro (Fracastorius) (c.1478–1553) is one of the first to describe the remains of ancient organisms, now called fossils.
1546	German scientist Georgius Agricola (Georg Bauer, 1494–1555) uses the word "fossil" for the first time.
1669	Danish geologist and anatomist Nicolaus Steno (Niels Stensen, 1638–1686) realizes that good-luck charms sold by the native of Malta were actually sharks' teeth turned to stone; he suggests the fossils were the remains of ancient creatures.
1600s	English scientist John Ray (1627–1705), the "father of naturalists," includes fossils in his work *The History of the World*; like Steno, he believes they are the remains of ancient creatures.
1695	English geologist John Woodward (1665–1728) proposes that strange organic-looking shapes found in rocks were the remains of creatures that failed to get on Noah's Ark before the great biblical flood; he believed they were preserved as fossils.
1725	Swiss geologist Johann Jakob Scheuchzer (1672–1733) publishes a work on fossils, titled *Homo diluvii testis* (*The Man Who Witnessed the Flood*); it suggests that the remains of a fossilized skeleton found in a German quarry were that of a man, which were later proven by George Cuvier in 1825 to be the bones of a salamander.
1812	French naturalist and anatomist Georges Léopold Chrétien Frédéric Dagobert Cuvier (1769–1832) concludes that fossils of certain animals represent forms that had died out completely.
mid-1800s	English naturalist Charles Darwin (1809–1882) proposes his theory of evolution, explaining why the most recent fossils more closely resembled living forms than the oldest ones.

Date	Event
1841	Scottish geologist Hugh Miller (1802–1856) notes the Devonian deposits of the Old Red Sandstone formation in Scotland, one of the most important fossil vertebrate-bearing sediments ever discovered; he also believes the fossil record confirmed the biblical account of creation.
1870s	The rivalry between Edward Drinker Cope (1840–1897), a U.S. paleontologist and herpetologist, and Othniel Charles Marsh (1831–1899), a U.S. professor of paleontology at Yale University, leads to the discovery of numerous dinosaur fossils in the western United States.
1891	French anthropoligist Eugene Dubois (1858–1940) describes the first-discovered fossilized remains of our immediate ancestor, *Homo erectus*.
early 1900s	The invention of radiometric dating methods enables scientists to accurately determine the age of rocks and the fossils within them.
1925	Australian paleontologist Raymond Arthur Dart (1893–1989) is brought a human skull he hypothesizes was the missing link between humans and apes; he later announces that fossil remains found in a Botswana quarry were those of an *Australopithecus*—the earliest known hominids.
1950s–1960s	British anthropologist Louis Seymour Bazett Leakey (1903–1972) and his wife, British paleoanthropologist Mary Nichol Leakey (1913–1996), discover the fossil remains of *Homo habilis* and *Australopithecus* in East Africa.
1974	Anthropologist Donald Johanson (1943–) discovers the fossilized skeleton of the earliest known human ancestor, nicknamed "Lucy," an *Australopithecus afarensis* approximately 3 to 3.5 million years old.
1994	The allegedly earliest-known land life is found in Arizona—fossil tubular microorganisms dating from around 1.2 billion years ago; more data must be found to confirm the discovery.
1996	Scientists believe that structures found in a Mars meteorite may be fossil remains, indicating that life has existed on planets other than our own; this "evidence of life" continues to be a hotly debated subject.

History of Fossils

Early Fossil Studies

Fossils have been around since the first animals and plants died in the oceans billions of years ago. Most of these organisms were made only of soft parts that quickly decayed after death, thus their true numbers will probably never be known. It is the hard parts—bones, teeth—that are most often found as fossils in the geologic record. Over millions of years of life, fossils filled many rock layers, which scientists now use to interpret the record of life over the natural history of the earth.

The first record we have of humans interacting with fossils comes from groups of Stone Age people who buried their dead in graves with carefully arranged fossils. But most scientists believe that this burial method was due to the peoples' cultural beliefs and rituals, not because they understood the meaning of fossils.

The Greek philosophers Xenophanes and Pythagoras were the first to record their ideas about fossils sometime around 500 B.C.; they felt fossils were the remains of once-living animals or plants that were no longer present in the known world. About a century later, the Greek philosopher and historian Herodotus studied the tiny fossils embedded in the sandstone of the Egyptian pyramids; he concluded that the ocean had once covered the present desert. Another century went by, and Aristotle, the best known of the Greek philosophers, stated that fossils were the failures left behind when life spontaneously generated from mud.

Despite these early theories, the actual meaning of fossils was still unknown around 2,000 years ago, with most theories as to their origins and uses being connected to superstition and religion. Records from China show that the culture attributed healing powers to fossils and crushed them to make medicines.

About A.D. 400, the Western church stated that fossils were placed in the ground by the devil to tempt people into questioning the literal truth of the Bible. Any fossils resembling human bones were thought to be the victims of the Great Flood and were reburied with Christian rituals.

By the fifteenth century, scientists took another look at fossils. The great scientist, inventor, and artist Leonardo da Vinci studied fossils and suspected their actual origin. He found fossils of sea creatures in the high Apennines, leading him to speculate that this area had once been covered by water and later uplifted.

By the mid-1600s, certain discoveries about layers of the earth would lead to changes in the notion of the true nature of fossils. Nicolaus Steno (Niels Steensen), a Danish scholar, realized that "tongue-stones," sold on the Mediterranean island of Malta as good-luck charms, were actually fossilized sharks' teeth. He also deduced that in a given rock layer the deepest rocks had been formed first and were the oldest; this is known as the Principle of Superposition, or Steno's law. Around the same time, John Ray, the "Father of Naturalists," included fossils in his work, *The History of the World*. It was the beginning of the true exploration of the fossil record.

Modern Fossil Studies

The scientific study of fossils, along with the knowledge of their true origins, has only occurred fairly recently. We now know that fossils are the remains of plants and animals that have been preserved in the Earth; the word "fossil" itself comes from the Latin *fossilis*, meaning "something dug up." The different types of fossils that are found in the Earth depend on the remains and conditions present at the time the organism died. Fossils are most commonly formed from the hard parts of an organism, such as teeth, shells, bones, or wood. Recently, some soft parts, such as the internal organs of a dinosaur, have also been found, but such fossils are much rarer. And it isn't always stone that fossils are found in; animals and plants have also been preserved in ice, tar, peat, and the resin of ancient trees, though not in the same numbers as found in stone.

The actual process of **fossilization** was only truly recognized by the nineteenth century. Baron Georges Cuvier, a French scientist at the National Museum of Natural History in Paris, who is considered to be the founder of modern paleontology (the study of ancient life), was the first to propose the idea, in the early 1800s, that fossils of certain animals represented forms that had died out completely—that is, went extinct. Cuvier also suggested a version of the "catastrophe theory" (see Evolution) to reconcile this finding with the interpretation of the Bible. Charles Darwin, an English biologist and geologist, later proposed his theory of evolution, which eventually superseded the catastrophe theories. Evolution explained why the most recent fossils more closely resembled living forms than the oldest ones.

The discovery of various fossils continued throughout the 1800s, making several important additions to the scientific understanding of fossils. By the 1870s, the great rivalry between Othniel Charles Marsh and Edward Drinker Cope, both U.S. paleontologists, led to the discovery of numerous dinosaur fossils in the western United States. Even **trace fossils** were found and interpreted in many places—such as the fossilized footprints of dinosaurs in the Connecticut River valley—traces that are indirect evidence of the animals. And by 1891, Eugene Dubois described the first-discovered fossilized remains of our immediate ancestor, *Homo erectus*.

As important as fossils themselves are, their age, both relative to each other and in absolute terms, is also of significant importance. Today, a number of methods are used to find the age of fossils. The first method of dating fossils began in the 1800s—an indirect method—in which the age of the soil or rock layer in which the fossils are found are dated, not the fossils themselves. In many cases, the age of the rock can be determined by other fossils within that rock.

One of the most important events in the interpretation of fossils occurred in the early 1900s. The invention of radiometric **dating techniques** enabled scientists to accurately and absolutely determine the age of rocks and the fossils within them.

Some of the most exciting fossil discoveries of the twentieth century are those that are thought to be human. Although scientists knew that humans bones could also be fossilized over time (bones, teeth, and hard parts if the conditions were right), no one thought that the line of humanity would reach that far into the past. In 1974, the fossilized skeleton of the earliest known human ancestor was found in Ethiopia by anthropologist Donald Johanson. Nicknamed "Lucy," the fossil was of a 3.5-foot (1-meter) high, approximately 20-year-old female. The fossil was about 3 to 3.5 million years old—making Lucy's species, *Australopithecus afarensis*, one of the oldest links to our past.

The Good and Bad of Fossil Fuels

Our modern world depends on plants and animals that died millions of years ago—organisms that died in just the right place at the right time to eventually turn into the organics we use for power. The term for these products, "fossil fuels," is familiar to most of us—they refer to coal, oil, and gas deposits we use for fueling our modes of transportation and heating our homes and workplaces.

These deposits were not always in the ground but were created by a process similar to fossilization. First a deposit of organic material is buried by sediment over thousands to millions of years. Coal layers, for example, come from many epochs, but some of the most abundant (and best) seem to come from the Carboniferous period, 345 to 280 million years ago. These deposits were made in warm, swampy river deltas, and underwent a complex process. Initially, the plants' carbohydrates and waxy materials were attacked by swamp bacteria and fungi; this process caused gases such as carbon dioxide and methane to be released. As time went on, the mix became rich in sulfides and hydrocarbons—perfect to eventually form petroleum and coal products.

As the layers built up, and the organics were buried deeper into the Earth, water was squeezed from the layers, and the temperature increased about 2° F (1° C) for every 100 feet (30 meters) in depth—causing various chemical reactions to turn the peaty, organic matter into coal. In general, the top quality coal usually has to be buried at a depth of about 3 miles (5 kilometers); the variations in the temperatures created the various coals—the best being the anthracite and bituminous (over 92 percent carbon) coals, and the worst, the lignites (at 60 percent carbon).

Gas and oil formed in a similar fashion. Layers of small animals and plants called plankton were also buried under thick layers of sediment and as the heat and pressure increased with depth, the fats and oils from marine organisms combined into a thick compound called kerogen. As layers were added, temperatures continued to increase, and hydrogen and carbon atom links broke down in the kerogen, creating a heavy oil; even more heat produced natural gas and light oils. The resulting oils tended to accumulate in sandstones—rocks that are so permeable they "soak" up the oil like a sponge. And usually, this layer was contained by less permeable rock—especially layers of shale.

Even though humans depend a great deal on fossil fuels, their use is cause for concern. For one thing, even though it took millions of years for the fossil fuels to form, it only takes centuries to consume them. And because fossil fuels are nonrenewable—meaning that once they are extracted from the earth, they are not replaced by other fossil fuels—the supply is finite. Some estimates are that our coal reserves, at the pace we are using up the fuel today, will last only for another 300 years; the world's oil reserves seem to have an even bleaker future, with estimates of only about 50 more years of reserves.

The second major concern is related to the refinement methods, and the uses of the oils and gases—which more often than not cause pollution. Some of these pollutants can be reused (such as oils); but others are released into the atmosphere, land, or water. And the dangers differ: The effect of a spill of light oil in a hot climate differs from a heavy crude oil spill off Alaska—the spread of oil and amount of destruction rely on how quickly the volatile elements in the oil evaporate, or how fast the thicker residues break up. But any way it occurs, or how much oil is spilled, a spill can cause major damage to marine and shore organisms alike.

Because of all these concerns, scientists are searching for renewable methods of power, such as solar, hydrogen, or hydro-power. (Nuclear power is still a nonrenewable source of energy, and it also has the potential to cause problems, such as where to store nuclear wastes.) In addition, replacement (artificial) fossil fuels—such as synthetic oils—have been and continue to be produced, helping to slow down the depletion of natural resources. And although progress has been made, such endeavors receive little support.

All the energy alternatives have a long way to go before they match the power potentials of fossil fuel. Part of the problem is people's attitudes toward the use of these alternatives. Most people have always used certain mass consumptive resources, and they feel they have no reason to trust, or need to understand, the energy alternatives that could save, or at least slow down the consumption of, certain resources. In addition to using alternate energy sources, appliances could be made to use less energy, at a lower cost, and yet be easier to use—thus, making these low energy consumption appliances more appealing to the public. But it is a difficult paradigm shift—and until energy supplies reach a critical point, or more leadership accepts the conservationist view, few people will even think in terms of alternative energy or of trying to conserve energy.

Today, the hunt for fossils centers around looking for ones that reveal the details of the evolution of organisms and discovering the oldest fossils that represent certain major steps in the earth's natural history. For example, in 1994, the earliest-known land life may have been found in Arizona: fossil tubular microorganisms dating from 1.2 billion years ago. With better detection techniques, more such ancient fossils may be found—confirming this and other finds, and determining what type of life existed so long ago. *See also* AMPHIBIANS, ANIMALS, ARTHROPODS, BIRDS, CLASSIFICATION OF ORGANISMS, DINOSAURS, EARTH, EVOLUTION, EXTINCTION, FISH, FUNGI, HUMANS, LIFE, PLANTS, PROTISTA, and REPTILES.

Topic Terms

dating techniques—Numerous dating techniques are used to determine the age of rock. Most of these techniques use the rate of change of elements into isotopes. In most cases, fossils themselves are not dated; rather, the rock around the fossil is dated, and the fossil's age extrapolated from these data.

Here are several dating techniques used by scientists to determine the age of rock and fossils:

isotopic (radiometric) dating techniques—The most well-known way to determine the age of rock is through the use of radiometric dating. Radioactivity within the earth continuously bombards the atoms in minerals, exciting electrons that become trapped in the crystals' structures. Using this knowledge, scientists use certain techniques to determine the age of the minerals, including electron spin resonance and thermoluminescence. (Both methods measure the number of excited electrons in minerals found in the rock; but spin resonance measures the amount of energy trapped in a crystal, while the thermoluminescence method uses heat to free the trapped electrons.) By determining the number of excited electrons present in the minerals—and comparing it to known data that represents the actual rate of increase of similar excited electrons—the time it took for the amount of excited electrons to accumulate can be calculated. In turn, these data can be used to determine the age of the rock—and the fossils within the rock.

Other variations are also used. For example, uranium-series dating measures the amount of thorium-230 present in limestone deposits. Limestone deposits form with uranium present and almost no thorium. Because scientists know the decay rate of uranium into thorium-230, the age of the limestone rocks and the fossils found in them can be calculated from the amount of thorium-230 found within a particular limestone layer.

carbon dating techniques—Carbon dating techniques can be used to date organisms that range from several thousand years old to ones that are relatively young in terms of geologic history. Carbon has two stable isotopes, C-12 and C-13, the first slightly lighter than the second. Living organisms tend to select the lighter isotope, as it is easier to absorb (it takes less energy). Thus, when ancient sedimentary rock is found to have more than the usual ratio of C-12 to C-13, scientists know that some form of life has altered the amounts—thus there was probably life in the ancient rock.

Carbon dating is used to determine the age of more recent fossils and even human-made implements (such as pottery)—especially those objects younger than about 45,000 years—because the decay of the carbon isotopes does not occur over a longer period. In this dating process, scientists use carbon-14 (C-14). C-14 is produced when nitrogen-14 (N-14) is bombarded by cosmic rays in the atmosphere. It drifts down, and plants absorb it from the air. Animals eat the plants (or other animals that eat the plants) and take the C-14 into their bodies. When a living organism dies, it no longer takes in C-14. By measuring how much C-14 has disintegrated, and how much is left, scientists can determine the object's approximate age. (C-14 decays to N-14; C-14 has a half-life, or rate of decay, of 5,730 years. This means half the C-14 will change to N-14 in 5,730 years, and half of that half in another 5,730 years, and so on.)

fossilization—The process for fossilization is, in general, very similar for animals and plants; the hard parts of animals, such as bones, teeth, and shells, as well as the seeds or woody parts of plants, are covered by sediment, such as sand or mud. Over millions of years, more and more layers of sediment accumulate, burying these remains deep within the earth. The sediment eventually turns to stone, and the animal or plant remains are chemically altered by mineralization, becoming a form of stone themselves. This same process can produce petrified wood, coprolites (petrified excrement), molds, casts, imprints, and trace fossils.

But not all organisms become fossils—many of them completely decay away or are chewed apart by other animals. Because of this fact, some paleontologists estimate that very few of the billions of flora and fauna that have lived on the earth have become fossils. Thus, the fossils we do find on Earth represent only a fraction of the animals and plants that ever lived on our planet. Taphomony is the name given to the

These trace fossils represent theropod tracks from the Triassic period. Scientists don't know what type of dinosaur left these tracks— no corresponding bones were found around the site of the trace fossils.

study of the processes that go into the formation of a fossil, which, it turns out, is a very difficult and chancy process. The general steps in the fossilization process are outlined below, using an animal as an example.

After the animal dies, first comes scavenging and decay; it does not take long after death for scavengers to remove the soft fleshy parts of the animal's body. Those parts that are not eaten decay at a fast or slow rate, depending on the prevailing climate. In any case, within a relatively short amount of time, only a skeleton of the once living animal remains. But even the remaining hard body parts are not immune to change. They are often weathered by the action of wind, water, sunlight, and chemicals in the surroundings, which round the bones or reduce them to small pieces.

A crucial step in forming a fossil is location; if the animal's skeleton is in an area in which rapid burial does not take place, then the chances of fossilization are slim. The bones can be broken and scattered, often moved by changing river courses or flash floods. Occasionally, this transport increases the chance of fossilization by moving the bones to a better area for preservation, like a sandbank in a river.

The most crucial step in becoming a fossil is burial: the sooner the burial of the bones, the better the chance of a good fossil being created. If the bones are covered by mud or sand, the amount of further damage is lessened; in addition, the amount of oxygen is reduced, thus slowing additional decay of the bones. Some damage can still occur, however, primarily from the pressure created by the increasing amount of sediment on top of the bones, or even from acidic chemicals that dissolve into the sediment.

The fourth step in becoming a fossil is the actual process of fossilization. Here, the sediments surrounding the fossil slowly turn to stone by the pressure of the overlying sediment layers and loss of water. Eventually, the grains become cemented together into the hard structure we call rock. The bones fossilize, as the spaces in the bone structures fill with minerals, such as calcite (calcium carbonate) or other iron-containing minerals; the actual mineral component of the bone itself, apatite (calcium phosphate), may recrystallize also.

Lastly, the deeply buried and fossilized bones must be exposed on the surface where they can be discovered. This involves the uplift of the bone-containing sedimentary rock to the surface, where erosion by wind and water can expose the fossilized skeleton. If the fossils are not found in time, however, the action of wind, ice, and water can destroy the precious record of this ancient animal species.

In fact, so many fossils have been lost to the elements in this way that there are gaps in the fossil record and in our understanding of the past. Gaps can also be caused by mountain uplift, which destroys fossils, and volcanic activity, with the hot magma physically changing any rock, and thus any fossils, it touches.

trace fossils—Trace fossils are the traces a creature left behind, usually in soft sediment like sand or mud. For example, small animals could have bored branching tunnels in the mud of a lake bed in search of food, and dinosaurs could have hunted for meals along a river bank, leaving their footprints in the soft sand. Similar to the fossil formation of hard parts, the tunnels and footprints were filled in by sediment, then buried by layers of more sediment over millions of years, eventually solidifying. Today we see the results of this long-ago activity as trace fossils. Many originators of trace fossils are unidentifiable—in other words, the creatures left no hard parts in the area, just their tracks. Some of the most famous trace fossils are those of dinosaurs tracks and human-like footprints— all found in hardened sediment.

17. Fungi

Introduction

Fungi (or the singular, fungus) are the great decomposers of the world, causing certain organisms to decay. Since they first evolved nearly 900 million years ago, fungi have played a vital role in the natural cycle of life, decomposing dead animal and plant material and releasing nutrients for a new generation of organisms. In addition, fungi interact with almost every other organism—as a parasite, in a symbiotic relationship, or directly and indirectly, causing certain diseases in plants and animals. And it should not be overlooked that fungi have played an important role in human history, as their fermentations give us bread, wine, beer, and many antibiotics.

Timeline

(note: mya=million years ago)

Date	Event

Prehistoric Events

~900 mya	The first fungi evolve in the oceans.
~440–410 mya	The first fungi evolve on land in association with smaller arthropods; they are the ascomycetes, the largest division of modern fungi.
~410–360 mya	The first fungi that cause wood decay evolve.
~325–286 mya	All the modern classes of fungi evolve.

Modern Events and Discoveries

mid–1500s	The "late blight of potato" caused by the fungus *Phytophthora infestans* is discovered in Europe, having been introduced there from South America; it would later cause many potato epidemics.
1804	Scientists discover that barberry hosts a fungus that is easily transferred from the barberry to cereal plants.
1821	Swedish mycologist Elias Magnus Fries (1794–1878) writes his *Systemia mycologicum*, the first standard work on fungi.
1830	A potato blight, caused by the fungus *Phytophthora infestans*, is first reported in the United States.
1845	A potato blight, caused by the fungus *Phytophthora infestans*, causes a severe epidemic in Europe; it is the major cause of the Irish potato famine.
1873	One of the most destructive fungus diseases to citrus plants is detected; it invades the root system of citrus trees.
early 1900s	A major fungal disease of chestnut trees, *Endothia* canker or chestnut blight, destroys most of the chestnut trees in the Appalachian range of the eastern United States.
1928	Scottish bacteriologist Sir Alexander Fleming (1881–1955) discovers penicillin; it uses a fungus to be produced.
1976	About 54 percent of all elm trees die from a fungus in the United States; it is known as the Dutch Elm disease.
1992	A 1,500-year-old fungus, an *Armillaria bulbosa*, is found in Michigan; it covers 30 acres underground and is thought to the oldest organism known to exist.

History of Fungi

Early Fungi

Scientist have found some fossil evidence that suggests **fungi** evolved from the protista, dividing into several evolutionary lines, but the actual lineage is still questioned. Thus, classifying the organisms is a confusing task. Depending on the classification, fungi are sometimes placed with the Kingdom Protista or within the Kingdom Plantae. But the majority of scientists give fungi its own kingdom, Fungi, because of the organisms' different structures, growth, and feeding habits.

Based on the fossil record, these eukaryotic organisms evolved just under a billion years ago, in the late Proterozoic era. Apparently, the first land forms of the ascomycetes—today, the largest division of modern fungi—were often found with the smaller arthropods of the Silurian period (440–410 million years ago). Fossils that show wood decay, or **decomposition**, from the action of fungi are often associated with plants in the Devonian period (410–360 million years ago). Fungal fossil diversity increased throughout the entire Paleozoic era; and by the Pennsylvanian period (325–286 million years ago), all the modern classes of fungi had evolved.

Most of the fungi evolved into land-dwelling organisms, but all of the major groups have invaded marine and freshwater habitats. One exception is the phylum Chytridiomycota, which probably had an aquatic origin and still prefers water—although some species of this phylum do live in terrestrial environments, primarily as plant fungi.

Modern Fungi

Although fungi have been around for millions of years and have been used by humans for thousands more, they weren't studied in great detail until relatively recently. Centuries ago, various cultures discovered ways to make wine, beer, and bread—all of which use fungi. At that time, they did not truly know the reason behind the process—only that the process of fermentation worked or that bread rose because of yeast. Other fungi, especially mushrooms, were also used in cooking. And even some slightly toxic mushrooms were used in certain cultures for religious ceremonies.

Gradually, scientists and others began to discover the true nature of these organisms. By the 1500s, Europeans knew about the "late blight of potato." This fungus, the *Phytophthora infestans*, was introduced from South America, an unwanted "export" in the continual exploration of the New World. Details of some fungi species became known two centuries later, as Elias Magnus Fries wrote his *Systemia mycologicum*, the first standard work on fungi.

The actual discoveries of certain fungi were directly tied to the organisms' effects on foodstuffs around the world. By the nineteenth century, many fungi were discovered because they caused disease—and even famines. For example, in 1804, scientists studying fungi discovered that barberry plants host a fungus that is easily transferred from the barberry to cereal plants. And in 1873, one of the most destructive fungus diseases to citrus plants, *Armillaria mellae*, was detected invading the root system of citrus trees.

In fact, fungi produced some of the worst blights and destruction of the environment in the past few centuries. Probably one of the worst fungi in the 1800s had been "discovered" three centuries before—the fungus *Phytophthora infestans* that caused the potato blight. It was first reported in the United States in 1830; by 1845, the spread of the fungus had reached epidemic proportions in Europe—and was the cause of the Irish potato famine. In the early 1900s, a major fungal disease of chestnut trees, *Endothia* canker or chestnut blight, destroyed most of the chestnut trees in the Appalachian range of the eastern United States. And in 1930, when the first Elm disease (Dutch Elm disease) was reported in Ohio, more than 77 million elms were located around the populated areas of the United States. By 1976, approximately 54 percent of all the elms in the United States had died. By the late 1990s, the loss was even greater. This elm disease, caused by the fungus *Ceratocysis ulmi*, was carried by several types of beetles, and grew in the new ring of the wood adjacent to the bark.

Besides these fungi that attack timber and cause its destruction, other fungi have caused extensive damage. They produce potent toxins and carcinogens that contaminate or spoil the food of birds, fish, humans, and other mammals. They attack fabrics and other materials, which they mildew and ruin. But although these processes appear negative and destructive, it should also be pointed out that in nature, they are necessary to prepare for the new growth of certain organisms.

In addition to their contributing to necessary natural processes, today, a number of fungi are used in the processing and flavoring of foods. For example, in the making of foodstuffs, baker's and brewer's yeasts—both types of fungi—are used, and fungi are used in certain cheesemaking. They are also used to produce organic acids. In addition, one edible fungus is commercially important and well-known—the mushroom *Agaricus campestris*.

But perhaps the most important fungus is *Penicillium chrysogenum*, known for its production of the antibiotic penicillin. Other antibiotics now exist that are produced by a wide variety of organisms, but penicillin was the first and is the most well known. In fact, when Sir Alexander Fleming discovered penicillin in the late 1920s, he kept the culture of the original penicillin; amazingly, this culture was auctioned off by Sotheby's of London to a pharmaceutical company for about 23,000 pounds—almost double the amount in U.S. currency. This price seems insignificant when compared to how many lives penicillin has saved over the past half-century—and how many more people it will save in the future. However, we have to be careful with such cultures in the future: According to many scientists, penicillin and other antibiotics have been misused and overprescribed, resulting in many resistant microorganisms. This overuse has also led—and will continue to lead—to untreatable bacterial infections and diseases.

Modern studies of fungi focus on the organisms' **lifecycles**, distribution, and various new species found in the more rugged regions of the world, such as Antarctica or the Amazon Rainforest. The number of modern fungal organisms is known to be immense. These include mushrooms, toadstools, smuts, rusts, molds, mildews, puffballs, truffles, morels, yeasts, and many less well-known organisms. About 70,000 species of fungi have been described, but some scientists believe that the total number may reach up to 1.5 million. *See also* ANIMALS, ARTHROPODS, BACTERIA AND VIRUSES, CLASSIFICATION OF ORGANISMS, CYTOLOGY, EVOLUTION, FOSSILS, GEOLOGIC TIME SCALE, HUMANS, LIFE, PLANTS, and PROTISTA.

Topic Terms

decomposition—Because of their ability to live in varied natural habitats, fungi are usually the primary decomposers (often with

---

Notable Fungi

Many fungi have made a hit in the biology world. Some are huge, covering large areas. Although many of the larger fungi mentioned below are probably fragmented and are no longer continuous bodies, they are still considered by most scientists to be one body. Here are some of the more notable fungi of the world:

What We Breathe—Fungal spores fill the air we breathe. On many days in some localities the number of fungal spores in the air far exceeds the pollen grains. Fungal spores also cause allergies; however, unlike seasonal pollen production, some fungi produce spores all year long. The largest number of fungal spores ever sampled was over 5.5 million per cubic foot in Wales.

What's This Fungi Helping?—One yeast-like ascomycete lives in the gut of cigar beetles and is essential to the beetle's health. These fungi detoxify the plant material—the beetles would die from poisoning without it.

Largest Basidiocarp—The largest basidiocarp known is that of a *Rigidioporus ulmarius*—and is listed in the *Guinness Book of Records*. The record-holding specimen is found in a shady, hidden corner of the Royal Botanic Gardens, in Kew, Surrey, England, where at the beginning of each year, an Annual Mensuration Ceremony of the basidiocarp occurs. By 1996, the basidiocarp was about 67 inches (170 centimeters) in length and 57 inches (146 centimeters) wide (the year before, it had been 62.5 inches [159 centimeters] long and 55 inches [140 centimeters] wide). It also grew 1.6 inches (4 centimeters) taller from the soil level, now measuring 21 inches (54 centimeters) high. Scientists have even estimated the fungus's weight—about 625 pounds (284 kilograms).

Contending basidiocarps—Other large basidiocarps include a puffball almost 9 feet (2.7 meters) in circumference in Canada; it is estimated to weigh just over 48 pounds. Another basidiocarp in England weighs more than 100 pounds.

Almost the Largest Fungus Body—In 1992, scientists made an astounding discovery in Michigan: a 1,500-year-old fungus, an *Armillaria bulbosa*. The fungus covers about 30 acres underground and is thought to the oldest organism known to exist. In this case, the fungus body is made of tubular filaments (hyphae) that clone themselves-and is estimated to weight as much as a blue whale—about 110 tons! The Michigan fungus clone grows in tree roots and soil.

Perhaps the Largest Fungus Body—An even larger fungal clone of *Armillaria ostoyae* was found in the state of Washington, covering over 1,500 acres. Scientists believe that each clone started from the germination of a single spore more than 1,000 years ago.

More on the Basidiomycetes—The basidiomycetes are the most familiar—especially those with large basidiocarps. But they also affect the ecosystem on the microscopic level. For example, the secret sex life of a yeast-like ascomycete human pathogen, *Coccidioides immitis*, is the cause of Valley Fever, which is endemic in parts of the southwestern United States.

the help of certain bacteria) of plant and animal organisms. Many fungi species are free-living saprobes (users of carbon fixed [a chemical process] by other organisms) found in woody substrates, soils, leaf litter, dead animals, and animal excrement. In many ways, these decomposers improve the living conditions of organisms within certain habitats. For example, in forests, the large cavities eaten out of living trees by wood-decaying fungi provide nest holes for a variety of animals; and they are also master decomposers, adding nutrients to the soil.

Where and why fungi decompose varies. Some fungi grow only on simple sugars, using them as sources of carbon, getting their nitrogen from inorganic nitrates and ammonium compounds. Other fungi release enzymes that continually digest the cells of dead plant and animal material, turning the decaying matter into a solution of simple nutrients the fungi can readily absorb. And still other fungi are parasitic or symbiotic, extracting their nourishment from the living host plant or animal. These fungi are called biotrophs, or organisms that derive nutrients from the tissues of the living host. For example, lichens are biotrophs and exist in a symbiotic relationship between the fungus and a certain type of green algae. They are often found on bare rock in various environments, including some of the harshest regions of the world.

fungi—Fungi have diverse habits and characteristics—thus, generalizations are difficult. And as for the conditions necessary for growth and reproduction, each species has its own preferences. Fungi grow in almost any habitat that contains organic substances. Wherever adequate moisture, temperature, and organic substrates are available, fungi are present. Normally, we think of fungi as growing in warm, moist forests, and the most diverse number of groups do tend to live in the tropical regions. But many species live in cold and periodically arid regions—or even inhospitable regions such as Antarctica. Numerous species are found in salt- and freshwater conditions; others are adapted to life on the surface of the earth or in the ground.

It is easy to see why fungi are considered neither animals nor plants. Fungi exist mostly in multicellular form, but they can go though an amoebae-like (single-celled) stage; others, such as yeast, exist only in single-celled form. Fungi do not use photosynthesis like plants and certain algae, which gather energy from the Sun. Fungi share with animals the ability to absorb food through their cell walls and plasma membranes. Instead of a stomach that digests food like an animal, fungi live in their own food supply and simply grow into new food as the nutrients are depleted in the local environment. In addition, the fungi cell walls contain chitin, the substance found in insects' exoskeletons; to compare, plant cell walls are made of cellulose and animal cells do not have cell walls at all—just a membrane.

Fungi reproduce sexually, and prior to mating, individual fungi communicate with other individuals chemically via pheromones. In every phylum at least one pheromone has been characterized, and they have remarkable names, ranging from sesquiterpines and derivatives of the carotenoid pathway in chytridiomycetes and zygomycetes to oligopeptides in ascomycetes and basidiomycetes.

Plants have a close connection to fungi. About four-fifths of all vascular plants have a symbiotic relationship with fungi, called fungus roots, or *mycorrhizae*. Because the fungi are able to absorb minerals in the soil more efficiently than the plant's roots, the plant "allows" the fungi to grow—and uses the minerals accumulated by the fungus. In exchange, the fungus receives energy by absorbing carbohydrates that the plant produces through photosynthesis.

By observing dense filaments of fungal colonies growing on nutrient-rich agar plates, biologists have discovered most fungi's fascinating food gathering process. When one of the filaments contacts a food supply, the entire colony mobilizes and relocates around the new food. When the food is depleted, spores are often sent out in search for more food. Although the filaments and spores are tiny, the entire colony can be very large, with individuals of some species reaching close to the mass of larger animals or plants.

The following is a list of the major classes of the modern fungi:

Phycomycetes—These fungi are quite similar to several different groups of green algae. The simpler members have only a single cell; others are multicellular. There are about 1,500 species of this fungi, and most of the species grow in water (such as the water-molds; and molds that grow on dead animals in water).

Ascomycetes—These fungi include some 40,000 species. The majority of the organisms are small, often minute, while others can reach 3 to 4 inches (7.5 to 10 centimeters) in length. Members of this class have a major economic impact on humans: They are often destructive to food and fiber crops, and include many parasites. But among the ascomycetes are also many beneficial species, such as yeasts and truffles.

Basidiomycetes—This class is the most well-known fungi and includes the toadstools, mushrooms, puffballs, and other forms. These fungi have a fruiting body called basidium (or mushroom or club), a club-shaped structure that bears four spores (reproductive cells) at its tip (apex). Some are poisonous to humans and other animals; some are not.

Imperfects (Fungi Imperfecti)—These fungi include those forms in which the sexual stage is not known, and thus cannot be assigned to any of the three major classes. They represent about 25,000 fungal species; members of this group are responsible for ringworm and athelete's foot, infecting people without sprouting fruiting bodies.

lifecycle—The fungi lifecycle varies, depending on the group; but the overwhelming majority reproduce by spores, a single cell surrounded by a protective coat of chitin. For example, the lifecycle of the basidiomycetes, such as mushrooms, is as follows:

First, the fungi disseminate their spores into the wind. Each spore has a haploid nucleus (containing only one copy of chromosomes). The spores germinate if they are lucky enough to land on a suitable surface and the conditions are favorable for growth, usually warm and wet. Fungal spores are actively or passively released through several different methods, such as wind, rain splash, or flowing water dispersal. Thus, the air we breathe is filled with spores of many fungi species—especially those that produce a large number of spores. These include species pathogenic to agricultural crops and trees, or even those that live within or on the surfaces of animals—particularly arthropods.

Next, if two sexually compatible "germlings" grow nearby, their hyphae (which look like fine threads or filaments) fuse—forming two different kinds of nuclei together. The fungus exists in this state for most of its life. Finally, when the organism has had enough to eat, or the conditions change for the worse, the fungus produces a fruiting body that will again go through sexual reproduction and dispersal.

18. Genetics

Introduction

All of the flora and fauna on Earth have, as one of their basic instincts, the urge to reproduce. This urge is necessary for the survival of the species—with limited lifespans and no new members, a species would soon die out. In addition, reproduction results in mutations, varying the traits within individuals while still keeping the species intact. These mutations, or changes, are thought to be one of the necessary components of evolution. But how does the code or instructions get passed along to the new species members? How does a rose get its characteristic odor, or a bird develop feathers instead of fur? The answer is found in genetics, or the study of heredity—which over the natural history of the Earth has had a profound effect on all organisms.

Timeline

Date	Event
Modern Events and Discoveries	
1866	Gregor Johann Mendel (1822–1884), an Austrian monk, publishes the results of his cross-breeding experiments with garden peas in an obscure journal.
1866	German biologist Ernst Heinrich Haeckel (Häckel, 1834–1919) hypothesizes that the nucleus of a cell transmits its hereditary information to other cells.
1869	English scientist, explorer, and Charles Darwin's cousin, Sir Francis Galton (1822–1911), claims heredity alone is responsible for a person's character traits. He later develops the idea of eugenics, the breeding of human beings for evolutionary improvement.
1881	German physician, bacteriologist, and chemist Paul Ehrlich (1854–1915) discovers staining of bacteria with methylene blue allows internal structures to be better seen; this discovery would increase the knowledge of genetics considerably.
1882	German cytologist Walther Flemming (1843–1905) discovers chromosomes, but no one connects the findings to heredity.
1887	Belgian biologist Edouard van Beneden (1846–1910) discovers organisms of the same species have the same number of chromosomes.
1888	German anatomist and physiologist Heinrich Wilhelm Gottfried von Waldeyer-Hartz (1836–1921) names the chromosome.
1892	August Weismann (1834–1914), a German biologist, publishes an essay on heredity, stating it was transmitted by a substance with a "chemical and molecular constitution."
1898	Flemming estimates the human chromosome count to be 24 pairs.
1900	Dutch plant physiologist Hugo De Vries (1848–1935), German botanist Karl Franz Joseph Correns (1864–1933), and Austrian botanist Erich Tschermak von Seysenegg (1871–1962) rediscover Gregor Mendel's work independently, studies that had been ignored for nearly four decades.
1909	Russian-born U.S. biochemist Phoebus Aaron Theodor Levene (1869–1940) discovers the chemical nature of ribonucleic acid (RNA).
1911	U.S. geneticist Thomas Hunt Morgan (1866–1945), with U.S. geneticist Alfred Hunt Sturtevant (1891–1970), demonstrates that genes linked along the chromosomes are the instruments of heredity, by experimenting with *Drosophelia* (fruit flies). Sturtevant later develops the first genetic map.
1944	U.S. bacteriologist Oswald Theodore Avery (1877–1955) and his colleagues discover the "instructions" (now called deoxyribonucleic acid [DNA]), or the material responsible for heredity in almost all living organisms.
1953	English molecular biologist Francis Crick (1916–) and U.S. biochemist James Watson (1928–) make a model of the "double-helix" structure of DNA, and prove genes determine heredity.
1950s	U.S. scientists Maurice Wilkins (1916–) and Rosalind Franklin (1920–1957), together with Crick and Watson, discover the chemical structure of DNA, starting a new branch of science called molecular biology.

Date	Event
1956	Java-born scientist Joe Hin Tjio (1919–) and Johan Albert Levan revise Walther Flemming's 1898 estimate of the human chromosome count from 24 pairs to 23 pairs.
1957	U.S. scientist Arthur Kornberg (1918–) produces DNA in a test tube.
1961	U.S. biochemists Marshall Warren Nirenberg (1927–) and J. Heinrich Matthaei report certain codes that open the way to identification of the genetic code.
1961–65	Several laboratories identify the genetic code words for the amino acids.
1967	British biologist John B. Gurden successfully clones a vertebrate—a South African clawed frog.
1972	U.S. scientist Paul Bergh (1926–) produces the first recombinant DNA molecule.
1979	DNA from malignant cells is used to transform cultured mouse cells, permitting cancer genes to be studied in cell culture.
1980	Construction is begun on the first industrial plant designed to make insulin using recombinant DNA methods.
1981	A U.S. court rules that genetically engineered bacteria can be patented.
1982	Human insulin, produced by recombinant DNA methods, is approved for use and marketed under the name Humulin.
1983	Researchers at Harvard University develop the world's first artificially made chromosomes.
1983	The complete 48,502 base pair sequence of the DNA of bacteriophage *lambda* is published.
1984	Sheep are successfully cloned.
1984	Through DNA, geneticists in the United States discover that chimpanzees are more closely related to humans than to gorillas or other apes; the genetic difference is about 1 percent.
1986	The United States Department of Agriculture grants the first license to market an organism that is produced by genetic engineering—a virus to vaccinate against a herpes disease in swine.
1987	Altered bacteria are released into the environment for the first time; they are released to save a California strawberry field from frost damage.
1989	Rennin, an enzyme used in the production of cheese, becomes the first genetically engineered edible product for human consumption.
1989	The oncmouse is the first genetically engineered animal to be covered by a U.S. patent.
1990s	The United States begins work on a 15-year project to map the 100,000 genes found in humans—the entire human genome.
1993	Researchers "clone" a human embryo for the first time.
1995	Geneticists sequence the full genomes of two different bacteria.
1997	The first artificial human chromosome is created.
1997	Dolly, a lamb cloned from the cells of an adult ewe, is born.

History of Genetics

Early Genetics

One of the most important people in the study of early genetics was an Austrian monk, Gregor Johann Mendel. In 1857, Mendel began his famous study of the pea plants in his garden; by 1866, he would publish the results of his cross-breeding experiments, the first step in understanding heredity and how it is passed from one generation to another. But the study was not as famous then as it is now. Mendel published his results in an obscure journal, and they went unheeded until about four decades later.

Mendel's story is a classic in scientific study. He had limited resources as a monk, and he based all his conclusions on his observations, not any experimentation with the internal workings of the plants. In particular, he was interested to discover whether or not the seeds produced by a certain plant resembled the seed from which the plant itself grew. His massive collection of detailed notes showed that he made hundreds of different crosses of a huge number of seeds—and he kept track of all of them. He theorized that the organisms did inherit characteristics from their ancestors—the basic foundation for the study of heredity.

Mendel knew that something carried the characteristics of an organism to each successive generation, but he did not know (and could not know) what scientists discovered much later: that each organism has coded biological instructions called **genes**, which carry an inherited trait from each parent to offspring; for smaller, more primitive forms of life that reproduce asexually, such as bacteria, a single identical set of genes is carried to the offspring.

The work Mendel completed seems all the more impressive because some of the most important parts of a cell involved in genetics weren't discovered until after his study was published. **Chromosomes** were discovered by Walther Flemming in 1882 and reported in his book *Cell Substance, Nucleus, and Cell Division*, in which he described chromosomes and mitosis. The existence of **amino acids** was discovered even earlier in 1806, but their importance to genetics was unknown until the late 1800s, when biochemist Thomas Osborne discovered that some **proteins** contain amino acids as building blocks essential for life. It would take until 1950 before U.S. biochemist William Cumming Rose established the protein-building role of the essential amino acids.

Mendel's work, now called **classical genetics**, appeared again in 1900, when Hugo Marie De Vries, Karl Franz Joseph Correns, and Erich Tschermak von Seysenegg independently rediscovered Gregor Mendel's work after four decades of neglect. And with their rediscovery came a wider interest in how heredity is carried from one generation to another. Another step towards understanding heredity occurred in 1907, when Thomas Hunt Morgan used fruit flies—organisms that have a short lifecycle that allowed the effects of generational changes to be quickly seen—to prove that Mendel's laws worked. He also found that certain inherited characteristics seemed to be strung together—and by 1911, demonstrated that fruitfly genes linked along the chromosomes were truly the instruments of heredity.

Eventually, scientists discovered a substance that every living organism carries within its cells—**DNA** (deoxyribonucleic acid). In 1944, Oswald Theodore Avery, Colin MacLeod, and Maclyn McCarthy discovered that DNA was the special hereditary material that almost all living organisms needed to reproduce and survive. With its long string of bases, DNA is the essential way cells "speak" to the next generation—without the influence of the genes and the specific combinations of DNA bases within the genes, no species, from primitive to complex, would survive. (Another kind of gene-bearing molecule, known as ribonucleic acid or **RNA**, was discovered by Phoebus Aaron Theodor Levene around 1909.)

Other discoveries about DNA continued. In the early 1950s, when Linus Pauling determined that the structure of a certain class of proteins was a helix (a three-dimensional spiral), many scientific eyes turned to another large biological molecule, DNA. Many players were involved in the hunt for DNA's structure. But the two who came out on top in 1953 were British molecular biologist Francis Crick and U.S. biochemist James Watson. They discovered DNA's famous "double helix" structure—which helped explain how the giant molecule was able to transmit heredity in all living organisms. At the same time, English scientists Maurice Wilkins and Rosalind Franklin produced the X-ray diffraction calculations that confirmed the structure.

Modern Genetics

Several major breakthroughs have been made in genetics in the past half-century—especially in the field of **genetic engineering**. The list of accomplishments in genetic engineering in just the past few decades is staggering—and will continue into the future. They range from special medicines for humans to the discovery of certain genes that control disease or other characteristics. A few of these highlights are mentioned here, but they are just a small sample. For example, in 1981, genetically engineered bacteria were officially sanctioned to be patented; in 1982, genetically engineered insulin was approved for human use, and a gene for growth was injected in a mouse embryo, resulting in an animal twice the normal size. Continuing with the altered mice theme, in 1989, the oncmouse was born, genetically altered to carry the gene that causes cancer—to use it to understand the disease in humans. And by 1994, scientists found a fat-regulating gene shared by mice and people.

Modern genetics has also allowed us to discover more about evolution, particularly of the human race. For example, in 1984, U.S. geneticists examining the DNA of chimpanzees and humans found that chimpanzees are more closely related to humans than either species are to gorillas or other apes. Based on this evidence—which also showed that the genetic difference was about 1 percent—scientists believe that humans and chimpanzees diverged from a common ancestor some 5 to 6 million years ago. Another human-gene connection began in late 1990s, when a government-financed program began to map the human genome (the full set of genes—thus traits; there are 100,000 genes on the 23 human chromosomes)—a feat that will take 15 years and includes the task of locating all the 100,000 genes.

Still other genetic engineering feats include altering genes to improve conditions within the environment. For example, in 1987, a gene-altered bacteria was sprayed on a strawberry field in California to prevent frost damage—the first officially approved release of altered bacteria into the environment. In 1989, rennin used in the production of cheese became the first genetically engineered edible product approved for human consumption. Since that time, the actual production of genetically altered foodstuff has increased—along with the controversy—as the general public continues to have doubts about the safety, in the present and future, of these genetically altered products.

But the most interesting, scary, and controversial—depending on whom you talk to—feats in all of genetic engineering is cloning. Without a doubt, it is one of the more controversial fields within genetic engineering, beginning in about the mid-twentieth century. British biologist John B. Gurden was the first to successfully clone a vertebrate in 1967—a South African clawed frog. In 1993, researchers "cloned" human embryos (they were not true clones as we think of them) for the first time. And by 1997, Scottish researchers, using a special technique, cloned the first mammal, a lamb, from the cell of an adult. The resulting sheep, named Dolly, is the adult animal's identical copy—something that had not been accomplished before. Just over a year later, fused cells from an adult cow's uterus and cow eggs (in which the DNA had been removed) produced several embryos, two of which were born to a surrogate mother cow—and are thought to be clones; there are even rumors, but no confirmation, of mice being cloned from adult mouse cells.

These cloning results continue to trigger a host of ethical questions—and will continue to in the future. In particular, it brings up questions of who or what should be cloned; what will such cloning do, if allowed to snowball, to the natural process of evolution? Right now, cloning seems to be controlled, and many countries agree to such restraints on this type of genetic production. But who decides what or who can or should be cloned? In 1998, a couple whose dog had died donated millions of dollars to a large university—to clone their dog they loved so much. The questions of how to cope with such decisions will continue to be highly debated in the future. *See also* CLASSIFICATION OF ORGANISMS, CYTOLOGY, EVOLUTION, HUMANS, LIFE, and PLANTS.

Topic Terms

amino acids—Amino acids are the components that make up the protein chains (large molecules composed of one or more chains of varying amounts of the 22 amino acids). Amino acids are water-soluble organic compounds composed of the ever-present carbon, hydrogen, nitrogen, and oxygen in varying configurations. (These four elements, incidentally, make up almost 96 percent of the weight of most life forms.) It takes a sequence of three bases on the DNA chain to specify an amino acid. This three-base sequence is sometimes referred to as a "codon." Their relationship to genetics is indirect; amino acids are the building blocks of proteins; and proteins make up some parts of nucleic acids (the major two types being DNA and RNA).

chromosomes—Chromosomes are the long, thin, thread-like structures within the nucleus of a cell that are composed of deoxyribonucleic acid (DNA) and protein. They contain genes, or the individual units of information that tell the cell what to do—and how and when to do it. All the instructions concerning life processes of the cell come from the chromosomes. A normal human cell has 46 chromosomes, with 23 of the chromosomes inherited from each parent. One of the primary tasks of genetic researchers is determining what genes are on what chromosomes.

classical genetics—Classical genetics is attributed to nineteenth-century monk, Gregor Mendel. Later, in 1900, Hugo Marie De Vries, Karl Franz Joseph Correns, and Erich Tschermak von Seysenegg rediscovered Mendel's work independently. In general, Mendel knew some traits were inherited from plant to plant. For example, if a white-flowered pea plant is bred with a red-flowered pea plant, the offspring would be pure red or pure white, not a combination of the colors such as pink. (See table below.) Although heredity is usually more complex than breeding flowers, they still exhibit a simple way of viewing genetics in action.

Classical genetics proposed that every individual has two copies of each gene, one of which is more dominant than the other—thus the dominant (R) and recessive (r) genes. For example, the red-flowered peas normally have two copies of a gene for color that are dominant; or they may have a dominant gene and a recessive gene. The white-flowered peas have two recessive genes that produce the white color. The different forms for a particular gene are called alleles; in the case of the pea plants, the genes are for colors, with the alleles giving the code for a particular color.

The following chart shows a very simple rendition of how the colors come out using dominant (R allele) and recessive (r allele) combinations (note: for any different mating, the offspring has a 75 percent chance of coming out red):

Classical Genetics—Parents and Offspring	
Parents (two plants bred with each other)	*Offspring*
Plant with R + Plant with R	RR (red)
Plant with R + Plant with r	Rr (red)
Plant with r + Plant with R	rR (red)
Plant with r + Plant with r	rr (white)

DNA—DNA is actually deoxyribonucleic acid. We now know that every living creature carries within its cells, or cell, a tiny concentration of DNA that carries the genes required for that creature's survival and reproduction. For higher order life forms, the individual will have two sets of genes, one contributed by each parent.

The actual DNA itself is made up a long strings of organic rings—composed of nitrogen, hydrogen, oxygen, and carbon—called bases. DNA has only four bases: adenine (A), guanine (G), cytosine (C), and thymine (T). Each of these bases provide the information needed by individuals to survive and reproduce. Bases are found in all lifeforms—from complex organisms such as mammals to elemental organisms such as bacteria. For example, DNA in bacteria may contain about 4,200,000 bases; a human being's DNA contains about 6,600,000,000 bases in each cell.

The actual structure of DNA is composed of pairs of bases, but they are not arranged in all the possible combinations: For example, the A base is always paired with a T base, and a G base with a C base—with each of the bases physically interlocked with only their specific pairing base. These paired structures form the famous "double-helix" of DNA that was discovered by Francis Crick and James Watson in 1953.

genes—Genes are the coded biological instructions found in every cell that control the transmission of inheritable traits from parents to their offspring; specifically, genes control the numbers and types of proteins. The genes are contained within chromosomes. Thus, one of the primary tasks of genetic researchers is determining what genes are on what chromosomes.

The long, continuous "thread" of base pairs along a DNA segment that gives information such as to how to create a certain bodily substance is called a gene. For example, a single human gene can vary in length from less than 1,000 base pairs to as many as two million. The substance that the gene specifies may define some recognizable trait of the individual—such as green eyes over brown eyes, or red hair instead of brown hair. From either human parent, as many as 50,000 to 100,000 genes may be present.

genetic engineering—Genetic engineering (or transgenics) is the deliberate alteration of the genetic makeup of an organism. It is accomplished by manipulation of the DNA (deoxyribonucleic acid) molecules within the organism's cells to cause a change in the hereditary traits of the organism—in particular, by introducing a foreign, or altered, gene into the cells. For humans, genetic engineering can be used to develop possible cures for certain genetically controlled diseases and to make medicines that attack viral diseases; in nonhuman applications, it offers a way to change some characteristics of plants and animals for useful purposes, such as plants engineered to be drought resistant or animals engineered to be heavier or hardier.

proteins—The formation of proteins is one of the most important functions of the genes in a cell. Proteins are absolutely essential to all life forms and are present in all living cells. They help to develop such necessary structural materials as skin or feathers, and they are found in cell coats and walls; they also transport mechanisms throughout the bloodstream and regulate certain hormones, such as those that control a body's metabolism. Some proteins even are poisonous, such as snake venoms that are composed of proteins.

Proteins are made up of amino acids, and their actual size varies greatly; they may have from 100 to up to as many as 1,800 amino acids. The way the proteins are sequenced, and the three-dimensional shape that they fold themselves into (called protein folding), is critically important to their function. Most bodily processes depend upon the ability of various substances to interlock with each other; thus, their shape is vitally important.

RNA—Ribonucleic acid (RNA) is the chemical that translates the genetic code of DNA into a protein. (The physical process of producing protein in the cell involves this other gene-bearing molecule, RNA.) RNA is very similar to DNA. It contains four bases (one of which differs slightly from its counterpart in DNA); has a single strand (not a double helix); and has a shorter chain of bases than DNA.

RNA works as follows: The DNA will synthesize a strand of RNA when a protein needs to be made, and one helix thread of the DNA will serve as a template to impress a corresponding base sequence on the shorter RNA strand. The RNA strand will then carry its message from the DNA to the production area of the cell (the ribosome), where the protein will be manufactured from the appropriate amino acids. Because of its message-carrying role in protein synthesis, RNA is often referred to as messenger RNA or simply "mRNA."

19. Geologic Time Scale

Introduction

To comprehend Earth's long natural history, scientists have developed the geologic time scale. In many ways, it represents something that is almost incomprehensible in its scale—about 4.55 billion years of history. This geochronology lets us know where rocks and organisms stand in the evolution of Earth. And it is only through the geologic time scale that we can understand more about evolution of all the organisms on the planet.

Timeline

Date	Event
Modern Events and Discoveries	
~1650	Archbishop James Ussher, Irish Protestant theologian and scholar (1581–1656), determines the age of the Earth based on his reading of the Bible, suggesting that the planet is close to 6,000 years old.
1695	English geologist John Woodward (1665–1728) proposes that strange organic-looking shapes found in rocks were the remains of creatures that failed to get on Noah's Ark before the great biblical flood; he believes that they were preserved as fossils.
1835	English geologist Adam Sedgwick (1785–1873) and Scottish geologist Roderick Murchison (1792–1871) publish a joint paper that outlines their studies of the Silurian and Cambrian divisions; the two men later dispute where the two systems overlap as further studies are made—studies that eventually led to the divisions of Silurian and Cambrian on the geologic time scale.
1846	British theoretical and experimental physicist Sir William Thomson, later Lord Kelvin (1824–1907), estimates (incorrectly) that the Earth is about 100 million years old, based on his determination of the planet's temperature.
1850	The geologic time scale, with its three major time divisions (Paleozoic, Mesozoic, and Cenozoic) and certain smaller time scale divisions called periods and epochs, are established by this time.
1896	French physicist Antoine Henri Becquerel (1852–1908) accidentally discovers that a photographic plate placed next to some mineral salts containing uranium had blackened; it proves that uranium gives off its own energy.
1902	New Zealand–born British physicist Lord Ernest Rutherford (1871–1937) collaborates with British chemist Frederic Soddy (1877–1966) to discover that the atoms of radioactive elements are unstable and decay to a more stable form.
1907	U.S. chemist Bertram Borden Boltwood (1870–1927) argues that knowing the decay rate of uranium and thorium into lead will make it possible to date rocks.
1911	English geologist Arthur Holmes (1890–1965) begins work on a geologic time scale based on the uranium-lead dating method.
1917	Geologist J. Barrel publishes a paper, "Rhythms and the Measurement of Geologic Time," presenting his own time scale; his dates are very close to those still used today.
1927	Holmes estimates the age of the Earth from the relative abundance of radioactive elements and their decay products; he suggests the crust is about 3.6 billion years old.
~1940	A.O. Nier uses uranium-lead dating based on mass spectrometric measurements; it is considered the start of modern geochronology.
1940	Canadian-American biochemist Martin David Kamen (1913–1992) discovers carbon-14, a radioactive isotope of carbon; it will play a significant part in determining the age of more recent fossils.
1947	U.S. chemist Willard Frank Libby (1908–1980) invents the technique of carbon dating; it can be used to determine the age of archeological objects as far back as about 45,000 years ago.

Date	Event
1956	C.C. Patterson compares the abundance ratios of lead isotopes in meteorites with terrestrial lead and determines the age of the Earth is about 4.55 billion years; this date is still the standard age, accepted by the majority of scientists today.
1960s	New isotopic dating techniques are developed throughout the decade, such as potassium-argon dating techniques.
1971	British geochronologist Stephen Moorbath dates some of the oldest rocks known, 3.75-billion-year-old rocks from Greenland.

History of Geologic Time Scale

Early Geologic Time Scale

The early attempts to create a **geologic time scale** took many centuries. After all, by the late seventeenth century, most European Christians (including many scientists who rejected other more reasonable suggestions on the age of the Earth) believed in the literal biblical rendition of the Earth. The Book of Genesis that states the planet (in its current form) was created in six days. In 1650, Archbishop James Ussher added up the lifespans of all the descendants since Adam and Eve, reasoning that the world was created in 4004 B.C.—which then made the planet close to 6,000 years old.

It was not until the late eighteenth century that scientists truly began to question this estimate of Earth's age—although change came slowly. In 1846, Sir William Thomson, later Lord Kelvin, estimated the planet's temperature, based on his observation that as coal miners tunneled deeper into the earth, they encountered rocks hotter than those at the surface. He believed that this heat was left over from the planet's formation, and by determining how fast the Earth would cool down, he resolved that the Earth was about 100 million years old. We know today that this estimate was wrong—Thomson's mistake was that he believed the earth cooled mainly by conduction, like a hot brick. What he did not, and could not, know was that the earth's interior is in constant motion and is convecting (giving off heat)—both factors making the determination of the rate of the cooling a complex problem.

By the nineteenth century, Thomson's age of the Earth was still accepted, but many scientists believed 100 million years was a very short period of time. Toward the end of his life, Charles Darwin began to doubt his theory of evolution, as it would take so much longer for evolution to take place than the then current estimate of the planet's age.

Modern Geologic Time Scale

Historically, scientists have used two major ways to determine the age of a rock, and thus know where it falls in the context of the geologic time scale. The classical way is by identifying fossils and the rock surrounding the fossils and using that information to put the target rock in its relative timeframe; this method, often referred to as the **relative time**, was readily used through the early twentieth century. It was then that new developments in physics allowed scientists to develop absolute numeric time boundaries based on radiometric age measurements, usually referred to as the **absolute time**.

Determining the divisions in Precambrian time was difficult using relative time. The biggest problem was the lack of fossils. Unlike the post-Precambrian time, when animal and plant fossils became more abundant, the Precambrian was almost devoid of any life—and thus, fossils. (Even today, although many fossils have been found that date back to the Precambrian time, they are still difficult to find. The main reason for this lack of fossil record is most life at the time consisted of soft parts and thus, usually left no trace of their existence.)

The immensity of the Precambrian time and the determination of absolute time on the time scale were not fully understood until the radiometric age measurements became available. The first break came in 1896, when Antoine Henri Becquerel discovered that a photographic plate left next to some mineral salts containing uranium had blackened—a serendipitous discovery proving that uranium gave off its own energy. By 1902, Ernest Rutherford had collaborated with Frederic Soddy to work out the process of radioactive decay—in particular, they found the atoms of radioactive elements are unstable and decay to a more stable form at a regular rate. The biggest break in determining geologic time came in 1907, when chemist Bertram Boltwood proposed that lead was the final product of the **radioactive decay** of uranium and thorium. He made the profound next step, arguing that by knowing the decay rate, absolute dating of the rock was possible. In other words, a rock could be dated by determining how much lead it contained—the more lead, the older the rock.

One of the foremost scientists involved in the development of the geologic time scale using the new radiometric techniques was British geologist Arthur Holmes, whose publications on the subject spanned almost a half-century. In 1911, Holmes began to produce a geologic time scale based on the uranium-lead dating method. Holmes's first extended time scale for the Phanerozoic eon was published in 1947, which included estimations of certain divisions, not single specific dates. (For example, his first chart in 1933 stated that the Jurassic period was 158 million years ago; his second chart stated it was 152–167 million years ago.) By 1917, geologist Jospeh Barrel published a paper, "Rhythms and the Measurement of Geologic Time," presenting his own time scale—also based on uranium-lead dating techniques.

Since World War II, progress has been made in geochronology with the introduction of other dating techniques, such as potassium-argon and rubidium-strontium techniques. In addition, improved uranium-thorium-lead techniques have allowed scientists to reach back further in time, testing certain minerals in the rock, such as zircon, that reveal ages of billions of years. Currently, the oldest rocks yet found on Earth are a staggering 3.75 billion years old—sedimentary rocks in the Isua region of Greenland, deposited in a volcanic environment; the oldest individual minerals, caught up in younger rock, are found in Acasta in northern Canada—crystallized during volcanic activity more than 4 billion years ago.

Talking Time

The geologic time scale is not just an arbitrary listing of the Earth's natural history. Each division represents a change or event that divides it from the next or previous set of events. Sometimes, the division is made to indicate a major catastrophe; other times, a division distinction is made due to evolutionary changes in animals or plants.

The divisions of the eras tell the stories. The following lists the eras, and the reasons for the divisions, from oldest to youngest:

Precambrian time—The Precambrian is the longest span of time, lasting from 4.55 billion years ago to 544 million years ago—about seven eighths of the earth's history. It includes the formation of the planet's crust; the formation of the first tectonic plates and their movement; the first of life on the planet in the form of eukaryotic cells; and the evolution of an oxygen-rich atmosphere. At the very end of the Precambrian, the first multicelled organisms, including the first animals, evolved in the oceans.

Paleozoic era—The Paleozoic era, 544 to 245 million years ago, is bracketed by two of the most important events that occurred in the animal world. At one end, the multicelled animals experienced an explosion in numbers and diversity, called the Cambrian Explosion; within that time, almost all modern animal phyla appeared within a few million years. At the other end of the Paleozoic, the largest mass extinction in the planet's history took place, wiping out between 90 and 96 percent of all species. In the middle of the Paleozoic era, animals, plants, and fungi began to colonize the land, and insects began to occupy the air. Thus, many divisions in this era are loosely based on these events; for example, at the end of the Permian period is when a great extinction took place.

Mesozoic era—The Mesozoic era, or "middle animals," is a period from 245 to 65 million years ago when the planet's terrestrial plants changed drastically from what they had been in the Paleozoic era. In particular, early in the era, ferns, cycads, ginkos, and other unusual plants dominated; in the middle of the era, gymnosperms (such as conifers) first appeared. Toward the end of the era, in the middle Cretaceous, the earliest angiosperms (flowering plants) took over. As for animals, this era brought out—and took away—one of the most famous animals of all time—the dinosaurs. This era was also the time of great animal diversity. In general, the amphibians evolved, then the reptiles from the amphibians, and even mammal-like reptiles evolved during this 180-million-year span of time. At the end of the Mesozoic era, another mass extinction took place, eliminating more than 50 percent of the species on the Earth, including the dinosaurs.

Cenozoic era—The Cenozoic era is the most recent of all the eras, and includes only the last 65 million years. It has two main divisions: the Tertiary and Quaternary periods. These divisions are also often divided into other periods by other charts (Paleogene [65 to 23 million years] and Neogene [23 to 2 million years] of the Tertiary; and the Anthropogene [2 million years to present] representing the Quaternary). The Cenozoic era is sometimes referred to as the "Age of Mammals" for a number of reasons. For example, after the mass extinction at the end of the Mesozoic era, mammals exploited empty niches, growing in numbers and diversity. Another reason is that the largest land animals have all been mammals during this time. But in reality, the nickname could have been the "Age of the Birds, Fish, Insects, or Flowering Plants" just as easily—as all these creatures have also diversified and grown in numbers over the past 65 million years.

While these radiometric techniques are excellent for determining the age of extremely old rocks, they are less useful for determining more recent dates, especially for fossils. To determine the age of fossils and other younger rocks, scientists had to come up with another technique—carbon dating. This method of determining the age of younger rocks—and even fossils—was discovered by Martin Kamen in 1940—it dates younger material because of carbon's shorter half-life. (See Fossils and Humans.) Carbon dating would become one of the most useful of all the radioactive tracers, not only to paleontologists, but to anthropologists—especially to date ancient human burial sites.

The modern geologic time scale is usually standard up to the period **time units**. But epochs, ages, and stages are usually different in specific places around the world. This is because not all areas contain the same rock layers or fossils. The various time units are not precise or found in consistent spans, but overall, the time periods are so large, there is no reason for such precision. *See also* AMPHIBIANS, ANIMALS, ARTHROPODS, BIRDS, CLASSIFICATION OF ORGANISMS, CYTOLOGY, DINOSAURS, EARTH, EVOLUTION, EXTINCTION, FOSSILS, LIFE, PLANTS, and REPTILES.

Topic Terms

absolute time—Absolute time is determined by radiometric means. As the name implies, it is the determination of the (approximate) absolute time when the rock formed. In comparison, relative time is based on determining the age of a rock based on where it sits in comparison to other layers of rock—thus, it is only relative, not absolute, time.

geologic time scale—The geologic time scale is a representation of the Earth's long history, from the formation of the Earth about 4.55 billion years ago to the present.

One of the most difficult ideas about the geologic time scale is to envision the scale. Author John McPhee, in his book *Basin and Range*, gives a way of understanding the vastness of the Earth's natural history—and geologic time. Stand with your arms held out to each side; the extent of the Earth's history is represented as the distance between the tips of your fingers on the left and right hands. If someone were to run a file across the fingernail of your right middle finger—and that represented time—it would erase the amount of time humans have been on the planet!

In general, the major divisions of the scale occur just after the Cambrian period, when life began to flourish on the planet. There is a good reason for this bias: Much of our knowledge of the geologic past is based on the fossils we find in rock—fossils of once living organisms. And those fossils did not show up in the rock layers in large numbers until after the great Cambrian Explosion, when life grew exponentially in the oceans.

The following lists the divisions and names on the geologic time scale:

The Geologic Time Scale (U.S. Version)

Era   Period Epoch

Holocene epoch—11,000 years to present
Pleistocene epoch—1.8 million years to 11,000 years ago
Quaternary period—1.8 million years to present
Pliocene epoch—5 million years to 1.8 million years ago
Miocene epoch—23 million years to 5 million years ago
Oligocene epoch—38 million years to 23 million years ago
Eocene epoch—54 million years to 38 million years ago
Paleocene epoch—65 million years to 54 million years ago
Tertiary period—65 million years to 1.8 million years ago
Beginning of the Cenozoic era
Cretaceous period—146 million years to 65 million years ago
Jurassic period—208 million years to 146 million years ago
Triassic period—245 million years to 208 million years ago
Beginning of the Mesozoic era
Permian period—286 million years to 245 million years ago
Carboniferous period—360 million years to 286 million years ago
Pennsylvanian—325 million years to 286 million years ago
Mississippian—360 million years to 325 million years ago
Devonian period—410 million years to 360 million years ago
Silurian period—440 million years to 410 million years ago
Ordovician period—505 million years to 440 million years ago
Cambrian period—544 million years to 505 million years ago
Beginning of the Paleozoic era
Precambrian time—544 million years ago to present
Vendian period—650 million years to 544 million years ago
Proterozoic era—2.5 billion to 544 million years ago
Archaean era—3.8 to 2.5 billion years ago
Hadean time—4.55 to around 3.8 billion years ago (not a geological period as such—no rocks on the Earth are this old—except meteorites; it's when the Earth was still mostly molten)

radioactive decay—Atoms of each element have the same number of protons in their nuclei, but atoms don't have to have the same number of neutrons. In nature, these elements will be in various forms depending on the different numbers of neutrons—with these different forms of the same element called isotopes. Simply put, if the isotope is unstable, it will decay into another isotope. For example, an atom of the radioactive element uranium-238 decays first into an atom of thorium-234; it will further breakdown into lead. The amount of time it takes for half of the isotope to break down is called the half life—which can range from a few seconds to millions of years depending on the element. Because some half-lives take so long to decay, scientists use these elements to determine the age of certain rock layers and minerals. (See Fossils.)

relative time—Relative time is based on where a rock is located within other rock layers. Thus, the relative time is usually a comparison of one rock layer's age to those layers around it; to compare, the absolute time is determined using radiometric dating of the rock.

time units—The geologic time scale uses about six major time units. These units are not precise but are merely ways of trying to keep track of the earth's long history.

eon—A large part or grand division of geologic time that represents the longest geologic unit on the scale. It is sometimes defined as 1 billion years on some geologic time scales.

era—This includes two or more periods.
period—This is a subdivision of the era.
epoch—This is a subdivision of the period.
age—This is a subdivision of an epoch.
subage—This is a subdivision of an age, but is rarely used.

20. Humans

Introduction

Modern humans are currently the dominant animal species on Earth. But this supremacy has only occurred in the last 30,000 or so years—a very short period of our planet's natural history. Currently no one theory of modern human origins is generally accepted; this lack of consensus is not surprising, because the evidence is similar to a large jigsaw puzzle with most of the pieces missing. New findings change some of the details and resolve some of the controversies, but the complete story may never be known. What follows is an overview that most, but not all, scientists feel is closest to the true story of the origins of modern humans. It is amazing to think that we are the creatures who have affected the natural balance of the Earth the most—but have lived here for only a short time in terms of geologic history.

Timeline

(note: mya=million years ago; ya=years ago)

Date	Event
Prehistoric Events	
~4 mya	The earliest links to *Homo sapiens sapiens* evolves, although no fossils have yet been found to confirm this age; the oldest known fossils are about 3 to 3.5 million years old.
~3.5–3 mya	Australopithecines, ancestors of early humans, may have evolved at this time; the bones of "Lucy," an *Australopithecus afarensis* and one of the oldest human fossils known, have been dated as being this old.
~2.5 mya	*Homo habilis* is thought to have evolved from an australopithecine ancestor.
~2–1 mya	Groups of *Homo erectus* begin moving out of Africa, spreading to the Middle East, Europe, and Asia.
~1.5–1 mya	*Australopithecus boisei* becomes extinct.
~120,000–30,000 ya	*Homo erectus* begins to evolve toward the anatomy of *Homo sapiens sapiens*. (Although, as with all these dates and fossils, this theory is highly debated.)
~90,000 ya	*Homo sapiens sapiens* are thought to have evolved at this time.
~30,000 ya	The only remaining species of humans on Earth are the *Homo sapiens sapiens*.

Modern Events and Discoveries

1852	Even before Charles R. Darwin (1809–1882) and Alfred Wallace (1823–1913) announce their theory of evolution, English scientist and sociologist Herbert Spencer (1820–1903) proposes the phrase "survival of the fittest"; he refers to the process by which organisms that are less well-adapted to their environment tend to perish, while the better-adapted organisms survive.
1850s	English naturalist Alfred Russel Wallace, a contemporary of English naturalist Charles Darwin, independently develops the theory of evolution by natural selection at the same time as Darwin; Wallace's theory is the result of his study of the flora and fauna of Brazil and Southeast Asia in the 1840s.
1856	Fossils of a Neanderthal (or Neandertal) are found in a cave in the Neander Valley of Germany.
1858	Alfred Wallace and Charles Darwin make public the theory of natural selection in a joint paper.
1871	Charles Darwin publishes *The Descent of Man* and includes man as an evolved animal.
1891	Dutch paleontologist Marie Eugene Dubois (1858–1940) describes the first-discovered fossilized remains of our immediate ancestor, *Homo erectus*; it is also called the Java man, named after the area in Indonesia (then the Dutch East Indies) where it was found (Kedung Brebus, Java).

Date	Event
1924	Australian-born South African paleontologist Raymond Arthur Dart (1893–1988) is brought a human skull that he hypothesizes was the missing link between humans and apes; he later announces that these fossil remains found in a Botswana quarry were those of an *Australopithecus africanus*—thought to be the earliest known hominids.
1925	The Scopes "Monkey Trial" begins in Dayton, Tennessee, as John T. Scopes (1900–1970) is brought to trial for teaching evolution, his opponents believing it was unlawful to teach in any public school any theory that denied the divine creation of man; he is convicted and sentenced, but the decision is later reversed and the law repealed in 1967.
1927	Canadian anthropologist Davidson Black (1884–1934) discovers the molar of *Sinanthropus pekinensis*, or Peking man, in a cave near Peking (now Beijing), China; it is an extinct relative of humans, and is considered an example of *Homo erectus*.
1950s–1960s	British anthropologist Louis Seymour Bazett Leakey (1903–1972) and his wife, British paleoanthropologist Mary Nichol Leakey (1913–1996) discover the fossil remains of *Homo habilis* and *Australopithecus* in East Africa; *Homo habilis* is the earliest known hominid species to make stone tools.
1974	U.S. anthropologist Donald Johanson (1943–) discovers the fossilized skeleton of the earliest known human ancestor, nicknamed "Lucy," an *Australopithecus afarensis* approximately 3 to 3.5 million years old.
1984	Paleontologist Alan Walker, with Richard Leakey, uncovers a 1.5-million-year-old, nearly intact fossil of *Homo erectus* that they call "Nariokotome boy."
1986	Paleontologists Tim White and Donald Johanson discover the first known limb bones of a female *Homo habilis*, dating back about 1.8 million years ago.
1988	Fossil of early humans are found in Israel and bear many characteristics of *Homo sapiens sapiens*; they date back 90,000 to 100,000 years ago—twice as old as the previously known specimens of modern humans.
1988	A charcoal dating at a site in Monte Verde, Chili, points to ancestors of Native Americans arriving in the Western Hemisphere at least 33,000 years ago, three times earlier than the accepted date.
1989	Richard Erskine Frere Leakey (1944–), Kenyan paleontologist and son of Louis and Mary Leakey, is appointed director of the Wildlife Service in Kenya; he once found some of the oldest known humanoid fossils in Kenya.
1992	U.S. paleontologists Andrew Hill and Steven Ward announce that a skull fragment from Kenya's Lake Baringo Basin is 2.4 million years old—half-a-million-years older than previously known specimens; it also proves that stone toolmaking hominids did emerge this long ago.

History of Humans

Early Humans

For millions of years, mammals, of which humans are but one species, were small in number as well as in stature. During the Mesozoic era from 245 to 65 million years ago, the large dinosaurs dominated the landscape. Their extinction almost 65 million years ago, along with many other forms of life, left an ecological void that mammals were able to fill. In turn, a wide range of diverse mammal species proliferated over the planet. One of the species that eventually evolved was a type of hominid— that eventually evolved into what we are today, *Homo sapiens sapiens*.

What is the story of **human evolution**? How did we spread across the earth, adapting to a wide variety of geography and climate? We can trace clues left by ancient civilizations and the bones of primitive humans, but eventually we reach a point where the records stop and only incomplete fossil pieces and bits of old stone tools remain. We have reached the time of the early hominids.

Our evolution was not straightforward or linear. There were many branches of hominids that arose and many evolutionary dead-ends. The number of fossil fragments is small, and complete fossils are extremely rare; so too are other remains, such as tools and artwork. Given the lack of fossil evidence, the limitations of dating techniques, the overlap of ranges, the large time scales involved, the ensuing geologic changes, and the worldwide distribution of some ancient populations, it is no surprise that paleoanthropology is a scientific field filled with opposing theories and controversies.

Paleoanthropologists have continued to theorize from the existing human fossil record—incomplete and contradictory as it is. As in the study of most species, scientists always search for the common ancestor, and the study of *Homo sapiens sapiens* is no exception. We are part of the kingdom Animalia, phylum Chordata, subphylum Vertebrata, class Mammalia, and order **Primates**. This means our ancestors were mammals, and the modern mammals that resemble us the most are the primates, which include modern apes and chimpanzees.

This theory of human evolution did not develop until relatively recently, and even then, not without controversy. Charles Darwin speculated in the late nineteenth century that we had descended from primitive primates; modern evidence also seems to lean in this direction. Creationists of Darwin's century, on the other hand, felt that modern humans were created many years ago by divine intervention—and that the discovered fossils represented primitive forms of primates, both extinct and living.

Fossil discoveries abounded in the twentieth century, and as Darwin's ideas on evolution became more accepted, fossils that might have once been classified as animals were instead being classified as hominids. For example, in 1924, the fossilized remains of what appeared to be a young girl were discovered in South Africa and brought to Australian-born South African anthropologist Raymond Arthur Dart. He determined that the cranial capacity of the skull was that of a gorilla, but the teeth resembled those of modern humans. Here was a likely ancient ancestor to modern man, and Dart gave it the name *Australopithecus africanus* ("southern ape of Africa"). His findings were highly controversial at the time, but today his theory is generally accepted.

Other fossil discoveries were made, and the confirmation of the age of the fossils was improved by new rock and fossil **dating techniques**. For example, carbon dating, developed in 1947, helped to date archaeological objects as far back as 45,000 years ago. Other dating techniques of rock allowed fossils found within certain layers to be accurately dated—including hominid fossils that were millions of years old.

Modern Studies of Humans

In the latter part of the twentieth century, the study of human evolution has centered not only on the lineage of humans, but also on their origin. Over the years, many places in the world have been thought to be the site where humans originated. At present, most anthropologists feel that the evidence points to Africa as the cradle of human evolution. The fossil evidence also appears to indicate that about 1 to 2 million years ago groups of *Homo erectus* began moving out of Africa, spreading to the Middle East, Europe, and Asia. But, by about 30,000 years ago, the only species of humans on Earth were the *Homo sapiens sapiens*. What happened in between?

Currently, paleoanthropologists are divided into two main groups of thought—multiregionalism and "Out of Africa." The multiregionalism theory holds that the effect of geography during these migrations millions of years ago produced isolated pockets of *Homo erectus*, with only an occasional interbreeding between neighboring groups. Some anthropologists go even further, theorizing that the various groups, from *Homo erectus* to *Homo sapiens*, were not separate species—but were all the same species derived from *Homo erectus*. Occasional interbreeding between neighboring groups kept the genetic traits of the species intact, while the geographical isolation of the groups led to the formation of distinctive regional characteristics that are still present today.

The best evidence for this theory comes from Asia, where fossils from China, Indonesia, New Guinea, and Australia show, when placed in chronological order, a smooth transition from *Homo erectus* to modern humans. Characteristics found in 700,000-year-old fossils can still be seen today in humans living in this region. Opponents of this theory hold that the evidence of regional characteristics suggests very limited contact and interbreeding between widely separated groups. But how then did the human species remain intact, if not by interbreeding across continents? The isolation that led to the distinctive differences should have led to separate species according to these critics.

The "Out of Africa" theory contends that modern humans evolved from *Homo erectus* as a distinct species approximately 100,000 to 200,000 years ago in Africa, then migrated out into the far-flung reaches of the planet. *Homo sapiens* did not interbreed with other human species they encountered during their migration, rather, they replaced them. By about 30,000 years ago, the only human species left was ours.

The best evidence supporting this theory comes from more recent technology—especially the work of molecular geneticists who have studied DNA found in the parts of human cells known as mitochondria. This mitochondrial DNA is extremely useful for evolutionary studies for two reasons: First, it is passed intact from the mother to the children, with no mixing of genes from the father. Second, because this form of DNA quickly picks up mutations and keeps them, it can be used as a kind of "clock" that shows when populations of *Homo sapiens* split.

Studies of mitochondrial DNA from African, Asian, Australian, Caucasian, and New Guinean ethnic groups established that 133 variants are present, and these data could be arranged into a chronological evolutionary tree. The results of this mapping showed that the "trunk" and longest branch of the human tree were composed of Africans; the other branch was composed of a subgroup of humans that had left Africa and migrated throughout the world. Also, all of the DNA from the far-flung groups was very similar, indicating that our species is young. But the African DNA had the most mutations in the study, showing that the African lineage was the oldest and, therefore, that all modern humans can trace their roots back to Africa.

Using the known rate of mutations in primates, geneticists were able to extrapolate backward in time to determine how long it took for the 133 variants in *Homo sapiens* to occur by mutation; they found that the time period was approximately 140,000 to 290,000 years ago. This estimate is consistent with the time frame for the origin of our species in the "Out of Africa" theory. And, from this data, they determined that the splitting off of the evolutionary tree, caused by the migrations, occurred approximately 90,000 to 180,000 years ago. Subsequent genetic studies have confirmed and sharpened the time estimates for the rise of our species in Africa and its migration out into the rest of the world. The earliest, female, common ancestor of our species was symbolically named "Eve," and new studies show that she was alive approximately 143,000 years ago. From one group of DNA, "Adam," a common male ancestor, was found to have been alive approximately 270,000 years ago.

Opponents of the "Out of Africa" theory argue that the original study that determined the evolutionary tree was flawed by the misuse of a computer program; in addition, they say that the fatal flaw in these studies is the assumption that different populations of humans would not interbreed. According to these scientists, there was no mass migration out of Africa, just a long period of moving about by human species and interbreeding, which eventually led to the evolution of *Homo sapiens*.

The true story may never be known, even if more fossils are found and dating techniques improved. Because scientists and others are seeking knowledge of our own species, it is hard for them to be objective and aloof. Our hopes, fears, and biases all color our interpretations of the data, leading to widely diverse theories about the evolution, origin, and future of *Homo sapiens sapiens*. See also ANIMALS, CLASSIFICATION OF ORGANISMS, DINOSAURS, EARTH, EVOLUTION, EXTINCTION, GENETICS, GEOLOGIC TIME SCALE, LIFE, and REPTILES.

Topic Terms

dating techniques—This term refers to the methods used by scientists to determine the age of fossils. Accurate dating is essential when trying to put together the jigsaw puzzle of human evolution; most dating techniques in use today are indirect. They do not measure the age of the fossil remains themselves, but the material, such as soil or rock, in which the remains were discovered.

Two techniques, electron spin resonance and thermoluminescence, measure the number of excited electrons present in minerals. Radioactive elements in the earth constantly bombard the atoms in minerals, exciting some electrons out of their normal orbits. A fraction of these excited electrons do not decay back to their normal state, but remain permanently trapped in the mineral structure, or crystal. Over time, the number of excited electrons increases; if the rate of increase is known, the age can be determined by measuring the total number of excited electrons in a material.

In electron spin resonance, the electrons are counted by measuring the amount of energy trapped in a crystal; the thermoluminescence technique heats the object to free the trapped electrons. When these excited electrons decay to their normal state, they release a specific amount of light, which can be measured to determine the number of excited electrons.

A third technique, called uranium-series dating, works for remains found in limestone deposits, which are typically present in caves. Uranium atoms decay at a known rate into lighter elements, including the isotope thorium-230. When the limestone deposits were formed, they contained uranium, but almost no thorium. Over time, the uranium atoms decay into other elements, including the thorium. By measuring the amount of thorium-230 present, the age of the limestone deposits, and thus the fossil remains associated with them, can be dated.

human evolution—Many primitive primate forms were living in Africa more than 25 million years ago; and therefore, the evolutionary tree has many branches. (This is based on the "Out of Africa" theory.) As time went on, several hominid precursors such as *Ranapithecus*, *Aegyptopithecus*, and *Proconsul* arose and evolved. Presented here are the dates and theories on the subsequent human predecessors—and remember, this is only one of many theories on human evolution:

Australopithecus—About four to six million years ago, from a currently unknown common ancestor, the evolutionary split between the apes and humans occurred. It was in this period that the australopithecine forms, ape-like primates that are the earliest known modern human predecessors, began their rise in South and Central Africa.

Discoveries of australopithecines over the decades in Africa produced similar fossils with different traits, which were divided into different species by scientists. A larger species was named *Australopithecus robustus*; another species, found in Tanzania by Mary Leakey, was named *Zinjanthropus boisei*, now called *Australopithecus boisei*. And a more ancient fossil, nicknamed Lucy, was found in 1974 by Donald Johanson in Ethiopia; subsequently, similar remains and traces of footprints were found, and this species, *Australopithecus afarensis*, was proven to have been bipedal (walking on two feet) and erect, though only approximately 3 feet tall.

Based on these fossils and dating techniques, some paleoanthropologists feel that Lucy and her species, *Australopithecus afarensis*, were the common ancestors to the following australopithecines and eventually the humans. *Australopithecus boisei* went extinct one to one-and-a-half million years ago, and it is not clear whether *Australopithecus africanus* also went extinct or was another human branch. However,

---

Famous Human Finds

The number of hominid (or human-like) fossils found over the years has not been immense. Like other fossils of organisms, the hominids would have to be at the right place at the right time—and not chewed on by predators or damaged by a sweeping river carrying them away. Here are a few of the more famous human fossil finds within the last two centuries:

Heidelberg Man—In 1907, Otto Schoetensack discovered the mandible of a primitive hominid in a sandpit in Mauer, Germany. This specimen has been referred to as "Heidelberg Man" and is a less archaic form of *Homo erectus* than Java Man.

Java Man—The popular name given to the hominid fossil remains discovered by Eugene Dubois in the late nineteenth century on the island of Java, Indonesia. "Java man" was named *Pithecanthropus erectus*, or "erect ape-man," because Dubois felt it was the missing link between apes and humans. It was subsequently determined to be the remains of a *Homo erectus*.

Lucy—Found by Don Johanson in 1974, this specimen of *Australopithecus afarensis* was uncovered in the region of Hadar, Ethiopia—known as the Afar Depression, hence the *afarensis* designation. The skeleton is popularly known as "Lucy" and is thought to have been a short (approximately 3.5-feet [1-meter] high) female about 20 years old at the time of her death. The fossil remains have been dated at 3.5–3 million years old, making it one of the oldest, complete links to our distant past.

Nariokotome boy—In 1984, paleontologist Alan Walker, with Richard Leakey, uncovered a 1.5-million-year-old, nearly intact fossil of *Homo erectus*, on the west side of Lake Turkana in Kenya. They called the fossil "Nariokotome boy."

Peking man—The popular name given to the fossil remains from what is thought to be *Homo erectus* that have been missing since the Japanese invasion of China in 1941. It was originally discovered in 1927 by Canadian anthropologist Davidson Black. In that year, he first discovered the molar of *Sinanthropus pekinensis*, or Peking man, in a cave near Peking (now Beijing), China. It is now thought to be an extinct relative of humans and is considered an example of *Homo erectus*.

Taung Baby—In 1925, Raymond Arthur Dart presented the "Taung Baby" fossil (originally brought to him in 1924), now classified as an *Australopithecus africanus*. The find was especially significant because it included a rough cast of the individual's brain.

what is clear from the evidence is that the australopithecines, being erect and bipedal, were early hominid ancestors to humans.

Homo habilis—The next step in the evolutionary process is thought by some to be *Homo habilis*, or "handy human." Paleoanthropologists Louis and Mary Leakey, working in Tanzania, East Africa, during the early 1960s, found fossil remains of a hominid that were subsequently dated at 1.8 to 2 million years old. Further finds have been made in southern Africa as well and have been dated to the same period. These remains, though similar in many respects to the australopithecines, had some crucial differences. The skeletal bones resembled those of modern humans, save for the long, ape-like arms, and smaller rear teeth. And, most importantly, the cranial cavity had more volume, indicating a larger brain than the australopithecines. For comparison, the cranial capacity of *Australopithecus africanus* ranged from approximately 22.9 to 29.6 cubic inches (375 to 485 cubic centimeters); the *Homo habilis*' cranial capacity was just under twice that amount.

Associated with the fossil remains were stone tools that were thought to have been made and used by these early hominids, hence the name "handy human." Studies of the inside of *Homo habilis* skulls seem to indicate that at least one region of the brain associated with speech was developed in this species. Based on these findings, *Homo habilis* is considered to be the earliest member of the genus *Homo*, and the transitional form between the australopithecines and the next species to evolve, *Homo erectus*.

Homo erectus—In the late nineteenth century, Eugene Dubois discovered some hominid fossil remains on the island of Java, in Indonesia. This find, termed "Java man," was originally named *Pithecanthropus erectus*, or "erect ape-man" because Dubois felt it was the missing link between apes and humans. Subsequently, this species was renamed *Homo erectus* because of its human-like physical characteristics and evolutionary closeness to *Homo sapiens*. As more and more fossil remains were found of this species, it became clear just how close it truly was to modern humans. In 1984, Alan Walker and Richard Leakey found an almost complete fossil on the west side of Lake Turkana in Kenya; they termed it "Nariokotome boy" and dated it at 1.5 million years old. These and other fossils were extensively analyzed by a variety of scientists to discover the characteristics of our most recent ancestors.

The time span when *Homo erectus* was present on Earth ranged from 1.6 million years ago, when the species first appeared, to approximately 250,000 years ago, when it gradually disappeared from the fossil record. The cranial capacity increased over this time; early remains have approximately one-half the capacity of modern humans, while later remains have approximately two-thirds. Its physical size increased also, close to that of modern humans, and it walked erect. The stone tools found with the remains were more complex, specialized, and required more manual skill to create and use them than was the case with the tools of *Homo habilis* or the australopithecines. In addition, *Homo erectus* discovered the use of fire, and some evidence points toward their use of cooperative group hunting, as opposed to earlier, theorized scavenging by earlier ancestors.

All of these factors made the migration of this species possible. They could cope with varied climates and geographic localities using their tools, hunting skills, and fire. Unlike its ancestors, *Homo erectus* has been found in parts of the world other than Africa. In addition to Olduvai gorge in Tanzania, Africa, fossil remains have been found in Java, Indonesia; Peking, China; Tautavel, France; Vertesszollos, Hungary; Heidelberg, Germany; and Petralona, Greece. Some scientists have noted different characteristics between fossil remains found in Asia and Africa, prompting the naming of the African species as *Homo ergaster*, a sister species to the predominately Asian *Homo erectus*. There is also some evidence that *Homo ergaster* is more closely related physically to modern humans than *Homo erectus*. These details aside, the evolution, migration, and skills of *Homo erectus/ ergaster* laid the groundwork for the emergence of modern humans, *Homo sapiens*.

As we have seen, the fossil record shows a general increase, over millions of years, in cranial capacity, bipedal ability, physical size, and the ability to make and use tools and fire by the australopithecines, *Homo habilis,* and *Homo erectus*. This trend continued with the emergence of the next species, *Homo sapiens*, and has brought us to the present time. Again, the path to the present was neither linear nor direct. Side-branches, other species of *Homo sapiens,* became extinct for one reason or another. We could look upon ourselves, *Homo sapiens sapiens*, not as the final apex of thousands of years of evolution, but as the species of *Homo sapiens* that was best able to adapt to the world in which we found ourselves.

Archaic Homo sapiens—Approximately 750,000 to 500,000 years ago, hominids appeared that were so advanced compared to *Homo erectus/ergaster* that they were designated *Homo sapiens*, a new species. This group had a number of subspecies; the fossil record, however, is incomplete and fragmented, and scientists debate which of these subspecies was our direct ancestor, and whether or not the others went extinct, interbred with, or were replaced by modern humans. What is clear is that our direct ancestors were not the only advanced hominid subspecies on Earth at the time. This group, with anatomical characteristics different from modern humans, is referred to as *Archaic Homo sapiens* or *Homo heidelbergensis*.

Homo sapiens neanderthalensis—One subspecies thought to be part of the archaic *Homo sapiens* group was the Neanderthals (or Neandertals), or *Homo sapiens neanderthalensis*. A fossil skeleton of a hominid was discovered in 1857 in the Neander Valley, near Duesseldorf, Germany; it was subsequently named Neanderthal. Other fossils and remains have been found since, and this evidence gives rise to some startling conclusions. The Neanderthals were very advanced, walked erect, used stone tools, and had warm clothing. They buried their dead with artifacts, created primitive art, lived in tribes, and used fire for warmth and light. To survive in the harsh, cold climates created by the Ice Age, they had to be resourceful and adaptive. In short, they displayed almost all of the traits exhibited by modern humans. Neanderthals lived in the areas of North Africa, the Middle East, and eastern and central Europe. Their latest remains have been found at Zaraffaya, Spain, and are dated approximately 26,000 years ago.

Because of these advanced characteristics, some anthropologists and other scientists have long regarded Neanderthals as a previous ancestor of modern humans; that they were a subspecies is reflected by their scientific name, *Homo sapiens neanderthalensis*. Opponents of this theory, however, point to the anatomical differences, such as the low skull, small forehead, and heavy brow ridges. Neanderthals were also bigger than modern humans and had larger brains. This debate has continued for a number of years and has given rise to different theories about the disappearance of the Neanderthals from the fossil record.

One theory is that the Neanderthals were wiped out by modern humans, *Homo sapiens sapiens*; another says that Neanderthals gradually evolved into modern humans, and still another says that the Neanderthals were an evolutionary dead-end that went extinct. Until recently, these were just hotly debated theories. Recent tests by geneticists of the mitochondrial DNA extracted from the first discovered Neanderthal skeleton found an average of 27 differences in the DNA sequences when compared to that of modern humans. The Neanderthals, therefore, were not our direct ancestors, but an evolutionary dead-end. They were a different species that appear to have split from our common ancestor approximately 500,000 years ago and were present when modern humans appeared on the scene. Whether they went extinct, interbred with, or were wiped out by *Homo sapiens sapiens* is still a matter of conjecture, but they were not our direct ancestor.

Anatomically Modern Humans—The first appearance of a subspecies of *Homo sapiens* that was anatomically like modern humans occurred approximately 130,000 to 100,000 years ago, meaning we have been around, in our current form, for an extremely short period in terms of geologic time. These modern humans had brain sizes of 79.3 to 82.4 cubic inches (1,300 to 1,350 cubic centimeters). Fossil remains have been found and dated from this time in numerous Sub-Saharan African sites, such as Omo-Kibish, Laetoli, Dar-es-Soltan, Florisbad, Djebel Irhoud, Border cave, and Klasies River Mouth. Though there is still debate over the origins of this subspecies, known as *Homo sapiens sapiens*, recent genetic studies have indicated that all modern humans are descended from a single common ancestor known as "Eve" and that she lived in Africa around this time frame.

After developing in Africa, these modern humans migrated throughout the old world. Fossils have been found in Israel dated at approximately 96,000 years ago; modern humans moved into Asia, Australia, and other areas of the Far East approximately 75,000 years ago. This species first appeared in Europe about 40,000 years ago. The term "Cro-Magnon" is sometimes applied to the European branch of these modern humans, from a discovery of fossils and artifacts made in 1868 at a site near Les Eyzies, France. But Cro-Magnon is just a regional term for *Homo sapiens sapiens* that lived in Europe during this time and does not indicate a different species.

As time when on, the modern humans of the world increased their knowledge, sophistication, language, and technical and social abilities. They developed cave art, agriculture, written language, transportation, conflict, laws, religions and religious structures. They domesticated more animals, built cities, and discovered the secrets of the world and universe around them; they explored the Earth, the oceans, and outer space.

primates—The order of mammals that contain modern humans. Humans are part of the class Mammalia, having, among other characteristics, mammary glands, hair, and embryonic development. The order Primates has two suborders; the shorter snouted animals are in the Anthropoidea, which includes the families Callitrichidae (marmosets), Cebidae (New World monkeys), Cercopithecidae (Old World monkeys), Hylobatidae (gibbons), Pongidae (gorillas, chimpanzees, orangutans), and Hominidae (human beings).

21. Life

Introduction

The origins of life on Earth are an integral part of our natural history—not only from the standpoint of understanding its beginnings and its overall effect on the planet, but also in terms of life elsewhere in the universe. The word "life" raises such questions as how did organisms first begin? How did cells begin to replicate? Does life come in different forms? And perhaps the most interesting and philosophical question of all: Are we the only life in the universe? This important subject in natural history is also one of its greatest mysteries.

Timeline

(note: bya=billion years ago; mya=million years ago)

Date	Event

Prehistoric Events

Date	Event
~4 bya	Something sparks the first inorganic chemicals to develop into primitive "cells."
~3.75 bya	The earliest form of life is thought to have evolved around this time; recent unconfirmed findings show it may be closer to 3.85 billion years ago.
~3.5 bya	Biomolecules begin to gather and somehow begin to self-replicate.
~3.5–3.2 bya	Primitive bacteria may have lived in ancient hot springs around volcanoes.
~2.1 bya	The Earth's atmosphere becomes sufficiently oxygen-rich to support an ozone layer; this layer would have a profound effect on the evolution of life on the planet, as it would protect organisms from harmful ultraviolet radiation from the Sun.
~2 bya	The first cells begin to form in greater numbers and develop into simple unicellular microorganisms.
~1.5 bya	The first eukaryotic cells form; they have a nucleus and complex internal structures; they are the precursors to protozoa, algae, and all multicellular life.
~544 mya	The Cambrian Explosion occurs, in which life on Earth experiences explosive growth; the ancestors of most modern animal groups "suddenly" appear in the oceans.
~520 mya	Representatives of most of the main groups (phyla) of animals have appeared on Earth.
~500 mya	The first life moves on land, in the form of mats of algae, lichens, and bacteria; they seem to only colonize the edges of shallow pools.
~440 mya	The first animals, probably tiny arthropods, crawl on land, and a second burst of growth and diversification takes place on Earth.

Modern Events and Discoveries

Date	Event
~600 B.C.	Greek philosopher and astronomer Thales of Miletos (c.625–c.547 B.C.) suggests that all life came from water.
1739	Swiss scientist Abraham Trembley (1710–1784) discovers the hydra, a freshwater polyp; it is interpreted by some to be the missing link between the animal and plant kingdoms.
1754	French scientist Denis Diderot (1713–1784) erroneously suggests, similar to an earlier theory by Empedocles of Acragas, that animals' organs at one time existed independently and eventually joined together, forming the creatures we see today and those we see from the past.
1877	Italian astronomer Giovanni Schiaparelli (1835–1910) discovers what he believes is a network of linear markings on the planet Mars; the results generate speculation on the possibilities of Martian life—especially intelligent inhabitants.

Date	Event
1936	Russian biochemist Alexander Ivanovich Oparin (1894–1980) suggests that life began from a random chemical process in the ocean—the biochemical soup theory that is still popular today.
1954	U.S. chemists Stanley Miller (1930–) and Harold Urey (1893–1981) discover a way to produce the conditions that existed in the Earth's early atmosphere.
1972	U.S. geologist Preston Cloud observes that no red beds older than 2 billion years have been found, rocks that show that oxygen did not exist in abundance in the Earth's atmosphere until that time; oxygen is thought to be evidence of plant life.
1977	The research submarine *Alvin* discovers a new type of ecosystem based on chemosynthesis (converting chemicals to food) at volcanic vents in the Pacific Ocean's Galapagos Rift.
1996	Scientists study a meteorite collected from Antarctica thought to have come from the planet Mars; they discover what they believe to be signs of life—although the findings continue to be debated.
1996	Paleobiologists Maarten de Whit and Frances Westall discover what is thought to be the oldest evidence of fossil life on Earth—the fossilized remains of primitive bacteria in 3.5–3.2-billion-year-old rock from southern Africa.
1997	Paleontologists uncover evidence that sponges are the most primitive type of multicellular animal on Earth.

History of Life

Early Life

No one knows the precise time that **life** began on Earth. The main reason is because the single-celled organisms that made up early life all had soft parts—and because these are the first to decay and disappear after death, it is almost impossible to find the remains.

In addition, the organisms' small size (most were the size of bacteria) make it difficult to detect them in ancient rocks—especially rock that has been exposed to the natural heat and pressure of geologic activity. To compare, some modern viruses are only about 18 nanometers (18-billionths of a meter) across; modern bacteria typically measure 1,000 nanometers across. Still, it is estimated that the first life began about 4 billion years ago (we only have indirect evidence from 3.8-billion-year rock)—although the organisms did not survive on oxygen, but carbon dioxide.

Because we have found so little fossil evidence, it is difficult to know the true shapes of the earliest life. Scientists believe that early life was composed of primitive single cells and started in the oceans. The reason is simple: Life needed some form of filter to protect it from the incoming ultraviolet energy from the Sun—and the ocean waters gave life that protection; the atmosphere at this time wasn't able to block out the Sun's harmful radiation.

Stromatolites were some of the first simple unicellular microorganisms on the Earth—at least that we know about—and have changed little since their beginnings almost 2 billion years ago. Various stromatolites eventually developed the ability to photosynthesize, producing oxygen as a waste product, and over time, creating higher levels of oxygen (instead of the more prevalent carbon dioxide) in the atmosphere.

Although **photosynthesis** seems to have appeared about 2.5 billion years ago, it took a long time for its oxygen-producing effects to become established on a global scale. This is because the oxygen produced did not get into the atmosphere; instead, it was absorbed by rocks, minerals, or organics to form substances such as iron oxides and carbonates. By about 2.1 billion years ago, the Earth's atmosphere had become sufficiently oxygen-rich to support an ozone layer. This layer would eventually have a profound effect on the evolution of life—the blanket of ozone creating a shield to protect organisms from the ultraviolet radiation from the Sun.

The next "stage" of life evolved as more highly developed cells (more complex than the simple cells of bacteria) called **eukaryotic cells** appeared about the same time as the oxygen in the atmosphere—just a little less than 2 billion years ago. And from these cells, single-celled protozoa and algae evolved, and eventually, all multicellular life.

Perhaps the most amazing progression of life (besides the formation of life itself) occurred about 544 million years ago, during what is called the Cambrian Explosion. At the end of the Precambrian era and beginning of the Paleozoic era (or end of the Vendian time and beginning of the Cambrian period), multicellular life experienced a rapid growth in diversity in the oceans. In particular, invertebrate animals with hard parts began to appear at this time; it is unknown if a spike of growth in soft-bodied invertebrates occurred at this time (such as jellyfish, segmented worms, or sea pens), as only very rare traces have been left by these soft-bodied organisms.

The progression of life seemed to proceed rapidly after the Cambrian Explosion. Approximately 520 million years ago, representatives of most of the main groups (phyla) of animals had appeared in the Earth's oceans. By about 500 million years ago, the first life moved on land, in the form of mats of algae, lichens, and bacteria. But they did not go very far, preferring to colonize only the edges of shallow pools and shorelines (much like modern stromatolites). About 440 million years ago, the first animals crawled onto land, and a second burst of animal growth—and diversification—took place. These tiny animals, probably tiny, millipede-like **arthropods**, developed shells to protect themselves from drying out. They also came ashore just after the first true land plants began to colonize the land. And during the late Cambrian period, the first animals with backbones—the fish—evolved in the oceans; eventually, one line would evolve a bony skeleton, air-breathing lungs, and "limbs" strong enough to support them on land—the first four-legged vertebrates, evolving around 360 million years ago (although this number is only an approximation).

The Search for Life's Origins

Theories of how life could have grown on early Earth abound—and in actuality, many of these conditions could have existed simultaneously and produced the planet's early life. One theory states that life grew from a "primordial soup" (or "prebiotic soup")—a thick stew of biomolecules and water. Chemical reactions were then triggered—either by the Sun's ultraviolet rays, lightning, or even the shock waves from a violent meteor strike. These reactions produced various carbon compounds—including **amino acids**, which make up the proteins found in all living organisms.

This theory was first postulated after a famous experiment performed at the University of Chicago in 1954, by then graduate student Stanley Miller and his advisor, chemist Harold Urey. They showed that chemicals thought to exist in the early Earth's atmosphere about 4 billion years ago could form amino acids when combined with water and zapped by lightning. In the experiment, a flask of boiling water provided heat and water vapor; this water was then mixed with hydrogen, methane, and ammonia—thought to be the chemical constituents of the early atmosphere. An electric discharge, simulating lightning, was sent through the mixture; the resultant material was then cooled to condense it—and the chemicals found in this liquid represented the building blocks of life: amino acids, nucleotides, sugars, and fatty acids. These simple chemicals could eventually form larger molecules (polymers)—which could have, after thousands or millions of years, become self-replicating molecules that would eventually lead to life.

One more recent theory developed in the 1970s came from the discovery of organisms living around deep sea vents, and primitive bacteria found in some of the oldest rocks on the planet. Hydrothermal vents are cracks caused by volcanic magma seeping through the deep ocean floor. In 1977, the research submersible *Alvin* discovered a new type of ecosystem based on chemosynthesis (converting chemicals to food) at volcanic vents in the Pacific Ocean's Galapagos Rift. This discovery revealed to scientists that these vents could hold an abundance of life—without the need for the usual light or photosynthesis processes. There is evidence of primitive bacteria in rock around 3.2 to 3.5 billion years ago—could they have evolved around such vents—away from the harsh conditions on the rest of the forming planet?

The early Earth probably had many more such hydrothermal vents—and perhaps primitive bacteria around them—as the crust was newer, and thus thinner, than today's cooled off, thicker crust. Thus, it is thought the organisms around these vents did not need to rely on photosynthesis or any other light process for energy. Today's volcanic vent organisms live off the bacteria around the vents, which in turn extract energy from the hot, hydrogen sulfide–rich water found around the cracks in the ocean floor. Early organisms could have survived in much the same way.

Other more recent theories have been proposed to explain how life progressed so slowly on early Earth. One theory has to do with idea that objects from space often strike our planet over time. Because scientists believe that the early solar system had a period of major bombardment of planet-sized or smaller objects (called the Late Heavy Bombardment), they also suggest that life may have started over and over on the Earth. They speculate that once life began—either around ocean vents and/or in the shallow seas—comets and asteroids would strike the planet, killing off all the beginning stages of life. This may have happened many times over millions of years—until life became stable enough to sustain and diversify itself.

Our record of early life is somewhat constrained by the age of rocks on the Earth. The oldest individual minerals, caught up in younger rock found in Acasta in northern Canada, crystallized during volcanic activity more than 4 billion years ago. The Earth's oldest preserved rocks are approximately 3.8 billion years old. The oldest rock so far that appears to have life—in this case, indirect evidence in the form of a peculiar chemical signature of living organisms—is an approximately 3.8 billion-year-old sequence of banded silica and iron-rich rocks from western Greenland. Scientists believe the oldest rocks that actually hold fossils were found in 1996 in the Barberton Greenstone Belt of southern Africa. These fossils, in the form of primitive bacteria—rice-grain-shaped creatures a few thousandths-of-an-inch across—are thought to be between 3.2 and 3.5 billion years old. *See also* AMPHIBIANS, ANIMALS, ARTHROPODS, ATMOSPHERE, BACTERIA AND VIRUSES, BIRDS, CYTOLOGY, DINOSAURS, EVOLUTION, EXTINCTION, FISH, GENETICS, GEOLOGIC TIME SCALE, HUMANS, PLANTS, PROTISTA, SOLAR SYSTEM, and VOLCANOES.

Topic Terms

amino acids—See GENETICS, Topic Terms

arthropods—See ARTHROPODS, Topic Terms

eukaryotic cells—See ANIMALS, Topic Terms

life—Scientists recognize something called the "life zone," a region around a star in which life can develop if the conditions, especially temperatures, are right. For the Earth, we had the right temperature that allowed water to circulate in all three states—solid, liquid, and water vapor—conditions that eventually helped to encourage life.

The Earth is the only known planet with an atmosphere that sustains life as we know it. The Earth's atmosphere maintains life for several reasons: (1) the atmosphere provides a combination of elements that all organisms need; (2) the atmosphere acts as a trap, collecting the Sun's heat and causing the weather patterns close to the surface; and (3) the atmosphere contains a layer of ozone, which protects organisms from harmful ultraviolet rays.

The overall definition of "life" is not agreed upon by biologists—or any other scientists whose fields even remotely touch on the study of biological life. But life has some generally agreed upon attributes: (1) life is highly organized, and body functions are directly linked together in most cases; (2) life can respond to surrounding stimuli; (3) life takes in energy from the surrounding environment and use it for its own energy—including those forms of life that seem to be on the border of being called life, such as viruses; (4) life is composed mainly of carbon-based molecules, although some forms can have other elements as part of their structure, such as some animals with silicon shells; (5) life is usually suited to its environment, unless there is a drastic change—and in most cases, it will adapt to those changes; and (6) life is able to reproduce itself, either asexually or sexually, and the offspring usually resemble the parents.

photosynthesis—See PLANTS, Topic Terms

stromatolites—Stromatolites ("stone mattresses") are microorganisms that form layered structures found in sedimentary limestones on every continent. They are easy to recognize, resembling giant layered mushrooms, flat corrugations, or simple branching pillars in shallow sea water; they can be hundreds of miles across and contain deposits over a half-mile (1-kilometer) thick. They existed as far back as 2 billion years ago; modern stromatolites are found in western Australia, Florida, and other warm climate regions. These often huge microbial mats secrete a tacky gel. The gel was primarily used to protect the organisms from the ultraviolet rays of the Sun and environmental contaminants; but in the meantime, the gel also stuck to the natural sediment floating in the shallow oceans. When the sediment grew too thick, the community reached skywards, and a new mat started to grow; then another, and so on.

Life and Space?

In the 1830s, a reporter on the New York newspaper, *The Sun*, led people to believe that the Moon was occupied by great civilizations. The reporter, Richard Alton Lock, was writing about what he thought were the discoveries of John Herschel, who was doing a star survey of the South African sky. Communications were slow and uncertain, but Lock wrote that the Moon was occupied by yellow quasi-humans and animals, and had crystal mountains.

Toward the latter part of the nineteenth century, when observations showed that the Moon was dead and airless, people turned to Mars. With its known atmosphere and polar caps, it seemed to be a perfect place for life similar to our own. The idea gained in popularity in 1877 when Italian astronomer Giovanni Schiaparelli drew pictures of the long, linear features on the Martian surface, which he called "canali," Italian for channels. By the turn of the century, astronomer Percival Lowell, whose specialty was Mars, incorrectly translated "canali" to mean artificial canals, and the idea that Mars had a planet-wide irrigation system developed by a great civilization was born. Today, we know the "canali" are actually the play of shadow and light on the many seasonal features of the red planet's surface.

Did life's precursors come from space? Comets and asteroids are known to have struck Earth in the past. These space objects contain certain organic elements that are thought to be the building blocks of life. Because of this, some scientists believe that life may have started on Earth as a result of comets and asteroids bringing organics to the planet. The idea that life on Earth came from space, or at least that the impact of comets and asteroids helped precipitate life—is not new. Sales-Guyon de Montlivault suggested such an idea in 1821, mentioning that the seeds from the Moon were the source of life on Earth.

Comets have been around since the beginning of the solar system. And because they carry great amounts of frozen water, comets have long been thought of as the source of water in Earth's oceans—periodically pelting Earth over the past 4.55 billion years. Another recent finding that reinforced this idea was the discovery of house-sized or smaller space objects, all containing a great deal of water, that strike the earth's surface every day. Still, many scientists don't believe that these smaller objects were responsible for the origin of ocean waters, sighting the release of water on Earth by volcanic eruptions and cracks in early Earth's surface.

In addition, life probably didn't succeed the first, second, or many times over. Many scientists believe that the massive bombardment of the early Earth by asteroids (and perhaps comets) left over from the formation of the solar system may have wiped out early life over and over again.

Can life exist elsewhere in our solar system? This is one of the major questions that have plagued astronomers for years. Places such as Mercury, with no atmosphere, or Venus, with a thick, superheated atmosphere probably have no life. But what about possibilities elsewhere? Here are a few speculations:

Mars—The red planet was once apparently covered with water, as numerous liquid-cut channels cover its surface. No one knows where the water is today, but several scientists speculate that it is held frozen underground. If water was prevalent on the early Martian surface, life may have developed, albeit in a primitive form. Another clue was discovered in 1996, the possible evidence of fossil life in a rock from Mars, found on Antarctica. The rock, ALH84001 (actually discovered in 1984, but not studied in such depth until 1996), is thought to have been blasted from Mars, then made its way to Earth, falling as a meteorite. The rock reportedly showed fossil evidence of minuscule bacteria, biogenerated materials, and organic compounds suggestive of life. If the tiny rounded features found in the rock are evidence of life on Mars, it will be the first time any other life has been discovered in the universe—and we will know we are not alone. But the findings are still highly debated. Some scientists believe the carbonates are artifacts of the techniques used to analyze the rock; others believe that it was so hot during the formation of the rock that organisms would not have survived.

Europa—Europa is one of the larger Galilean moons of Jupiter. Recent images sent back from the Galileo spacecraft have shown that the satellite's hard icy crust may harbor a vast ocean underneath. If so, many scientists point to the fact that on our own planet, hot volcanic vents have helped encourage life to grow. Thus, life may exist below the ice of Europa.

Titan—Of all the satellites of the solar system, scientists believe Titan, the largest moon around Saturn, may be the best candidate for harboring life. Not only does it have a thick atmosphere, but it also has some of the major necessities needed for life—especially hydrocarbons.

Outer Planets—Scientists have long thought that the outer gaseous planets of Jupiter, Saturn, Uranus, and Neptune do not hold life, as the atmospheric pressures are so great. But some scientists have speculated that these planets may have life that is very different than our own—perhaps flattened creatures that can tolerate the lesser pressures of the upper atmospheres. It is doubtful, though, as the compositions of the atmospheres are mostly toxic to life as we know it.

22. Moon

Introduction

The Moon plays an integral part not only in the natural history of our solar system, but also in the natural history of Earth itself. In particular, the tides that have such an effect on our coastal shorelines are controlled by the movements of the Moon. Other observed events, such as lunar and solar eclipses, occur on Earth because of the Moon—Earth's ever-present companion as it travels through space. The Moon is the only large, natural satellite orbiting Earth. And similar to all the planets and satellites of the solar system, it shines by reflected light from the Sun. Because the Moon travels around Earth—and thus also the Sun—with different portions of its illuminated side facing Earth at different times, the Moon displays phases. It is thought that the Moon may once have been a part of Earth; thus, the Moon and Earth are often referred to as a "double planet" system.

Timeline

(note: bya=billion years ago)

Date	Event

Prehistoric Events

Date	Event
~4.55 bya	The solar system forms from the solar nebula; the oldest Moon rocks have been dated from this time.
~4.3 bya	The Moon is strongly heated, causing the melting and formation of igneous (volcanic-type) rocks in its outer layer.
~4.0–3.8 bya	The Moon is intensely bombarded by space objects during what is called the Late Heavy Bombardment.
~3.8–3.2 bya	The Moon's large impact basins and other depressions are flooded by large basaltic lavas; they form the dark mare plains that cover close to 18 percent of the lunar surface.

Modern Events and Discoveries

Date	Event
~130–121 B.C.	Greek astronomer and mathematician Hipparchus of Nicea (c. 170–c. 120 B.C.) uses a total eclipse of the Sun and parallax to determine the distance and size of the Moon.
1178 A.D.	A group of English monks reports observing a major impact on the Moon.
1879	English astronomer George Howard Darwin (1845–1912) proposes that the Moon actually came from Earth; he calls his idea the fission theory.
early 1900s	Thomas Jefferson Jackson See proposes the capture hypothesis of the Moon's formation, stating that the gravitational pull of Earth captured the Moon as it came close after formation.
1966	The first soft landing of an unmanned spacecraft, the United States' Surveyor 1, takes place on the Moon.
1969	U.S. astronauts Neil Armstrong (1930–) and Edwin E. "Buzz" Aldrin Jr. (1930–) become the first humans to land on the Moon; they bring back the first samples of the satellite's surface.
1970	The first unmanned spacecraft, the Luna 16 (built by the former Soviet Union), returns to Earth with samples of the Moon; another craft, the Luna 17, is the first roving vehicle on the Moon that same year.
mid-1970s	The giant impact hypothesis is proposed, stating that a huge impacting body struck Earth and sent out a portion of the planet, creating the Moon.
1982	A meteorite is found in the Allan Hills of Antarctica and is originally thought to be the first lunar rock found on Earth; later, it is thought to be of Martian origin.
1996	Another possible small moon, about the size of a small asteroid, called Toro, is discovered moving around Earth; whether it can actually be called one of Earth's "official" moons is still controversial.
1998	The spacecraft Lunar Prospector orbiting the Moon discovers evidence of frozen water buried beneath the lunar poles.

History of the Moon

Early Moon

The formation of the **Moon** remains a mystery; it is thought to have formed about 4.55 billion years ago, similar to the planets and satellites of the inner solar system. Many scientists believe our Moon formed as a result of a major impact between a planetary-sized object and Earth. The Earth was forming at this time, too; as the body struck, it dug deep into the mantle, throwing off material that eventually began to orbit Earth. But this theory is just one of many theories on the Moon's formation. (See the next section, Moon Studies, for more theories.)

Hundreds of millions of years later, after the formation of the Moon, scientists believe the satellite became strongly heated, causing the melting and formation of igneous (volcanic-type) rocks in its outer layer. Between around 4 to 3.8 billion years ago, the Moon was intensely bombarded by space objects during what is called the Late Heavy Bombardment. (It is thought that all the planets and satellites in the inner solar system experienced this bombardment at the same time.)

By about 3.8 to 3.2 billion years ago, large impact basins and other depressions were heavily flooded by basaltic lavas. They were thought of as oceans by the first observers of the lunar surface. (Thus, they were given such names as Sea of Tranquillity and Ocean of Storms.) But later telescopic observations showed they were waterless. Today, it is thought that after the formation of the early Moon, space objects struck the surface, creating huge cracks that released molten rock. The molten rock formed the dark, flat, gray plains we see today called mare (plural maria) and covered close to 18 percent of the satellite's surface. Some of the volcanic basalts contain a highly radioactive rock called KREEP, composed of a high percent of potassium (chemical symbol K), rare Earth elements (REE), and phosphorus (chemical symbol P).

After the initial formation of the Moon, other features formed. For example, the lunar mountains are found in association with two major areas of the lunar surface: edges of craters and areas called the lunar highlands. The majority of mountains around craters formed during the Late Heavy Bombardment, when multiple impacts occurred. The highlands apparently also formed as a result of this major bombardment.

Probably the most spectacular and noticeable features on the Moon are the craters, which also show the best evidence of the continual bombardment during, and for some years after, the Late Heavy Bombardment. One of the largest craters on the Moon's surface is named Bailly, and measures about 183 miles (295 kilometers) in diameter. Craters are natural holes that vary in diameter from hundreds of miles to 1 foot, formed when a space object strikes the surface of a planet or satellite. The size and depth of a crater is dependent on the size, speed of entry, and composition of the striking object; in addition, the surface material of the body can be a factor in the characteristics of a crater. (Astronomers use crater counts to determine the age of the surface features on the planets and satellites of the solar system.)

After about 3.5 billion years, few geological processes appear to have occurred on the Moon. This makes our Moon a "fossil" of the early solar system, as it has virtually no atmosphere or water to erode the surface. The rate of meteorite impact on the satellite continues to be low even today.

One process that seems to have continued after the bombardment was the formation of the regolith, another name for the lunar soil seen in most areas of the surface as a dust-like covering. The regolith is composed mainly of anorthosites, rocks that contain mostly feldspar materials and anorthite. Most of this grayish, dusty soil developed in one of several ways: (1) from the pulverized rock from impacts; (2) from debris ejected from volcanic eruptions during the Moon's early history; or (3) from the daily intense heating of the surface. (The moon has virtually no atmosphere to protect its surface from the Sun's radiation.) Because of the weight of the Moon's upper layers of rock and soil, regolith an average of a few feet below the surface has hardened into rock.

Moon Studies

The actual origin of the Moon is still a mystery. Past theories include the capture theory in the early 1900s, which proposes that as the Moon swung by Earth, it was dragged into orbit by Earth's more powerful gravitational pull. Another theory suggests that the Moon formed in the same way most people believe Earth formed: by accreting material from the primordial solar nebula.

More recently, many scientists believe that the Moon came from Earth. Initially, this idea was called the fission theory, proposed over a century ago by George H. Darwin. His idea was that the Moon formed as the rapidly spinning Earth ejected material that eventually became the Moon.

The most recent theory states that the Moon did come from Earth, but through a different process. Some

The Case of Giordano Bruno

One of the more interesting questions in the study of the Moon has its roots in the ancient past: Have any major impacting bodies recently struck the Moon? Scientists currently believe that the formation of most of the larger craters occurred many millions of years ago; any huge impacts probably would not occur in recorded history. There just aren't as many large impacting bodies in the solar system to go around.

At least that is what scientists thought, but there is one caveat. In A.D. 1178, a group of monks in Canterbury, England, reported a strange sighting: The chronicles of Gervase of Canterbury stated that the monks were "prepared to stake their honor on an oath that they made no addition or falsification . . ." in reference to their observations on June 25, 1178. They believed they witnessed an impact on the Moon, but no one truly knows where on the Moon.

The matter lay dormant for centuries. Then, in 1976, several scientists showed a likely candidate: the crater Giordano Bruno, located on the far side of the Moon, on the face we cannot see. This crater is very bright, and has a very sharp rim; this indicates that it is relatively young and has not been worn down by other impacts. And although the crater cannot be seen from Earth, it is close to the "edge" of the far side. Thus, if the impact occurred under the best conditions, it would be possible for the ejecta (material splashed as the impacting body hit the surface) to rise to a considerable height and be seen from Earth.

Is such an impact possible? Two pieces of recent evidence seem to reinforce such a possibility. The first comes from the network of seismometers left on the Moon by the Apollo astronauts. In 1975, the instruments recorded a storm of meteorite impacts, around 10 to 15 per day, as opposed to the usual one per day. In fact, it is thought that the Moon orbits through this meteor path in late June of each year, and in deference to the early English monks, it is called the "Canterbury Swarm." It is estimated that these rocks range in size from small, weighing little more than a pound, to just less than a half-mile long, weighing tons.

Another event supporting the possibility of recent impacts on the Moon was the recent "show" on Jupiter in July 1994. As scientists watched, a string of fragments from a broken comet struck the giant planet, giving us a better idea that impacting is still alive and well—albeit not as frequent an occurrence as it was billions of years ago—in our solar system.

---

to buckle and collapse; like a splash of water from a rock dropped in a river, a huge chunk of Earth flew into space. The mostly molten chunk circled Earth, its momentum causing it to spin and congeal, the debris eventually creating the Moon. The result was the Moon we now see—a part of our planet in orbit around us.

What physical evidence do scientists have for such a scenario? Thanks to the Apollo manned missions to the Moon (starting with the Apollo 11 landing in July 1969 and lasting until the final landing in 1972 of Apollo 17), scientists have analyzed about 882 pounds (400 kilograms) of rock and soil from our satellite. Rocks found on the Moon and characteristics of the Earth's mantle seem to indicate that the Moon may really have come from deep within the Earth. If such a large planet-sized object did strike early Earth, it could have torn the material from the mantle.

This huge boulder was photographed on the *Apollo 17* mission, one of the many astronaut visits to the Moon. The boulder was probably moved here by a huge impact that ejected large chunks of the lunar surface and dropped them nearby. *(Photo courtesy of NASA)*

In recent years, scientists have been keeping better track of the movements of the Moon and its effects on us, such as **eclipses** and **tides**. And thanks to programs such as the Apollo missions, we know a great deal more about the **interior of the Moon** and rock types, and even about **moonquakes**.

Still other recent studies, using improved optics and technology, search for more moons around our solar system. In fact, in 1996, scientists believe they found another natural—albeit smaller—astronomical body close to Earth: Toro is a small "moonlet"—actually a tiny rock in space—that orbits the Earth at the L4 position along the Moon's orbit. (The L-points are points in which the gravitational pull of a planet or

scientists suggest that a Mars-sized object struck the Earth while it was still forming. The collision of the planetoid caused the not-quite-solid Earth's thin crust satellite is somewhat balanced.) Toro's origin is a guess, just like our Moon's—and so is its status as a "moon." Some scientists believe it was a part of the Moon that broke off, perhaps due to a collision with a larger space object, and was caught in the Moon's orbit. Another theory states that Toro was captured by the Moon's and Earth's gravity, caught in the gravitational well of the Moon's orbit. *See also* EARTH, EARTHQUAKES, SOLAR SYSTEM, and UNIVERSE.

Topic Terms

eclipses—The changing positions of the Moon, Sun, and Earth result in eclipses. In order for there to be an eclipse, the Moon must be at or near one of its nodes—or the point at which the Moon's orbit intersects the plane of the ecliptic, and where the Earth and Sun are lined up with the Moon.

Eclipses probably occurred for millions of years, but with the positions of the Moon, Sun, and Earth in different relationships. As the Moon continues to move farther away from Earth in the next hundreds of thousands of years, total solar eclipses will probably disappear, as the Moon's conal shadow will no longer reach Earth's surface.

There are two main types of eclipses: lunar and solar eclipses. The variations in lunar eclipses occur because the Moon travels through Earth's penumbra (lighter part of the shadow) or umbra (darker part of the shadow); the variations in solar eclipses occur because of the differences in the distance between the Moon and Earth. Eclipses of the Sun are only seen in specific regions; eclipses of the Moon can be seen in the entire nighttime hemisphere of Earth. The time of totality of a total solar eclipse is shorter than for a total lunar eclipse. A solar eclipse can only be seen during a new Moon; a lunar eclipse can only been seen during a full Moon.

lunar: Lunar eclipses occur when Earth is between the Sun and the Moon, resulting in the Moon falling within Earth's shadow. If the Moon's orbit takes it within the umbra, the resulting eclipse is called an umbral, or total lunar, eclipse. In this case, the Moon's disk is darkened, with reported highlights of a coppery-red color (caused by the refraction of the sunlight through Earth's atmosphere). If the Moon's orbit takes it within the penumbra, the resulting eclipse is a penumbral lunar eclipse. In this case, the Moon's disk will not be as dark as an umbral eclipse. Partial lunar eclipses also occur when the Moon is not exactly in line with Earth's umbra; only part of the Moon is covered by Earth's shadow during a partial lunar eclipse.

solar: Solar eclipses occur when the Moon comes between the Sun and Earth. There are several types of solar eclipses, depending on the position of the Moon from Earth. Annular solar eclipses occur when the Moon is farther away from Earth as it passes through the node, and the shadow cone does not reach Earth's surface; because the Moon is too small to create a total eclipse, the Sun's rim is visible as a bright ring around the Moon. Total or full solar eclipses occur when the Moon's shadow cone reaches the surface of Earth and the disk of the Sun is entirely obscured by the Moon; the total eclipse will be seen along a specific, narrow eclipse path. Partial solar eclipses occur if there is a partial blockage of the Sun by the Moon, usually observed on either side of the total eclipse path.

interior of the Moon—The Moon's interior differs greatly from Earth's interior, as the Moon is almost completely solid with a small amount of material near the core that may be hot enough to be molten.

core: The Moon probably has a thick metallic core, which may be partially melted with a diameter from 746 to 1,119 miles (1,200 to 1,800 kilometers). It is thought to be made of heavier elements such as iron.

mantle: The lunar mantle is thought to be mostly solid, with the main moonquakes occurring near the base. It is thought to extend from about 37 to 621 miles (60 to 1,000 kilometers) below the lunar surface. In general, the mantle is composed of dense volcanic rocks of basaltic pyroxene and olivine.

crust: The lunar crust is much thicker than Earth's, probably because the internal temperatures of the Moon are much lower. The layer just under the regolith is composed of fractured volcanic rock called basalt, which extends to about 12 miles (20 kilometers); underneath is a tier of anorthositic gabbro, another volcanic rock, which extends to the mantle at about 37 miles (60 kilometers) below the lunar surface.

regolith: The regolith of the Moon is very shallow, averaging just over 1 mile (2 kilometers) in thickness. It is composed of shattered bedrock, broken by the action of space objects striking the surface and the action of the Sun's radiation.

Moon—The Moon displays phases in response to its position as it orbits Earth. Because the Moon is lit by the Sun, the phases depend on where the Moon is in its orbit in relation to the Sun. The phases, in order of occurrence, are waxing crescent (or new), first quarter, waxing gibbous, full, waning gibbous, third quarter, waning crescent (or old) Moon. (The dark side of the Moon is actually the "far side" that constantly faces away from us.)

The composition of the Moon was discovered after the Apollo astronaut missions (1969–1972) brought back to Earth about 882 pounds (400 kilograms) of lunar rock. The following charts list the composition and location of the lunar rocks:

Composition of Lunar Rocks	
Chemical Composition	Percent
SiO_2 (silicon dioxide)	43%
FeO (iron oxide)	16%
AlO_3 (aluminum oxide)	13%
CaO (calcium oxide)	12%
MgO (magnesium oxide)	8%
TiO_2 (titanium dioxide)	7%
other elements	1%

Types and Locations of Lunar Rocks		
Name	Chemical Composition	Mainly Found In
Anorthite	calcium, aluminum, silicon, oxygen	regolith and crust of Highlands
Ilmenite	iron, titanium, oxygen	regolith of maria; mantle
Plagioclase	calcium, aluminum, silicon, oxygen	regolith and crust of Highlands
Pyroxene	calcium, magnesium, iron, silicon, oxygen	regolith of maria; mantle
Olivine	magnesium, iron, silicon, oxygen	regolith of maria; mantle

Characteristics of the Moon	
Average distance from Earth (miles/kilometers)	238,331/384,405
Diameter at equator (miles/kilometers)	2,160/3,476
Surface gravity (Earth = 1)	0.165
Day (rotation)	27 days, 7 hours, 27 minutes
Revolve around the Earth (from new moon to new moon)	29.53059 days
Mass (Earth = 1)	0.0123
Highest surface temperature (approximate)	273° F (134° C)
Lowest surface temperature (approximate)	–243° F (-153° C)

moonquakes—Moonquakes are similar to the smaller tremors felt during earthquakes on Earth. Quake monitors left behind by several of the Apollo manned missions to the Moon detected the presence of lunar quakes. Roughly 3,000 minor tremors occur each year, most of them so small, they are hardly noticeable. Most quakes occur when the Moon is closest to Earth (perigee), when Earth's gravitational pull on the Moon's crust is at its strongest. Another possible cause of quakes is the collision of small meteoroids on the lunar surface.

Over the billions of years that the Moon has been in existence, larger moonquakes no doubt have had a great effect on features on the lunar surface. For example, quakes can cause the unstable Moon's regolith to slide down steeper slopes (such as those along deep craters) in the form of landslides.

tides—See OCEANS, Topic Terms.

23. Mountains

Introduction

Some of the most striking land features on our planet are mountains. They tower above, seemingly immortal, wrapped in clouds and weather of their own making. The interplay of light and shadow on their slopes gives them an almost living presence. Indeed, some early humans thought of these lofty places as homes of the gods, sacred places that were not to be defiled by ordinary people. With the advent of scientific reason and a new spirit of adventure, mountains were studied, climbed, probed, and measured. Although no longer thought of as homes of the gods, they are still areas of great beauty and serenity. They still create unique weather patterns that influence the areas around them. And they do indeed have a lifecycle: one of birth, dominance of the horizon, and eventual death as their heights are ground down to flatness over millions of years by the forces of nature.

Timeline

(note: bya=billion years ago)

Date	Event

Prehistoric Events

Date	Event
~4.55–2.5 bya	The ancient cratons, or cores, of the early continents form, but the still hot Earth prevents any large continents from forming.
~3.96 bya	One of the earliest mountain orogenies occurs at this time.
~2.5–2 bya	The Earth's crust cools, and crustal movements slow, allowing the larger landmasses to form.

Modern Events and Discoveries

Date	Event
early 1500s	Leonardo da Vinci (1452–1519), Italian artist and inventor, is the first European to explore the high Alps of Europe; in the northern Italian Alps, he observes and records well-preserved fossils of marine creatures, from which he deduces that an ancient sea floor was uplifted to form the mountains.
1545	German geologist Georgius Agricola (Georg Bauer, 1494–1555) proposes that mountains originate by means of water erosion, atmospheric and subterranean winds, earthquakes, and the Earth's fiery interior.
1669	Danish geologist and anatomist Nicolaus Steno (Niels Stensen, 1638–1686) concludes from field studies in Tuscany that mountains form from the shifting of the Earth's strata, through either violent upthrusting, slippage, or downfall.
1740	Italian scientist (Antonio) Lazzaro Moro (1687–1764) describes the history of early Earth, contending that the reason for mountains is a central fire within the Earth.
1779–1796	Swiss scientist Horace de Saussure (1740–1799) publishes his four-volume series of the geology, meteorology, and botany of the Alps and other European mountains.
1830	French scientist Élie de Beaumont (1798–1874) suggests that each mountain range formed because of a catastrophe, and the mountains are due to the cooling, and thus contracting, of the Earth's crust.
1880	Swiss scientist Alphonse Favre (1815–1890) shows that mountains can form under lateral compression, using a model consisting of a layer of soft clay and a thick band of rubber; he believes the Alps, Appalachians, and Jura mountains were formed in this way.
1887	French geologist Marcel Alexandre Bertrand (1847–1907) suggests that Europe has gone through three major periods of intense folding and orogenies, creating the Caledonian, Hercynian, and Alpine mountain chains.
1889	U.S. geologist Clarence Edward Dutton (1841–1912) proposes the concept of isostasy—the rising of the land after the Ice Ages ended.
1998	The current theories of mountain building link uplift, climate, and erosion in a complex interaction of feedback, with mountain ranges having formative, steady-state, and erosional stages in their lifecycle.

History of Mountains

Early Mountains

The process of **mountain** building, called **orogeny**, has occurred many times in the past—and no doubt will continue in the foreseeable future unless the driving force behind plate tectonics comes to a halt (as the continental plates come together, a mountain range usually results). The first known orogeny on record took place in North America and is evidenced from gneissic rock that dates to about 3.96 billion years ago. Some of the early mountains that once towered over the planet have eroded away to hills or plains over millions of years. Others are still in the process of erosion. And still others are beginning and continuing to uplift today depending on their location in terms of continental plates. (See orogeny in the topic terms section for information on orogenies throughout Earth's history.)

Modern Studies of Mountains

Human beings have always had a fascination with mountains, but that fascination took many forms, depending on the culture and area. The ancient Greeks revered their mountains, believing them to be the homes of gods and goddesses. Olympia was where Zeus, Apollo, Hera, and others looked down on the mere mortals and influenced their lives. The Buddhists of China and Japan felt that mountains were associated with divinity and built shrines and temples upon the peaks. In contrast, most Europeans, until the early nineteenth century, regarded mountains from a superstitious viewpoint. Aided by religious beliefs, they felt that these apparitions were the tumors or wens of the Earth, populated by demons, goblins, and gnomes—and avoided them as much as possible.

In fact, it was not until the early fifteenth century that Leonardo da Vinci became the first European to explore the high Alps of Europe. In northern Italy, he found fossil seashells in the mountains; others felt the animals had been driven to that point by the biblical Deluge. But da Vinci felt the shells would have been broken by such action and instead deduced correctly that the mountains were actually ancient sea floors uplifted to their present lofty positions.

In 1899, the first comprehensive model of the evolution of mountains was published. Called the "geographic cycle," it proposed a lifecycle of mountains that included a fairly brief birth—caused by sudden, violent uplift—and an old age—caused by slow erosion. This model endured for almost 100 years, but is currently being modified as new research is done on the processes shaping our planet's mountains.

Currently, the mechanism of mountain-building is thought to be a complex interplay between the forces of **plate tectonics**, climate, and erosional processes. These are all linked together and influence each other in a feedback cycle that continually changes the height and topography of the mountains and nearby regions.

The uplifting of a region from the collision of continental plates is only the beginning of the mountain-building process. To craft the uplifted regions into the shapes that we are familiar with, the processes of erosion and climate must work their magic on the rocks. Erosional agents that work on a landscape include water, wind, glacial ice, and gravity. Erosion weathers the bedrock, turning it into smaller and smaller pieces of sediment. This sediment is then moved down the slopes to the bases of the mountains, and from there transported away in moving water, such as rivers. The rate of erosion is closely tied to the presence and extent of these erosion agents, the type of rocks that have been uplifted, the steepness of the topography, and the climate.

The climate is an integral part of this erosional process, affecting the amount of material that is removed and how fast this process occurs. The relationship between climate and erosion, however, is not straightforward. For example, mountains in climates that are wetter will generally erode faster, but this climate may also promote the growth of plants that anchor the soil and hinder erosion. Cold climates foster the development of glaciers, which can aggressively erode mountains. Examples of this type of process are found in the European Alps and the Sierra Nevada mountains of western North America. But the ice sheets in Antarctica and Greenland are normally frozen to the underlying rock, and little erosion occurs.

In fact, the interaction of mountains, climate, and erosion is best viewed from a system point of view, with a change in one factor leading to changes in the others. As mountains grow higher due to tectonic activity, they change the climate around them; it normally becomes colder and wetter. Winds that collide with mountains in their path dump their moisture as they climb higher, leading to rain storms and glacier formation. This increases the rate of erosion of the mountains, creating the craggy slopes that are familiar to us. On the lee side of the mountains, the now dry winds create a rain shadow, or desert-like area, resulting in a dry area with little erosion.

Another example that supports this feedback theory is the finding that erosion can actually lead to higher mountains. Mountains are akin to icebergs; they are supported on the denser, fluid-like mantle below by the strength of their underlying crust and by a buoyant "root" of crust that penetrates into the mantle. Similar to an iceberg, there is more material below than above; crustal roots tens of miles deep support mountains several miles high. The mountain "icebergs" float on the mantle in an equilibrium position due to buoyancy; as erosion removes more and more of the exposed material, the weight on top diminishes and the mountains rise in a process called **isostatic** uplift. If tectonic activity is still present, the erosional forces can actually increase the rate of that activity.

These findings have prompted geologists to theorize that there are three stages of development that occur during the lifecycle of mountains: the formative, the steady-state, and the erosional. The formative years occur when tectonic activity is dominant, and the rate of uplift is greater than the rate of erosion. As the mountains rise, however, they change the climate around them and increase the rate of erosion. The steady-state years occur when the rate of uplift, whether due to continuing tectonic activity or isostasy, equals the rate of erosion. As the uplift diminishes, erosion becomes dominant and the final stage is reached. The mountains slowly begin to round and diminish, as their uplifted rock is eaten away by the agents of change. They may eventually disappear; if tectonic activity reoccurs or if the climate drastically changes, they may linger on for many millions of years, or once again rise up and dominate the horizon.

Today, mountains along plate boundaries rise a certain amount each year. And new technology, such as the global positioning system (GPS), is allowing scientists to understand just how rapidly these mountains are rising. For example, with conventional surveying methods, scientists have estimated that Mt. Everest, the world's tallest mountain, rises an average of 1 to 2 inches (2.5 to 5 centimeters) per year as India's and Asia's continental plates collide. But the actual height is still disputed: Not only is the summit being constantly eroded, but the other data from the summit are imprecise because the peak is covered by a snowcap of unknown depth—that shrinks and grows differently each year. Scientists recently planted a GPS receiver about 60 feet (18 meters) from the summit on a rocky outcrop not covered by snow. It recorded altitude data for five days until it was retrieved; more measurements will be taken from the same spot in a few years to detect any changes. *See also* ATMOSPHERE, CLIMATE AND WEATHER, EARTH, EARTHQUAKES, FOSSILS, LIFE, MOON, OCEANS, SOLAR SYSTEM, and VOLCANOES.

Topic Terms

isostasy—Isostasy is the state in which blocks of Earth's crust are in equilibrium. The model of isostasy treats the blocks of the crust as if they were floating stationary in a liquid. To keep this balance, the mountains must have deep roots that are less dense than the mantle below. The best example of isostasy occurred after the Ice Ages. As the huge miles-thick ice sheets rolled over the land, the continents subsided; after the ice sheets retreated, the continents rebounded, and continue to rebound even today, to bring them back to balance. Today, many of the minor earthquakes in the Adirondack Mountains in New York are the result of the land—once covered by the ice—rebounding to reach equilibrium.

mountains—Mountains are a manifestation of the movement of the Earth's crust, which is broken into large plates that separate from, or collide with, each other over millions of years. Because of the shortness of our lifespans, these movements seem so slow as to be nonexistent; the Earth, to us, is essentially static. But on the geologic time scale, the Earth is a dynamic entity, as landmasses move into and away from each other, much like cars in a demolition derby. Each collision or separation leaves its mark on the landscape, and one of the most prominent marks are mountains. Mountains can also form by volcanic action. For example, the mountains of the Hawaiian Island chain have all been built up from volcanic activity.

These are only part of the long chain of Rocky Mountains that runs through the North American continent. The Rocky Mountains represent a more recent orogeny.

Mountain Ranges of the World

Name	Location
Great Dividing Range	Eastern Australia
Himalayas	Southern Asia
Altai Mountains	Western China
Greater Khingan Range	Northern China
Ural Mountains	Central Russia
Caucasus Mountains	Southern Russia
Carpathian Mountains	Eastern Europe
Alps	Southern Europe
Pyrenees	Between France and Spain
Atlas Mountains	Northwestern Africa
Ethiopian Highlands	Eastern Africa
Drakensberg Mountains	South Africa
Andes Mountains	Western South America
Appalachian Mountains	Eastern North America
Rocky Mountains	Western North America
Coastal Mountains	Northwestern North America
Alaska Range	Northwestern North America
Brooks Range	Northern Alaska, Northwestern North America

World's Highest Mountains

Name	Height (feet/meters)	Location
Everest	29,022/8,848	Nepal/Tibet
K2	28,250/8,611	Kashmir/China
Kanchenjunga	28,208/8,598	Nepal/Sikkim
Lhotse	27,890/8,501	Nepal/Tibet
Makalu I	27,790/8,470	Nepal/Tibet

orogeny—Orogenies are periods of mountain-building that occur as the result of plate tectonics—the movement of the Earth's crustal plates. The process of mountain building is called orogenesis. The results of the orogenies include folding, thrusting, and deformation of the rock. In addition, an orogeny is often accompanied by the invasion (intrusion) of igneous rocks or volcanics. The heat and pressure from an orogeny can also create metamorphic rocks.

Precambrian Orogenies

Name	Time Period (mya)	Location (referenced to modern day location)
Acasta gneiss (oldest rocks)	3960	North America
Kalahari craton (volcanic)	3300–2700	Africa
Baltic Shield (volcanic)	3200–2700	Baltica
Pilbara block (volcanic)	3050–2700	Australia
Canadian Shield (volcanic)	2700–2500	North America
Algoman	2600–2400	North America
Hudson	1820–1640	North America
Penokean	1750–1650	North America
Mid-Continent Rift (volcanic)	1200–900	North America
Grenville	1000–880	North America
Baikalian	600–550	Siberia

Paleozoic Orogenies

These episodes of mountain building occurred as the widely separated continental landmasses moved together to form a large northern continent, Laurasia, and a large southern continent, Gondwanaland.

Name	Time Period (mya)	Location (referenced to modern day location)
Adelaide	540–530	Australia
Appalachian (early)	480–460	North America, South America, Antarctica
Taconic	460–440	North America, Baltica
Caledonian	450–430	North America, Baltica
Acadian	410–380	North America
Antler	380–350	North America
Uralian	380–300	Baltica, Siberia, Kazakhstania
Tasman	380–250	Australia
Hercynian	350–245	Baltica
Ouchita	325–310	North America

Paleozoic to Mesozoic Era Transition

The delineation between the Paleozoic and Mesozoic eras occurred approximately 250 million years ago on the geological time scale. Laurasia and Gondwanaland were colliding to form the supercontinent of Panaega, with mountain building episodes occurring as Africa and eastern North America made contact.

Name	Time Period (mya)	Location (referenced to modern day location)
Allegheny	320–220	North America, Africa
Cape Folding	250	Africa
Siberian Traps (volcanic)	250	Siberia

Mesozoic Era Orogenies

No sooner did Pangea form, its seems, than it started to break back up into Laurasia and Gondwanaland. As the Mesozoic era continued, even Laurasia and Gondwanaland began to scatter into landmasses that vaguely resembled today's continents.

Name	Time Period (mya)	Location (referenced to modern day location)
Eastern N. America (volcanic)	200–190	North America
Rifting (volcanic)	170–160	Africa, Antarctica
Nevada	190–140	North America
Sevier	140–80	North America
S. Atlantic (volcanic)	135–130	South America, Africa
Ontong (volcanic)	122	Java
Kerguelen (volcanic)	112	South Indian Ocean
Rajmahal (volcanic)	110–100	India

Cenozoic Era Orogenies

The Cenozoic era also saw the movement of the continents continue into today's present positions.

Name	Time Period (mya)	Location (referenced to modern day location)
Laramide	84–50	North America
Andean	80–60	South America
Deccan Traps (volcanic)	65–63	India
Brito-Arctic	62–55	North America, Baltica
North Atlantic (volcanic)	57	North Atlantic Ocean

Cenozoic Era Orogenies (cont'd.)

Name	Time Period (mya)	Location (referenced to modern day location)
Pyreneen	55–45	Baltica
Alpine	40–5	Baltica, Africa
Ethiopa (volcanic)	24–18	Africa
Himalayan	24–0	India, China
Columbia River	18–15	North America

plate tectonics—The crust of the Earth is divided into large plates; at the boundaries between plates lie faults, or cracks. These plates "float" on the Earth's mantle and are moved about, scientists think, by convection currents in the Earth's interior. When two plates are slowly pushed together, one plate may be pushed downward into the interior of the planet in a process called subduction or the plates may also move horizontally to each other, as in the case of the San Andreas fault.

Another effect occurs when the movement is due to the internal heat energy of the Earth's mantle: The interfaces between plates are the areas where mountains form. In the simplest of terms, mountains are just masses of rock that have been lifted up. The crust is made up of plates floating on the mantle under it; the addition of mass makes the crust thicker. If the mass being added is molten, then the crust will be hotter. Geologists have found that thicker or hotter crusts will float higher on the mantle, uplifting that particular area into mountain regions.

There are a number of ways that the action at plate boundaries can increase the mass or temperature of the crust. When tectonic plates are colliding with each other, one of the plates may be subducted beneath the other, being driven down into the mantle and returning to a molten state. The crust at the subduction zone boundary will be thickened as a result of the compressive forces, and upwelling magma from the subducting plate will also add to the mass and increase the temperature, leading to an uplift. A well-known example of mountains formed by this subduction mechanism are those around the Pacific rim, known as the "Ring of Fire."

If neither of the colliding plates subducts beneath the other, then the action of the collision will add mass from both plates to the interfacial region, causing the uplift. An example of this type of spectacular mountain-building is the Himalayas and the Tibetan Plateau.

At an area where the plates are separating, the flow of magma and transfer of heat can also build mountains. Under water, for example, the Mid-Atlantic Ridge, where two plates have been separating for more than 200 million years, is the largest mountain range on the planet, extending more than 9,321 miles (15,000 kilometers) long, and rising 13,124 feet (4,000 meters) above the ocean floor. As these plates continue to separate, this mountain range continues to grow, as do other ridges, such as those in the eastern Pacific and Indian Oceans. On land, areas such as these can also be uplifted because the heat of the magma leads to a less dense, and hence more buoyant, crust.

24. Oceans

Introduction

Oceanography is the study of the oceans and seas—their compositions, currents, geology, and surrounding coastlines. The oceans play an important role in the shaping of the Earth's past, present, and future natural history. After all, about 70 percent of the Earth is covered by oceans; and about two-thirds of all animal and plant phyla are considered to be marine. These huge bodies of water affect the climate (the ocean interacting with the atmosphere), marine life (both flora and fauna in the oceans and along the coastlines), erosion and deposition of landmasses (along the coasts and on the ocean floor), and movement of the Earth through space (in the form of tides).

Timeline

(note: bya=billion years ago; mya=million years ago)

Date	Event

Prehistoric Events

Date	Event
~4.35 bya	The Earth's atmosphere and oceans are established; the Earth was also being almost continually bombarded by objects from space.
~3.8 bya	The heavy bombardment of the planet dies down, allowing the planet to continue to cool down; life probably began to form in the oceans around this time.
~500 mya	The large landmasses of Gondwanaland (or Gondwana) form to the south, and Laurasia (or Laurentia) forms around the equator, changing the ocean current regimes.
~250 mya	Another supercontinent forms called Pangea (or Pangaea), a large landmass composed of Gondwanaland and Laurasia huddled around the equator; this changes the ocean currents again.
~200 mya	Pangea breaks apart into again, Laurasia and Gondwanaland, and the continents and ocean currents begin to head toward the familiar continents and currents we see today.
~50 mya	The familiar continents we see today are relatively in place, with ocean currents in general in the same areas as today.

Modern Events and Discoveries

Date	Event
~1210	An Englishman named Wallingford (?–1213) is thought to be the first to record tide observations for the purpose of prediction, and tabulates the occurrences of floods at the London bridge.
1326–1335	Englishman Richard Wallingford (not the same person as in 1210; c.1292–1335) installs a mechanical clock near the abby of St. Albans in England to show the ebb and flow of the tide.
1513	Spanish explorer Juan Ponce de Léon (c.1460–1521) puts forth the first known written description of the Gulf Stream current in the North Atlantic Ocean.
1616	English explorer William Baffin (1584–1622) searches for the Northwest Passage, and records the existence of the open water region in the midst of pack ice in what is now called Baffin Bay.
1663	Oceanographer Isaac Vossius (1618–1689) correctly concludes that the North Atlantic circulation is clockwise; he incorrectly believes that tides and currents are caused by a rise in sea level at the equator that is tied to expansion brought on by the Sun's heat.
1672	Irish chemist and physicist Robert Boyle (1627–1691) rejects the idea that the salt in the sea is only at the surface, as was previously believed.
1683	English scientist Isaac Newton (1642–1727) formulates a mathematical theory of ocean tides, based on the gravitational attraction of the Sun and Moon.
1770	U.S. scientist, inventor, and diplomat Benjamin Franklin (1706–1790) publishes the first scientific chart of the Gulf Stream current in the North Atlantic Ocean; it is widely used by navigators and highly regarded by scientists.

Date	Event
1855	U.S. oceanographer Matthew Fontaine Maury (1806–1873) authors the first textbook on oceanography, *The Physical Geography of the Sea*.
1868	Scottish naturalist-oceanographer Charles Wyville Thomson (1830–1882) and Irish oceanographer William Carpenter (1813–1885) organize a deep water dredging expedition in the water north of Ireland and Scotland to search for life at great depths; the generally held belief at the time is that life cannot exist at the ocean bottom.
1869	Wyville Thomson and William Carpenter, along with Welsh scientist Gwyn Jeffries (1809–1885), further their search for deep sea life on another expedition, proving that life does exist at great depths.
1872–1876	Thomson and several other scientists begin the first orderly, detailed studies of the oceans on the H.M.S. *Challenger*; this is considered the birth of modern scientific oceanographic studies.
1889	German scientist Hermann von Helmholtz (1821–1894) suggests that the waves in the ocean are caused by the same instability that causes clouds to billow in the atmosphere.
1893–1896	Norwegian oceanographer Fridtjof Nansen (1861–1930) leads the *Fram* expedition to the Arctic Ocean; he later writes about his speculations regarding the Arctic circulation in a six-volume work published from 1900 to 1906.
1902	Nansen determines the first sea water salinities of the oceans; he also presents new ideas on ocean sea ice drift and dynamics.
1912	German oceanographer Gerhard Schott (1866–1961) presents the most comprehensive treatment of oceanography of the Atlantic Ocean to date in his *Geography of the Atlantic Ocean*.
1929	The first U.S. national park east of the Mississippi, and its first national park on an ocean, is established: The Acadia National Park, a 27,871-acre preserve in Maine.
1954	The nuclear powered submarine U.S.S. *Nautilus* is built, and becomes the first to cross the North Pole under the ice.
1959	The first submersible (diving saucer, or bathyscaph) is built by French oceanographer and author Jacques-Yves Cousteau (1910–1996), the *Soucoupe Plongeante*; it is only one of his many inventions, including the aqualung (with Emile Gagnan, called SCUBA, or Self-Contained Underwater Breathing Apparatus) and underwater photographic techniques.
1960	The bathyscaph *Trieste*, built for diving to deep depths, touches down at a depth of 6.8 miles (11 kilometers)—the deepest any vessel had ever reached.
1964	The three-man submersible, the *Alvin*, is built; it will be used to explore the wreck of the *Titanic* and discover undersea vents near the Galapagos Islands, among other discoveries, descending as deep as 13,000 feet (3,962 meters).
1970s	The CLIMAP project (Climate: Long-Range Investigation, Mapping, and Prediction) compares sea level changes throughout geologic history with those levels known today.
1977	Deep-sea vents at the Galapagos Rift are discovered, along with strange, unknown organisms living there.
1982	The oldest known shipwreck is found at Ulu Burun off the coast of Turkey, dating from the fourteenth century B.C.
1982–83; 1997–98	Two of the strongest El Niños, changes in the patterns of ocean waters off the continent of South America, are recorded.
1985	U.S. Oceanographer Robert D. Ballard and a team of U.S. and French researchers discover the sunken ocean liner, the *Titanic*.
1990s	It is estimated that more than 75 percent of the U.S. population lives within 50 miles (80 kilometers) of the seashore.

History of Oceans

Early Oceans

No one really knows how or even when the **oceans** and **seas** formed, although most scientists believe the Earth's atmosphere and oceans were established around 4.35 billion years ago. By about 3.8 billion years ago (or earlier) the first life developed in the oceans, and organisms began to increase in diversity and number in the oceans about 1 billion years ago. By around 544 million years ago, life was teeming in the oceans, during what is called the Cambrian Explosion at the beginning of the Paleozoic era. Over millions of years the continents changed from two great supercontinents into one larger landmass and back again, with corresponding changes in the ocean's currents. By about 50 million years ago, the familiar continents, and thus also ocean currents, we see today were relatively in place.

No one really knows how water filled the oceans. It is thought that early volcanic activity released water into the atmosphere in the form of gas; once the gas condensed and settled out of the atmosphere as liquid, it formed the oceans. Another theory states that the water from bombarding comets and asteroids during the early solar system formation created the oceans.

No matter how the oceans were created, several things are currently known about the oceans that have occurred over the Earth's long history. For example, it is known that **sea levels**—actually the "surface" of the oceans—have risen and fallen over time, and continue to do so to this day. It is also known that the early Earth's **seawater** differed greatly in terms of salinity and temperature, as the waters rose and fell from ice ages over time. And shallow and deep **currents** have changed due to the movement of the continental landmasses (the result of **plate tectonics**); and **ocean margins** (also called **coastal margins**) have changed over time—including dramatic changes in wetlands, **estuaries**, and sundry coastal features.

Oceanography

The oceans have been explored and used by humans for hundreds of years for travel and sustenance. Some of the first ocean studies were done in the early 1200s, when the first known recording of **tide** observations were carried out in England to predict the occurrence of the tides. And for the next 500 years, tide prediction was the major objective of oceanic observations.

The first ocean voyages were mostly for exploration and to find new trade routes. **Coastlines** were mapped, as were some major currents and **wave** actions along seacoasts. But the true field of oceanography would take centuries more to flourish.

By the mid-1800s, several expeditions were formed to explore the oceans. In 1868, Charles Wyville Thomson and oceanographer William Carpenter organized a deep-water dredging expedition on the H.M.S. *Lightning* in the water north of Ireland and Scotland to search for life at great depths. The overall belief at that time was that no life could survive at the ocean bottom—but the scientists were soon to prove that belief wrong, as they pulled up crustaceans, mollusks, sponges, and other organisms from the ocean floor. They also found differences in temperature as they tested the water at varying depths, crushing another belief of the time—that all of the water in the oceans was a constant temperature. Because of these findings, Wyville Thomson and William Carpenter, along with Gwyn Jeffries, furthered their search for deep-sea life on another expedition onboard the H.M.S. *Porcupine*. They dredged the ocean floor even deeper, definitely proving that life exists at great depths.

But it took another few years before an orderly, detailed study of the oceans was accomplished beginning with the voyage of the H.M.S. *Challenger* (1873–1876). This expedition gave birth to the science of oceanography and was led by naturalist-oceanographer Charles Wyville Thomson—experienced from his previous expeditions—and included numerous professional scientists. The expedition discovered thousands of new species, and evidence that the flat ocean bottom was actually filled with enormous mountains, plains, caverns, and cliffs. The resulting 50 volumes based on the findings of this voyage were published starting in 1895.

Marine plants and animals were not left out of these studies. Wyville Thomson was trained as a physician, but devoted himself from his 30s onward to the study of oceanography—in particular, ocean organisms. His work on all the expeditions was mostly to detail these marine organisms. Overall, the expedition collected 1,441 water samples, 13,000 plant and animal specimens, and hundreds of sea floor deposits.

In the years following the *Challenger's* voyage, almost every seafaring nation established their own oceanographic research efforts. The United States established the Scripps Laboratory in 1925 and Woods Hole Oceanographic Institution in 1930.

Prior to these developments, scientists continued to collect physical data on the oceans—especially studies in current flow. Oceanographer Matthew Maury, largely self-educated, was responsible for creating the first sea lanes for ships, developing charts based on trade winds and ocean currents compiled from the study of ships' logs. The charts made possible shorter passages using areas of the oceans that had the most favorable winds. He also did soundings of the Atlantic **ocean floor**, making way for the transoceanic telegraph cables; and was the author of the first textbook on oceanography, *The Physical Geography of the Sea* (1855).

One of the more interesting discoveries in the oceans occurred by serendipity. On the Grand Bank, Newfoundland, an earthquake in 1929 triggered a massive **turbidity current** along a canyon. The current, with its water, mud, and debris, swept out into the abyssal plain, tearing apart deep-sea telegraph cables and causing a great deal of erosion before it spread millions of tons of sediment over the Atlantic floor. Scientists estimate that the current dropped sediment in an area the size of Iceland—and about 3 feet (1 meter) thick. And all of these findings were discovered because of the snapping of the telegraph cables.

Studies of the oceans continued in the early twentieth century, as more observational oceanographers described the physical features of the oceans. For example, discoveries were made about the turbulent motion of the sea and the mixing of seawater; and about temperature, salinity, and the effect of currents. These studies tended to be focused in the Atlantic Ocean, because the main shipping lanes from Europe to the United States were the most frequently traveled. Even during World War II, when submarine warfare was predominant, such ocean studies were being conducted—and with many new devices, such as sonar, used not only for the war effort, but to collect data about the oceans.

In 1960, the special submarine (called a bathyscaph) *Trieste*, built for deep diving, touched down at a depth of 6.8 miles (11 kilometers). It was the deepest any vessel had ever reached and still holds that record today. Jacques Piccard and Donald Walsh were in the bathyscaph that Piccard's father (Auguste, famous for exploring ocean depths and the upper atmosphere) had designed and built. They set this record just off Guam on the floor of Challenger Deep in the Mariana Trench in the Pacific Ocean, the deepest **trench** known to exist. The two men were the first to record creatures at such depths and sighted, during the first few minutes of landing on the ocean floor, a footlong fish that resembled a flounder, with both eyes on the same side of its head.

Another significant study of the ocean floor came in 1977, as the *Alvin* submersible discovered deep sea **vents** on the ocean floor near the Galapagos Islands. These vents discharged very hot water and have since been found in other rift areas, including the Rivera Fracture Zone off the west coast of Mexico, near the mouth of the Gulf of California. Here, the seawater circulates through the newly formed oceanic crust, is heated by the underlying magma, and reacts chemically with the rocks; in addition, new types of organisms are found around the vents—all previously unknown. Significant marine expeditions continue even today—with the discoveries made available to the public in magazines, books, and on television shows. *See also* ATMOSPHERE, CLIMATE AND WEATHER, EARTHQUAKES, FOSSILS, GEOLOGIC TIME SCALE, LIFE, MOON, MOUNTAINS, PLANTS, SOLAR SYSTEM, and VOLCANOES.

Topic Terms

coastline—A coastline (also called seacoast or coast) is the edge of land that comes in contact with the oceans or seas along continents or islands. In less accurate terms, it is also used to describe the edge of land that comes in contact with larger lakes. Coastlines vary from low, muddy, or sandy to high cliffs; the type of coastline is dependent on the type of rock along the coastline and the amount of erosion it experiences from the ocean activity. Because of the nature of rock, coasts come in a wide variety of shapes, sizes, and lengths. Some coasts are irregular with rugged and steep cliffs, such as those along many of the coasts of the Hawaiian Islands—steep because the ocean waves have not yet eroded the volcanic island shores into flat, sandy beaches. Other coasts are smooth with long, sandy beaches leading out to the sea, such as the coast of the eastern United States. Still others are unique, for example, coasts of coral reefs, such as the Great Barrier Reef along the northeastern continent of Australia; and steep fjords caused by glaciers and melting glacial waters, such as the those along Norway's coast.

Coastlines—in the past and present—experience constant change because of their interaction with various natural processes; these processes affect many people, as nearly two-thirds of the world's population live in coastal areas. Various organisms, plant and animal, live within certain zones along the coastlines, with the organism variations dependent mainly on the composition of the beach, and the waves and tidal currents.

Each coast has a distinctive natural history: Coastal erosion due to wave and current action is prevalent in many coastal regions, including the changing barrier islands off the coast of North Carolina. Although certain measures to stop erosion along various coastlines have been taken, the process often continues slowly over many years, because the natural effect of the oceans is to change the coastlines. Years of tides and violent storms especially affect smaller island coasts, such as those along Haiti, the Dominican Republic, and Puerto Rico; and lowland coasts are especially susceptible to violent storms. Rivers also change the coastal boundary between the ocean and land by carrying sediment to the oceans, such as the Nile River in Egypt, which creates a fertile coastal land called a delta. These features are also susceptible to major storms, such as hurricanes and monsoons, which can cause large portions of a coast to be washed away.

Another reason for coastal change is the rise and fall of sea level over time (and thus, the continual erosion and deposition of sediment). Still another reason for coastline change has to do with the movement of the continental plates, as many coasts are located along a continental plate boundary. Two examples are the western coast of the United States, in which the Pacific plate and the North American plate move relative to one another, creating the rugged coast; and the western coast of South America, in which the Andes Mountains are the result of the Pacific plate subducting under the South American plate. The coastlines can be either the leading or trailing edges of the plates. For example, the western coast of the United States is considered the leading edge of a plate, while the Atlantic coast south of South Hook, New Jersey, is considered a trailing edge.

A coastal plain is a large area of low, flat-lying land adjacent to the oceans. If water encroaches onto a coastal plain, wetlands develop in the area in which the water pools; larger wetland areas include the Everglades of Florida and the area around the Amazon River. These wetlands are found on all the continents except Antarctica.

Coastal plains form in three ways, each method taking hundreds of thousands of years to develop: (1) they may begin as part of the continental shelf—as sea level naturally falls, the continental shelf is exposed, creating the coastal plain, and when sea level rises, the coastal plain may disappear underwater again; (2) they may begin as part of the continental shelf that has been raised by the movement of continental plates; and (3) they may form from the deposition of sediment from rivers into the oceans. If the sediment builds up high enough, it will eventually "emerge" from the ocean to form a coastal plain.

currents—Currents are the large scale movements of water in a specific direction. Currents can be seen as the circulatory system of the oceans; they are responsible for the vast movement of water around the planet. Winds can be seen as the atmospheric equivalent of currents. In fact, winds are responsible for surface currents, while density differences in the water, an effect of temperature and salinity, are responsible for the deep-ocean currents. Together, these currents comprise the oceanic circulation, which, like the atmosphere, is driven by the Sun's radiation and the Earth's rotation (the Coriolis effect). There are numerous important currents around the world, both surface and deep-ocean. Some, like the Gulf Stream, have important climatological effects.

Density, or the measure of a substance's mass per unit volume, has a great deal to do with the oceans. In

The Gulf Steam is a warm current that runs along the east coast of the United States and eventually reaches the British Isles. *(Photo courtesy of NOAA, Asheville, NC)*

the case of seawater, the density is affected by the temperature, the salinity, and, in a much smaller way, by the pressure. Cold water is denser than warm water, which is why the deep-ocean currents hug the ocean bottom and why water is warmer closer to the surface. Density differences also give rise to variations in the flora and fauna found at different depths. Salinity also contributes to water density—the higher the salinity of water, the higher the density. Where there is high temperature and low relative humidity, such as the Mediterranean Sea, the rate of evaporation is high, leaving the remaining water salt-rich and therefore denser.

Thermoclines also are connected to the currents. A thermocline represents a temperature difference in the water. Although water temperature normally decreases with depth, a thermocline is a zone in which the temperature of the water decreases more rapidly with depth than the surrounding areas. Some thermoclines are seasonal, such as those caused by the heating effect of the Sun in summer on the water near the surface, and others are permanent, a result of the "downward" flow of cold polar waters. In the oceans, the permanent thermal barrier separates the deep, cold waters from the relatively warmer surface waters and hinders their mixing. The deep waters make up about 98 percent of the ocean's volume and flow in deep-ocean currents; the remaining volume is found in surface waters, which receive the Sun's radiation and have currents driven by the wind.

And still another manifestation of currents are gyres, large-scale circular movements of the currents in the upper surface of the oceans. They occur when the surface currents created by the prevailing winds encounter continental barriers, modifying their paths into giant circulations of water. Although present in all oceans, one of the best defined is the North Atlantic gyre. This movement commences with the westward flowing equatorial current, is changed by the North American continent to a northeast flow along the coast of the United States (Gulf Stream), then flows east as the North Atlantic current and, to complete the circle, south as the Canaries current.

There are two major divisions of currents:

deep-ocean—These are also known as subsurface currents. They are the deep-water currents, which are created mostly by density differences in the water; these density differences are a function of the temperature and salinity. For example, deep beneath the northeast flowing Gulf Stream, a large, cold current is flowing in the opposite direction. The cold, dense water from around Greenland sinks and flows toward the equator, pushed up against the western edge of the Atlantic Ocean's basin by the Earth's rotation. Here it flows at approximately 0.5 miles (0.8 kilometers) per hour, hugging the bottom and displacing the warmer water upward.

surface—These oceanic currents are generated by the prevailing winds created as a result of the Sun's radiation and the Earth's rotation. For example, winds generated around the equator give rise to westward flowing surface currents in the areas north and south of the equator. With nothing to stop them, these equatorial currents would flow around the world; however, the continents are barriers that redirect their flow into circulation cells called gyres.

Surface currents also have a direct effect on the climate of regions nearby. For example, in the North Atlantic gyre, with its clockwise spin, the Gulf Stream is the fastest current, carrying warm water from Florida northeast along the coast of the United States, past Newfoundland and arriving at England, where it moderates the climate of this island. In New York, even though the state is the same latitude as warm southern Italy, the Labrador Current creates a much cooler climate.

Currents not only support marine organisms—plus contribute to the geologic changes along the coast and on the deep-ocean floor—but they also can have a profound effect on local economies—and sometimes on the world. One of the major events important to the fishery industries are the upwellings along some coastal regions. Usually thought to be cyclic in nature, upwellings occur where currents part or where winds blow surface waters away from a coast, causing the subsurface waters to rise and take their place. These upwellings, mostly on the west coasts of the continents, are prime fishing areas, because the upwellings bring up nutrient-rich waters from below. The result is hundreds of species of organisms—including fish—rushing into the feeding grounds. If the currents do not form, these fishing grounds are lost and local economies are affected.

Another more popularized periodic current is the El Niño, or "the young boy," in reference to the Christ child. (The current was discovered about 200 years ago and occurs around Christmas.) El Niño is a mass of warm water along the west coast of South America, just above the equator, that seems to peak every seven to 15 years or so. (Its counterpart is called La Niña, when the waters cool.) El Niño is known to change the climate around the world—sometimes for the worst. In bad El Niño years, such as in 1982 to 1983, this phenomenon caused tens of thousands of animals to die off in the Galapagos, and droughts were felt as far away as Australia and northern Brazil. In another peak in 1997 to 1998, the warmer waters caused several major hurricanes to reach the highest intensity category and go farther north than usual.

There are many major ocean currents around the world—some covering broad areas, some smaller areas—and are almost too many to count. Here is a list of some major currents:

Benguela Current: The Benguela current is a broad, shallow, slow-moving body of cold water flowing northward along the west coast of South Africa. It is part of the gyre of the South Atlantic Ocean. The area is dense with plankton and other marine life, as the prevailing winds displace the surface waters away from the coast; this causes an upwelling as the subsurface waters rise to the surface to replace the water.

California Current: The California Current is a shallow surface current, generally flowing southeastward along the west coast of North America, until it meets the North Equatorial Current. This current also affects the local climate, because moist air moving toward the coast moves over the cooler current waters, creating dense sea fogs along the coast, including the famous fog banks that roll into San Francisco.

Canaries Current: The Canaries Current is a branch of the North Atlantic Drift (or North Atlantic Current); the cooler waters of this current temper the climate of the Canary Islands, which would be tropical without them. This current is also responsible for the famous fogs seen along the western edge of the Pyrenees Mountains between Spain and France, as the current's southern direction meets the warm winds and waters of the Iberian peninsula.

Gulf Stream: The Gulf Stream is located just off the Atlantic coast, and brings a temperate climate to the coastlines it passes. It begins in the Caribbean Sea, an area near the equator that is greatly affected by the tropical heat. The heated water escapes into the Gulf of Mexico where it is warmed even more; it then moves though the Florida Straits and northward, just off the Atlantic coastline. Traveling about 80 miles (130 kilometers) a day, the current finally reaches the colder Labrador Current. The current then splits, creating the North Atlantic Drift, which flows northeast across the Atlantic; and the Canaries Current, which flows east, then southeast back toward the equator.

Labrador Current: A cold surface current located off the coast of Labrador and Newfoundland. It is slow moving and travels in a southward direction until it reaches and dips below (the colder water sinking because of its higher density) the northward-flowing Gulf Stream. In the warmer, summer months in the Northern Hemisphere, the current can extend down to about Cape Cod, Massachusetts; in the winter months, the current can extend down to about Virginia. The Labrador Current has been known to carry icebergs into northern shipping lanes during the winter months.

North Atlantic Drift (or North Atlantic Current): The North Atlantic Drift is a branch of the Gulf Stream, a northeastern-flowing shallow current that moves across the North Atlantic toward the Arctic Ocean. As with most larger currents, the North Atlantic Drift affects the climate of nearby landmasses; the warmer waters of the North Atlantic Drift are responsible for moderating the climate of western Europe. The result is more moderate winters than would be expected at this latitude. For example, this current generally keeps Norwegian ports free of ice during the winter.

estuaries—Estuaries are rivers and tributaries at the border between the ocean (salt water) and rivers (freshwater). They are usually affected by tidal action, the water levels rising and falling with the daily tides. Estuaries form in several ways: (1) the natural lowering of the sea level exposing a drowned river valley; and (2) the ocean rises, filling in lowland drainage areas with water.

Estuaries are prime locations for wetlands, a halfway point between water and land, often referred to as swamps or marshes. Because of this, estuaries are ideal places for a wide variety of birds and wildlife. Tall reeds, grasses, and other plants are perfect hiding places for a variety of waterfowl, songbirds, and other birds, fish, amphibians, reptiles, and mammals.

marine animals—The number of marine animals is almost exhaustive, and is beyond the scope of this book. Suffice it to say that marine animals have played an important part in the natural history of the Earth. In particular, fish were probably one of the most important stepping-stones that led to life on land; the first creatures to inhabit the land were fish that adapted by developing primitive lungs and fins for crawling. Also among the fish were other important marine animals—creatures that adapted and evolved in the watery environment. The oceans today are filled with fish, mammals, and other sundry vertebrates and invertebrate animals. Some examples of known marine animals include approximately 1,000 species of cephalopods (squids, octopi, and pearly nautiluses), 1,000 species of sea anemones, 7,000 species of echinoderms (starfishes, sea urchins, sea cucumbers, and sea lilies), 13,000 species of fish, and 50,000 species of mollusks.

The quantity of sea creatures is immense—and because of this, commercial use and exploitation of marine animals is common. For example, global fish production exceeds that of cattle, sheep, poultry, or eggs and is the biggest source of wild or domestic protein in the world. But this high rate of use is not without cost. It is estimated that by the turn of the century 15 of the world's 17 largest fisheries will be overfished or in trouble.

marine plants— It is difficult to comprehend the number and variety of plants that exist in the ocean. For example, brown algae alone has nearly 1,500 different species; and diatoms, tiny floating plants that live in the upper levels of the ocean waters, can number as many as 6 million in a cubic foot of sea water. The number and variety of marine plants is almost as exhaustive as the number of marine animals and similarly, is beyond the scope of this book. Suffice it to say that marine plants were perhaps one of the most important developments on the face of the Earth (or underneath the surface of the oceans)—they were the catalyst that allowed the growth of animals in the oceans and on land.

ocean—On Earth, an ocean is a large body of seawater. (This definition breaks down on other planets.) Approximately 139.4 million square miles (361 x 10^6 square kilometers) of the Earth's surface is covered by oceans, or about 70.78 percent; only 57.5 million square miles (149 x 10^6 square kilometers) is dry land, or 29.22 percent. Also, the proportion of ocean to land in each hemisphere differs. In the Southern Hemisphere, the proportion of seawater to land is approximately 4:1; in the Northern Hemisphere, it is approximately 3:2. In addition, the planet's overall water is not evenly distributed. The oceans include about 97 percent of the total water on Earth, amounting to 328 million cubic miles of water. The ice caps and glaciers hold about 2 percent; 0.6 percent is in groundwater; and the rest is in lakes, rivers, streams, and soil carried in the atmosphere in the form of clouds, rain, and snow.

Largest Oceans and Seas of the World	
Ocean or Sea	*Square Miles/Square Kilometers*
Pacific Ocean	63,980,000/165,720,000
Atlantic Ocean	31,530,000/81,660,000
Indian Ocean	28,360,000/73,440,000
Arctic Ocean	5,541,000/14,350,000
Mediterranean Sea	1,150,000/2,970,000
Bering Sea	876,000/2,270,000
Caribbean Sea	749,000/1,940,000
Gulf of Mexico	699,000/1,810,000
Sea of Okhotsk	591,000/1,530,000
East China Sea	483,000/1,250,000
Hudson Bay	475,000/1,230,000
Sea of Japan	405,000/1,050,000
North Sea	224,000/580,000
Black Sea	174,000/450,000
Red Sea	170,000/440,000
Baltic Sea	162,000/420,000

ocean floor—The ocean has an average depth of 1.8 miles (3 kilometers), but there are also trenches—chasms that are deeper than the highest mountains on land—flat plains, and rolling hills. Oftentimes, there are mountain ridges, volcanic mountains, or vents. And throughout the oceans, one of the most amazing areas—and one that is still mostly a mystery to us—are the abyssal plains.

The abyssal plains, the flat or gently sloping areas of the ocean basin floor, are often called some of the flattest places on Earth. The plains are adjacent to the continental rises, and are usually found at depths between 7,200 and 20,000 feet (2,200 and 5,500 meters). Because of their proximity to continental rises, scientists know that the locations of abyssal plains have changed over the Earth's natural history, as the continents (or crustal plates) have changed position. Today, the majority of abyssal plains are found in the Atlantic and Indian Oceans, as they receive the most sediment, mainly from erosion of the nearby continents. Abyssal plains also occur at the bottom of the deepest trenches, as fine clays settle out along the deep ocean floor.

The plains are usually covered with fine ooze blankets, a type of mud composed mainly of fine sediment and organic (plant and animal) remains. Some of the sediment settles out from the ocean water above. Turbidity currents from submarine canyons along the continental slope also deposit fine sediment on the plains. Many marine organisms that eat the remains of plants and animals, such as certain types of worms, fish, and bivalves, survive on the abyssal plains; the result of the mud feeders' activities, such as worm tubes and feeding tracks, are readily seen in the muddy bottom floor.

ocean (or coastal) margin—The ocean margin refers to the coastline that we can see, and the continental shelf, slope, and rise that lie underwater. The following lists the distinctions between these areas:

continental shelf—The continental shelf lies just off the coastline, a relatively shallow extension of the continent that ends at the continental slope. On the average, the depth of the continental shelf ranges from about 330 to 1,600 feet (100 to 500 meters). Its width varies greatly. For example, off the coast of Florida, the continental shelf is about 150 miles (240 kilometers) wide. Off the coast of Argentina, the continental shelf is about 350 miles (560 kilometers) wide; along some areas of the east Pacific Ocean, the continental shelf barely exists. It is thought that most continental shelves were built up by the deposition of sediment as rivers carried material from the continents out to sea; others, which were covered by ice sheets during the Ice Ages, may be the result of glacial deposition.

continental slope—The continental slope is an area just past the continental shelf that drops relatively steeply to the continental rise. The slope may be very steep or gradual. The continental slopes are often cut by deep V-shaped valleys and canyons. Scientists believe the majority of the deep canyons were carved during the last Ice Age; first the sea level decreased during the Ice Age, and then meltwater from the glaciers increased as the ice sheets retreated—both actions carving the canyons. Today, the canyons are further deepened by turbidity currents. These submarine canyons are single valleys, or systems of valleys cut into the ocean landscape, similar to those found on land. The majority of submarine canyons are V-shaped, narrow valleys along the continental slope or just beyond a river delta; others are broad valleys that look like river valleys on land. Other canyon-like features are the result of past glacial action, in which the sea levels were lower, allowing the canyons to be cut more readily, or volcanic activity, in which a volcanic magma or vents created a canyon.

continental rise—The continental rise is the area just below the continental slope and above the abyssal plains; it is also called the extreme edge of a continent. The rise is very narrow in most regions, but can be up to 370 miles (600 kilometers) wide. Depending on the area, the rise ranges in depth from 4,600 to 16,700 feet (1,400 to 5,100 meters). Much of the material found on the rise comes from sediments from the continental shelf that are carried down the continental slope by turbidity currents or else, more gently, under the influence of gravity.

plate tectonics—See MOUNTAINS, Topic Terms

sea level—Because the ocean is one continuous body of water, the level of the ocean surface attempts to seek the same level throughout the entire world. But many factors prevent the surface from being level, such as tides, landmasses, river discharges, waves, and even variations in gravity. A baseline measurement had to be made to determine the variations in these levels—and thus, the concept of local mean sea level was developed.

Sea level is the height of the ocean's surface at a certain spot and is dependent on varying conditions. In the United States, sea level was determined by taking hourly measurements of sea levels over a period of 19 years in various locations. The readings were then averaged for all the measurements, and sea levels were determined. The 19-year period is based on the Metonic cycle, a natural lunar cycle mostly determined by the Moon's declination, or the height of the Moon in the sky. Sea level height variation along a coastline is usually small, but there can be up to 65 feet (20 meters) of variance depending on conditions. The average water level at a certain place, through all tidal and wave conditions, is called the mean sea level (MSL).

Sea level is also used as a reference in measuring altitude—the height above or below the ocean's surface at a certain spot is described as its location above or below sea level, even if the spot is on land and far from any ocean.

Heights above and below sea level vary greatly from place to place as does the actual sea level. For example, between Nova Scotia and Florida, there is a difference in sea level of 16 inches (40 centimeters); and between southeast Japan and the Aleutian Islands, the difference is about 3 feet (1 meter). The highest point above sea level on land is Mt. Everest, which measures 29,022 feet, 7 inches (8,846 meters) above sea level. (This measurement was established in 1993.) One of the lowest elevations on land is Death Valley, California, which measures 280 feet (85 meters) below sea level. The greatest depth below mean sea level in the oceans is the Mariana Trench, measuring 36,198 feet (11,033 meters) below sea level.

Sea levels change over time. During the last Ice Age about 18,000 years ago, when much of the water was locked up in thick polar ice sheets, the oceans were some 330 feet (100 meters) lower than they are today. Thus, sea levels were lower during that time. And if the seas continue to rise over the next few thousand years, sea level measurements will also continue to change.

During the past century, the global sea level has apparently risen as much as 8 inches (21 centimeters). No one knows what caused, and continues to cause, the rise in sea level. Many scientists believe that global warming, caused by human pollutants and particulates released in the atmosphere, is responsible for the current rise in sea level. If this is true, worldwide warming will continue to raise global temperatures. This will result in both melting of the ice in the polar regions and expanding of the sea water due to the warmer temperatures—and thus, cause the sea levels to rise. Other scientists believe that the rise in sea level may be from rising landmasses. As areas rise—such as areas of Scandinavia and northern North America that are rebounding from the pressures of the past Ice Age ice sheets—it could cause sea level changes. In addition, other areas, such as around Miami and New Orleans, are sinking at a rate of several inches per year. Such movements will continue in various spots on the Earth in the future.

seawater—The composition, or content, of seawater is fairly uniform throughout the oceans of the world.

Studies of marine deposits on land from all geological time periods also show this consistency, leading to the assumption that the composition of seawater has been approximately the same throughout most of the Earth's long history. This would make sense, since the early oceans flowed unimpeded by continental barriers. There are, to be sure, local differences in temperature and salinity and also some differences in the concentrations of certain substances from ocean to ocean, but in general seawater contains mostly ions of salt. Other dissolved materials make up only 0.05 percent of the total content, but the trace ions from the dissolved materials represent almost all of the elements.

Dissolved sodium chloride (NaCl), or salt, is the most prevalent substance in seawater: The salinity of seawater is the amount of dissolved inorganic minerals (salts) in seawater, such as sodium chloride and magnesium chloride. Although salinity is about 35 parts per thousand by mass in the oceans, it varies greatly. For example, the salinity of subtropical seas such as the Mediterranean is higher because there is more evaporation; it is lower in areas that receive freshwater from rivers or melting ice.

Seawater is made up of water; sodium, calcium and magnesium chlorides; sodium, magnesium and calcium sulfates; and calcium carbonate. These ions account for more than 99.95 percent of the total weight of all the ions present in seawater; the rest are all found in trace (parts per million, or ppm) amounts. Salinity of the ocean waters has played a major part in the development of the Earth's natural history. The presence of salt in the oceans is thought to have come from the natural erosion of elements within rock. In more modern times, salt was known to humans and was even used by Neolithic man, as they harvested salt in salt pans, and used the product for trade.

seas—A general term used to designate a body of water. Seas may refer to a specific part of an ocean; the ocean as a whole; or large inland bodies of water that may, or may not, have an ocean access. Some land-locked "seas" are really lakes. Examples of seas include the Sea of Japan, the South China Sea, and the Norwegian Sea.

tides—Tides on Earth are influenced by the Moon and also somewhat by the Sun; without the Moon, Earth would experience much lower tides, as the Sun's effect is not great. In general, as the Moon rotates around Earth, its gravitational force raises an oceanic bulge on the side of Earth facing the Moon, and an opposite rise (due to "centrifugal" force) on the other side of the planet. Because Earth rotates on its axis, the surface sweeps beneath the two bulges (or troughs), forming the changing high and low tides.

Characteristics of Seawater

Most Common Ions in Seawater

Ion	Percent of total salt (by weight)
Cl^-	55.04
SO_4^{-2}	7.68
HCO_3^-	0.41
Br^-	0.19
H_3BO_3	0.07
Na^+	30.61
Mg^{+2}	3.69
Ca^{+2}	1.16
K^+	1.10

Trace Elements in Seawater

Elements	Concentrations
Strontium (Sr^{+2}) ion	8 ppm (parts per million)
Oxygen (as O_2)	4.6-7.5 ppm varies with depth; the greatest concentration is at the surface)
Silicon (Si^{+4}) ion	3 ppm
Fluorine (F^-) ion	1.3 ppm
Nitrogen (N_2, NO_2^-, NO_3^-, NH_4^+)	0.5 ppm
Argon (Ar)	0.5 ppm
Lithium (Li^+) ion	0.17 ppm
Phosphorus (HPO_4^{-2}, $H_2PO_4^-$) ions	0.017 ppm
Iodine (I^-) ion	0.06 ppm
Carbon (in CO_2)	traces

In general, tides are broken into spring and neap tides depending on the phase of the Moon and, to a lesser extent, the position of the Sun. The size of the tides mainly depends on the Moon's position in the lunar cycle.

   * *spring*—Spring tides occur at the new or full phases of the Moon, or when the Sun, Earth, and Moon are "in line" with each other. The average tides along Earth's shorelines are highest during the spring tides.

   * *neap*—Neap tides occur at the first- and third-quarter Moon phases, or when the Sun and Moon are at right angles to each other. The average tides along Earth's shorelines are lowest during the neap tides.

Because Earth is spinning "under" the spring and neap tides, the tides are further broken down into diurnal, semi-diurnal, and mixed tidal cycles, depending on the number of high and low tides per day. A diurnal tidal cycle is when a region experiences one high tide and one low tide a day, as in the Caribbean; in a semi-diurnal tidal cycle, as in most Atlantic coastal estuaries, there are two high and low tides a day. In parts of the Pacific and Indian Oceans, there are mixed cycles—tidal cycles that are even more complex. Many times, when high tides occur with the arrival of a major storm, the resulting tide can be even higher than usual.

The largest tidal ranges are found in bays and estuaries, on the average, ranging from a few feet to more than 10 feet; enclosed seas—or even lakes—experience tides, but the change in water level is negligible. The Bay of Fundy in Nova Scotia has the largest tidal range in the world, sometimes measuring more than 50 feet (15 meters).

The importance of tides through the Earth's natural history is threefold: The tides are instrumental in bringing sediment from the estuaries to the oceans and in eroding the shorelines (thus changing the coastlines of the world), and they have created an entire zone of organisms that count on the changing tides to survive.

trench—Trenches are very deep, V-shaped areas in the Earth's crust, with the deepest found underwater. These geologic features are a result of continental plate movement. As the plates are forced together, one plate is pushed underneath the other in a process called subduction, creating a trench. In the majority of cases, the resulting pushing and movement creates a zone of seismic and volcanic activity. The more well-known trenches include the Tonga, Puerto Rico, and Mariana Trenches. The Mariana Trench is located in the Pacific Ocean near the Mariana Islands. It was first measured in 1899; the latest measurement places its deepest point at 38,635 feet (11,708 meters)—or approximately 7.3 miles (11.7 kilometers)—below sea level.

Ocean Trenches

Name	Ocean	Approximate Depth (feet/meters below sea level)
Mariana Trench	Pacific	38,635/11,708
Tonga Trench	Pacific	35,505/10,822
Japan Trench	Pacific	34,626/10,554
Kuril Trench	Pacific	34,587/10,542
Mindanao Trench	Pacific	34,439/10,497
Kermadec Trench	Pacific	32,963/10,047
Puerto Rico Trench	Atlantic	30,184/9,200
Bougainville Deep	Pacific	29,987/9,140
South Sandwich Trench	Atlantic	27,651/8,428
Aleutian Trench	Pacific	25,663/7,822

turbidity currents—Turbidity currents are underwater avalanches that carry huge amounts of sediment from the continental shelf down the submarine canyons of the continental slope and then onto the continental rise. Most of the turbidity currents that originate on the continental shelf originate around areas of high sediment deposition, such as a river delta. The avalanches can be precipitated by a number of events: a sudden influx of river sediment, such as from a major storm or flooding; localized earthquakes; a buildup, then release of sediment within a canyon; or even organisms, such as whales, slamming into the sides of a canyon wall (although this is highly debated). No matter the mechanism, once the surge in water or disturbance occurs, it breaks up unconsolidated sediments, causing the turbidity currents. The hundreds-of-feet-thick, heavy sediment and water mixture can travel several miles in length and width, moving down the slope at speeds of up to 50 miles (80 kilometers) per hour. These avalanches are thought to be one of the major erosional factors along the continental slopes and rises—and the reason for the formation of new submarine canyons and the lack of sediments in already formed canyons.

vents—See VOLCANOES, Topic Terms

waves—Waves are the large scale movement of water molecules that manifest themselves in the form of crests and troughs; they are generated in water by a variety of sources. The crests and troughs are aligned perpendicular to the direction of the motion of the waves. The height of the crest above a neutral point is the amplitude; the measurement between the height of the crest and the bottom of the trough is the wave height. The distance from crest to crest (or trough to trough) is the wavelength.

Waves can be divided into two categories: Small waves with wavelengths less than 0.78 inches (2 centimeters) are called ripples. Longer waves, also known as gravity waves, are classified by the depth of the water through which they move. There are shallow-water waves such as tidal or river bores and deep-water waves. Deep-water waves are generated by storms or seismic activity, have wavelengths of several miles, and have large amounts of energy; they travel very rapidly and are deceptive since the surface is not changed much by their passage. But when these waves encounter shallow water, the enormous amount of energy in them translates into large, dramatic rises in the sea level that pound the shoreline.

25. Plants

Introduction

The world is filled with all types of plants—and almost all of them have something in common: They are one of the most important organisms on the planet, not only in modern terms, but also in terms of the natural history of the Earth. The first multicellular life on the planet began in the oceans in the form of plants. They produced the oxygen that eventually led to life on land. And now, they still provide one of the main links in the chain for the survival of all organisms on the Earth, by producing oxygen and taking up carbon dioxide, and by providing food—and even shelter and hiding places—for many other organisms. In other words, without plants to essentially bind the world together, the Earth would have no natural history.

Timeline

(note: bya=billion years ago; mya=million years ago)

Date	Event

Prehistoric Events

Date	Event
~3 bya	Photosynthesis probably starts at this time.
~1.5–1 bya	Oxidizable rock becomes saturated, allowing free oxygen to stay in the atmosphere; the ozone layer also forms around this time.
~470 mya	The first plants adapt to a land-water interface, such as tidal pools; it is the first step for plants to move from the oceans to land.
~430 mya	Vascular plants—those with roots, stems, and leaves—evolve.
~420 mya	The first true plants colonize land and include flowerless mosses, horsetails, and ferns.
~345 mya	Ferns eventually develop seeds.
~145 mya	The first flowering plants evolve on land.
~100 mya	Flowering plants begin to dominate the land.

Modern Events and Discoveries

Date	Event
~7000 B.C.	Agriculture, the planned cultivation of plants, begins in Mesopotamia.
~1500 B.C.	Egyptians produce a systematic arrangement of the known medicinal plants.
~300 B.C.	Greek scientist Theophrastus (c.372–c.287) publishes his botanical works, *Historia plantarum* and *De causis plantarum*, in which he describes, identifies, and classifies 550 different plants; he is known at the father of botany.
~40 A.D.	Greek physician Pedanius Dioscorides (c.20–c.90) writes *De materia medica*, about the medical properties of approximately 600 plants and nearly 1,000 drugs; he also makes the first recorded use of anesthesia.
1500	German-French scientist Hieronymus Brunschwig (c.1450–c.1512) publishes an account of the methods and apparatus needed to treat plants in order to obtain their medicinal benefits; he writes an expanded version in 1512.
1539	German scientist Hieronymus Tragus (Jerome Bock, 1492–1554) arranges plants by relation or resemblance, the first known attempt at a natural classification of plants.
1544	German botanist Valerius Cordus (1515–1544) leaves *Historia plantarum* for posthumous publication; it includes descriptions of 500 newly identified plant species.
1551–1565	Swiss naturalist Konrad von Gesner (also known as Conrad Gessner, 1516–1565) apparently distinguishes genus from species and order from class in his classification of plants, a practice that would not be introduced until 1623 by G. Bauhin; his works, *Opera Botanica* and *Historia Plantarum*, would influence such later taxonomists as Linnaeus and Cuvier.
1554	Italian naturalist Ulisse Aldrovandi (1522–1604) proposes a systematic study of plant classification in his publication *Herbarium*.

Date	Event
1583	Italian physician and botanist Andrea Cesalpino (also known as Andreas Caesalpinus; 1519–1603) suggests a plant classification in his publication *De plantis*; it classifies plants according to roots and fruit organs.
1596	Li Shi-Chen (?) writes the book, *Ben-zao Gang-mu*, in which he describes more than 1,000 plants and 1,000 animals.
1623	Swiss biologist Gaspard Bauhin (1560–1624) introduces the two-name system of classifying plants, one for genus, one for species, in his book *Pinax theatri botanici*; it is called binomial classification.
1667	English naturalist John Ray (1627–1705) classifies plants based on the number of seed leaves; he is known as the "father of English botany."
1694	German botanist Rudolph Jakob Camerarius (1665–1721) elaborates on the plant classification of Andrea Cesalpino; he classifies them based on the nature and number of stamens and pistils in the flower.
1694	Botany professor Joseph Pitton de Tournefort suggests a classification system, describing more than 8,000 plants; it would be the accepted classification system until Carolus Linnaeus in 1735.
1727	English botanist Stephen Hales (1677–1761) publishes a report, *Vegetable Staticks*, in which he explains his data on plants; he is one of the founders of the study of plant physiology.
1735	Swedish botanist Carl von Linné (Carolus Linnaeus) (1707–1778) introduces the first classification system to keep track of organisms on the Earth in his book, *Systema naturae*.
1737	Von Linné explains his method of systematic botany in the book *Genera plantorum* (*Genera of Plants*); he classifies 18,000 species of plants in the book.
1779	Dutch botanist Jan Ingenhousz (1730–1799) determines that the green portions of a plant give off oxygen, and with no light, the roots, flowers, and fruits give off carbon dioxide; he also determines that plants obtain carbon from the atmosphere, not the soil.
1789	French plant taxonomist Antoine-Laurent de Jussieu (1748–1836) sorts plants under a more natural classification system, classifying them into families such as grasses, lilies, and palms.
1793	German botanist Christian Sprengel (1750–1816) describes the plant pollination process, noting the importance of wind and insects in cross-pollination.
1805	German explorer and scientist Baron (Friedrich Wilhelm Heinrich) Alexander von Humboldt (1769–1859) publishes *Essays on the Geography of Plants*, concerning his five-year voyage to the Americas; it is the first reference to the field of plant geography.
1831	Scottish botanist Robert Brown (1773–1858) discovers the plant cell nucleus.
1838	German botanist Jacob Schleiden (1804–1881) announces his work on the development of plant cells, suggesting the existence of a cell nucleus where new cells develop and grow.
1868	French agricultural chemist Jean-Baptiste-Joseph-Dieudonné Boussingault (1802–1887) points out that plants require carbon dioxide for photosynthesis.
mid-1800s	German biologist Hugo von Mohl (1805–1872) develops the cell theory for plants, including finding the fibrous structure of the cell walls.
late 1800s	U.S. naturalist Luther Burbank (1849–1926) develops many new varieties of plants, including the Burbank potato, Shasta daisy, and new varieties of plums and berries.
late 1800s	U.S. scientist George Washington Carver (1864–1943), among other plant studies, works on plant diseases and soil and crop management practices that would eventually increase yields for farmers.

Date	Event
late 1800s	German botanist Julius von Sachs (1832–1897) determines that chlorophyll in plants is only found in small bodies (later named chloroplasts); he also discovers how plant chlorophyll works.
1878	Scientists show that microorganisms are responsible for changing ammonium in fertilizers into nitrites and nitrates needed by plants.
1898	U.S. biologist Charles Reid Barnes proposes the term "photosynthesis."
1901	Dutch plant physiologist and geneticist Hugo de Vries (1848–1935) discovers gene mutations in plants.
1906–26	German chemist Richard Willstätter (1872–1942) and coworkers discover the chemical structure of chlorophyll pigments.
1930	The parallelism between photosynthetic processes in bacteria and green plants is suggested.
1934	Dutch botanist Friedrich August Ferdinand Christian Went (1863–1935) discovers the existence of plant hormones.
1970s	Cloning plants from protoplasts becomes an active field of research.
1985	The first bioinsecticide is announced by the United States Department of Agriculture.
1990s	Chloroplasts' DNA and RNA are studied in order to further the classification of plants.
1998	Scientists discover a naturally cloned shrub thought to be 43,000 years old; if confirmed, it will be the world's oldest living plant.

History of Plants

Early Plants

Once photosynthetic organisms evolved, life started to have a major impact on the Earth's environment. These organisms took in the atmospheric carbon dioxide; at the same time, they started to produce oxygen through the **photosynthetic** process. For a long time, the oxygen was taken up by rocks and did not build up in the atmosphere. By about 1.5 to 1 billion years ago, the oxidized rock became saturated, allowing the free oxygen to stay in the air. This oxygen also formed the ozone layer, which made it possible for marine organisms to spread to land. Previously, the oceans acted as a block against the Sun's harmful radiation; the atmosphere's ozone layer later played the same role.

After the Cambrian Explosion of life about 544 million years ago, the oceans teemed with multicellular **plants** and animals. But the land remained empty of life, except for an occasional microbe—probably in shallow pools. Animals had no incentive to colonize the land, because there was nothing for them to eat there. So it was up to the plants to make the great leap from the oceans and become the first multicellular organisms to live on dry land.

The first primitive **nonflowering plants** (or gymnosperms) semi-adapted to a land-water interface, probably on the edge of tidal pools, about 470 million years ago, according to recent findings. But these plants were much different than modern plants: They were small and hugged the ground; they had no adaptations for saving or transporting water throughout their system (such as a xylem or roots); they had no leaves or stems; and they had no real protection against drying out. These characteristics had to evolve over the next millions of years in order for plants to survive and spread onto land. In particular, scientists believe liverworts—rootless patches of thin, leaf-like plants, were some of the first to adhere to rocks on semi-wet lands. (We still have liverworts in the world today, such as the common and braided liverworts in the United States; but so far, scientists do not know how the familial relationship among the more than 8,000 present species fits in with the lineage of the ancient liverworts.)

Vascular plants—those with roots, stems, and leaves—did not evolve until about 430 million years ago, but they were still mostly confined to the oceans at that time. Fossils reveal that the first true plants to colonize land appeared about 420 million years ago and included flowerless mosses, horsetails, and ferns. They reproduced by throwing out minute spores that carried the genetic blueprint for the plant. The ferns eventually developed reproduction by seeds, but that didn't happen until about 345 million years ago.

By about 220 to 130 million years ago, ferns and horsetails covered the ground; ginkgoes and tree ferns lined rivers and lakes; and cycads, conifers, and sequoias forested thousands of square miles of the drier lands. During the Jurassic period, there was an increase in abundance of the cycads, or short palm-like plants, (it is often referred to as the "Age of Cycads") and conifers, with most of the modern plant families evolving, such as cypresses, redwoods, yews, and junipers. The first truly modern ferns also appeared during the Jurassic period. In the early Cretaceous period about 145 million years ago, the cycads began their decline, leaving the conifers to dominate.

Not long afterward, in terms of geologic time, the angiosperms, or **flowering plants**, made their first appearance—considered the biggest environmental

This modern cycad is similar to those that evolved millions of years ago.

change during this time. Where flowering plants came from is still a mystery. One theory is that they evolved from a gymnosperm ancestor; another theory is that a seed-bearing fern was the immediate progenitor. One reason for the quandary is the lack of evidence: **Flowers** are very scarce as fossil remains until the late Cretaceous period.

The angiosperms' rise in numbers and diversity was explosive. About 110 million years ago, they were a small part of the world's vegetation, which was then dominated by gymnosperms; by 100 million years ago, many of the flowering plant families appeared—and the group was on its way to dominating the globe. The flowering plants diversified during the middle Cretaceous, taking over many areas that once held only the ferns and horsetails; by the end of the period 65 million years ago, angiosperm diversity had surpassed that of conifers. Of the 500 modern families of flowering plants, 50 appeared during the Cretaceous period, including the sycamore, magnolia, palm, holly, and trees of the willow and birch family.

The reason for the rapid spread of the flowering plants during the Cretaceous period may have been the dinosaurs. Some scientists suggest that as larger dinosaurs of the early Cretaceous period began to trample the low-growing angiosperms, the plants grew back very rapidly—similar to today's flowering plants that respond to cutting by growing even more profuse. In addition, insects, passing dinosaurs, flying or gliding reptiles, and wind probably carried the angiosperm seeds to other areas.

Modern Plant Studies

Most of the earliest studies of plants dealt with the connection between plants and medicine, and plants and agriculture. Agriculture began some 9,000 years ago in the fertile crescent of Mesopotamia. Written records of the medical use of plants also began to emerge about the seventh century B.C., in which some 700 medicinal and semi-medicinal plants were arranged by the Assyrians according to use and application. By 1500 B.C., Egyptians produced a systematic arrangement of the known medicinal plants.

The next large push in plants was to create a classification system to keep track of the organisms: Around 300 B.C., Theophrastus published a botanical work identifying, describing, and classifying 550 different plants—and is known as the father of botany. And even through the sixteenth century, systematic studies of plant classifications continued—with many of the classifications highlighting the plants around certain regions of the world.

By the late eighteenth century, botanists began to focus on the processes of the plants, such as the discovery of the plant **pollination** process in 1793, in which the importance of wind and insects in cross-pollination was noted. In addition, with better ability to explore the world, better plant classifications were determined as more scientists found and studied plants from around the globe.

In the late nineteenth century more discoveries were made concerning the mechanisms within plants. For example, Julius von Sachs determined that chlorophyll in plants is only found in small bodies (later named **chloroplasts**)—and that it was the chlorophyll that actually changed carbon dioxide and water into starch while releasing oxygen. The way microorganisms convert a fertilizer's ammonium compounds into nitrates and nitrites necessary for plant growth was also discovered. During this time, too, many new varieties of plants were found. One of the more well-known scientists who worked on developing a multitude of new plant varieties was Luther Burbank, who developed such varieties as the Burbank potato, Shasta daisy, and several types of plums and berries.

More recent plant studies deal mainly with genetic discoveries—or just the discovery of new plants altogether. For example, scientists know that the plants called liverworts were probably the first plants to colonize dry land because of recent research into plant genetics. For a long time, it was thought that plants took one giant step onto land, as opposed to the apparent repeated transitions of animals. Thus, they believe that the first plant on land is the ancestor of all living plants. Mosses and liverworts, both of which are related to the ocean's green algae, seemed to be the best candidates—but the fossil record was not good enough to come to any conclusions.

Enter genetic research, which focused on extraneous pieces of genes called introns, which are found in more than 300 modern plants. Over the course of evolution, introns have "pushed" their way into the genes of plants; they get "cut out" of the gene before it makes a protein. Scientists narrowed their research to three ancient introns, none of which are present in green algae; trees, flowers, mosses, and some other common modern plants have at least two introns. But the liverworts lacked all three ancient introns, making them the closest relatives to water-loving green algae. And because they knew green algae is one of the oldest types of plants (and organisms, for that matter), scientists concluded that liverworts were the first plants to colonize land. See also ATMOSPHERE, BACTERIA AND VIRUSES, CLASSIFICATION OF ORGANISMS, CYTOLOGY, DINOSAURS, EARTH, FOSSILS, GENETICS, GEOLOGIC TIME SCALE, LIFE, OCEANS, and PROTISTA.

The Strange of the Plant World

The plant world has many strange features—from the age of certain plants, and even to the behavior of plants. For instance, amazing as it may seem, many plant species seem to live a very long time—and many of these include tree species. For example, of the 850 different species of trees in the United States, the oldest is the bristlecone pine (*Pinus longaeva*), which can live probably more than 5,000 years. The oldest, called PNW-114, was found on Mount Wheeler, Nevada, and was thought to be 5,000 years old, but it is now gone; the oldest still living is a bristlecone pine nicknamed "Methuselah" in southern California's White Mountains—and is thought to be 4,700 years old.

Another long-living tree is the Redwood (*Sequoia sempervirens*), which can reach about 1,000 to 3,000 years old. Giant sequoias (*Sequoiadendron giganteum*) can live to about 2,500 years old; the Douglas fir (*Pseudotsuga menziesii*) to 750 years; and the bald cypress (*Taxodium distichum*) can live for about 600 years.

The tallest tree ever recorded was an Australian eucalyptus (*Eucalyptus regnans*) located at Watts River, Victoria, Australia. The claim was made in 1872, by a forester named William Ferguson. He measured the tree to be 435 feet (132 meters) tall, and it was probably originally 500 feet (150 meters) in height where it stood. The current highest tree still standing is thought to be the "National Geographic Society" coast redwood in Redwood National Park, Humboldt County, California—at 373 feet (113 meters) tall as of October 1990.

And there are even plants that are carnivorous, with between 450 to 500 species and 12 genera of carnivorous plants—all classified by their trapping mechanisms. For example, the Venus flytrap (*Dionaea muscipula*) and bladderworts (*Utricularia vulgaris*) are active traps, using a rapid motion to capture their prey. The two leaf-looking projections of the Venus flytrap snap together in response to an insect in its clutches. The butterwort (*Pinguicula vulgaris*) is a semi-active trap; it traps its pray by catching the insect in its adhesive fluid. As the insect struggles, the plant slowly tightens its grip. And finally, the passive traps usually lure insects by nectar—such as the pitcher plants that attract the insects into a reservoir of fluid where they drown.

Probably one of the strangest plants was discovered in Australia in 1998—a naturally cloned shrub thought to be 43,000 years old. If confirmed, it will be the world's oldest known living plant. The shrub, called King's Holly (*Lomatia tasmanica*), was found on Australia's island state of Tasmania. The plant, located in a 0.4-square-mile (1-square-kilometer) area in the rain forest, was found in 1930, but it wasn't studied in detail. The shrub is actually a self-propagating clone that doesn't produce seeds; instead, it reproduces by shedding cuttings of itself on the floor of the forest. And it is these cuttings that produce the genetically identical plants.

Topic Terms

chloroplasts—Chloroplasts are the microscopic functional units within a plant cell where photosynthesis takes place. They contain the green pigments called chlorophyll-*a* and -*b*; these pigments trap the light energy that the plant uses for photosynthesis. In the smallest, single-celled plant, there is only one chloroplast; in a plant leaf cell, there may be as many as 20 to 100.

flowering plants—The flowering plants belong to a distinct group known as the angiosperms. They probably had one evolutionary origin; or they may include two different lineages—but probably not any more (to compare, note the nonflowering plant lineage). The angiosperms include more than 80 percent of all living green plants; some consider these plants the pinnacle of plant evolution—but it is difficult to say, when one realizes that nonflowering plants have evolved and continued to survive for a longer period of geologic time.

flowers—Many parts make up a flower, the sexual reproductive unit of the plant. In general, they include the following:

nectar—The flower's organs secrete the nectar, a mixture that contains various amounts of sugar and proteins. In most flowers, it collects in the flower cup, or near the base of the cup formed by the flower parts. It is mostly "food" to attract pollinators.

petals—The often colorful petals are present mainly to attract pollinators such as insects or birds, and are usually dropped after pollination.

pistil—The pistil is the female part of the plant; it includes the stigma, style, and an ovary containing ovules. In most flowering plants, after the plant is fertilized, the ovules mature into seeds.

sepals—The sepals are usually found at the base of the flower, extending outward from the base when the flower is open, or from the bud. When the flower is in the budding stage, the sepals protect the bud from drying out; after the flower opens, the sepals often serve to ward off predators with spines or chemicals.

stamen—The stamens are the male part of the plant; they include the filament and anther (where the pollen is produced).

nonflowering plants—The nonflowering plants are an immense group—and can mean everything from a microscopic, one-celled algae to a giant redwood tree and everything in between, such as seaweeds, mosses, ferns, horsetails, ginkgos, cycads, and conifers. In other words, they are lumped together only on the basis of not having any flowers. This lack of flowers is significant, as nonflowering plants represented the first stages of plant evolution. They were more numerous in the past than now, and some were much larger. For example, giant clubmosses, relatives of today's species, made up the Carboniferous period swamps—most of which we see today as coal seams in rock layers.

Of the nonflowering plants, one of the more amazing (and difficult to distinguish from certain other organisms such as protists) are the algae. They include green, red, brown, golden-brown, and yellow-green algae, diatoms, dinoflagellates, and euglenoid protozoa. They all began as single-celled organisms, although some scientists believe the brown algae could be an offshoot of green algae. Some groups, such as the dinoflagellates, stayed single-celled, and formed loose colonies; others, such as green algae, evolved into multicellular forms.

It is clear that algae have several different lineages that evolved independently—with estimates from four to six, or even higher numbers. Some scientists believe the engulfing of photosynthetic bacteria by a larger host could have been the precursor to algae, with a different bacterium within each host. In fact, algae are usually classified by their color, or by the pigments involved in photosynthesis. And these pigments are believed to be the relics of those pigments that the original bacteria contained—such as some cyanobacteria that carried around red phycobilin pigment that would later evolve into certain red-colored algae.

photosynthesis—Photosynthesis is a process used by plants for food; the green plants use light energy from the Sun (or special, full spectrum artificial light) to synthesize sugar (carbohydrates) from carbon dioxide, water, and other elements with oxygen as a byproduct. The sugar is manufactured in the presence of chlorophyll, a molecule uniquely capable of turning light energy into glucose (sugar) that is stored as food.

plants—Plants (Kingdom: Plantae, with about 400,000 species) are multicellular organisms and are eukaryotic. They include chlorophyll-*a* and -*b*, which convert the Sun's energy to food for the plant; and they are diverse in numbers, shapes, sizes, and colors. They are found everywhere on the planet except in the polar regions, on the highest mountains, and in the deepest oceans.

Plants are necessary for life on Earth, primarily because they provide a critical link in the food chain and produce oxygen for the higher, more complex organisms on the planet. In fact, scientists estimate that almost 90 percent of the living mass on Earth is made up of plants. The earliest plants were the spore plants, or those that produced asexually or sexually with spores (microparticles) that grew into new plants. Spore plants still exist, but they are not as prolific as the seed plants.

Seed plants are divided into the gymnosperms and angiosperms. The gymnosperms, or "naked seeds," are the simplest seed plants. There are about 1,000 modern species of these plants, and they include the evergreen trees. The angiosperms, or "enclosed seeds," include about 250,000 species and are collectively called the flowering plants. They are some of the most advanced plants on the planet and include grasses, palms, orchids, and broad-leaved trees. The most common are wild and garden flowers and hardwood trees.

In general, the parts of the plant include the following (with the exception of the parts of a flower, which are listed above):

fruit—The fruit of a plant (and not all plants have a fruit) is a way for the plant to disperse its seeds. After the plant is fertilized (pollinated), the ovary begins to develop into a fruit, while the ovules develop into seeds.

leaf—Leaves act as light catchers for the plants; they also obtain and store water and food, and provide a place for photosynthesis to take place.

root—Roots range from a large, single, tap root to a mass of small, similar-sized roots; they grow in the soil by cellular division, usually at the root tip. The root has two major functions in most plants: to pull water and nutrients into the plant from the soil (or other medium); and to anchor the plant to the ground. The roots absorb mineral salts and continually take up water from the soil; after all, 90 percent of more of the water they take up is lost through transpiration (the evaporation of water and water vapor into the atmosphere from the leaves). Roots are also important to plants that reproduce asexually—they can easily split apart to form new plants.

seeds—After the plant is fertilized (pollinated), the ovary begins to develop into a fruit, while the ovules develop into seeds. These seeds are carried away by birds, the wind, and certain animals, and are usually dropped, eventually growing into a new plant.

stem—The stem of a plant is responsible for producing and supporting the new leaves, branches, and flowers; these structures are positioned at the best places to receive the optimum amount of light, water, and warmth.

Plants are often classified into 12 phyla based largely on their reproductive characteristics: by tissue structure into nonvascular (mosses) and vascular plants (all others); by "seed" structure into those that reproduced by naked seeds, covered seeds, or spores; and by stature, divided into mosses, ferns, shrubs and vines, trees, and herbs. The majority of these plants are flowering herbs. The general divisions of plants are as follows:

- mosses and allies (mosses, liverworts, hornworts)
- ferns and allies (horsetails, club mosses, whisk ferns)
- conifers and allies (cycads, ginko, herb-like cone-bearing plants)
- flowering dicot plants (trees, shrubs, vines, and flowers)
- flowering monocot plants (orchids, lilies, palms, grasses)

pollination—Pollination occurs when viable pollen is transferred to a plant's stigmas, ovule-bearing organs, or ovules (seed precursors). It occurs in most plants and is usually carried out in a multitude of ways. Although some plants are able to self-pollinate, most rely on other methods. These methods, including passing pollen by insects, wind, water, birds, or brushing on other animals, are necessary since plants are immobile organisms. The transfer of pollen by insects and by other animals or processes is called cross-pollination, in which one plant's pollen is moved to another plant's stigma. Flowers have developed many methods to attract such animals to the pollen. For example, certain orchids use color and smell to attract a certain type of bee and wasp species; certain insects will try to mate with the flower because the ruse is so convincing. Some flowers even attract only certain species of insects, which tends to keep the plant species pure.

26. Protista

Introduction

The single-celled protists, also called protoctist, represent some of the smallest organisms on Earth. Under the Kingdom Protista (although another classification lists the Kingdom Protoctista), scientists include the protozoans, some algae, and fungi (although some classifications only include the protozoans and some don't include fungi). For years, many of the protists were classified with other single-celled organisms, such as bacteria. But today, because of their specialized characteristics, scientists have given them a classification of their own. The organisms in the Protista kingdom are known to be extremely important to modern natural processes and to animals and plants, but they also have played a major role in the natural history of our planet. In particular, some of the first life on Earth may have been a form of protista—making these organisms some of the oldest known on our planet.

Timeline

(note: bya=billion years ago)

Date	Event

Prehistoric Events

Date	Event
~4 bya	Something sparks the first inorganic chemicals to develop into primitive "cells."
~3.75 bya	The earliest form of life is thought to have evolved around this time; although recent unconfirmed findings show it may be closer to 3.85 billion years ago.
~3.5–3.2 bya	Primitive bacteria may have lived in ancient hot springs around volcanoes.
~2.1 bya	The Earth's atmosphere becomes sufficiently oxygen-rich to support an ozone layer; this layer would have a profound effect on the evolution of life on the planet, as it would protect organisms from the solar radiation.
~2 bya	The first cells begin to form in greater numbers and develop into simple unicellular microorganisms.
~1.5 bya	The first eukaryotic cells form; they have a nucleus and complex internal structures and are the precursors to protozoa, algae, and all multicellular life.

Modern Events and Discoveries

Date	Event
1677	Anton van Leewenhoek (1632–1723) uses his compound lens microscope to discover protists; in 1696, he publishes *Mysteries of Nature*, noting his findings of these small creatures he called "animalculae."
1846	German zoologist and parasitologist Karl Theodor Ernst von Siebold (1804–1885) determines that protists are single-celled organisms; but he incorrectly assumes they are the basis for all other organisms.
1848	Von Siebold establishes protozoa as the basic phylum of the animal kingdom.
mid-1800s	German biologist and microscopist Christian Gottfried Ehrenberg (1795–1876) is a pioneer in studying invertebrates, but he incorrectly proposes that protists have organ systems similar to vertebrates.
1880	The first disease caused by a protist is discovered: Malaria is discovered in the blood of a soldier.
early 1900s	Several discoveries are made connecting protist to disease generation.

History of Protists

Early Protists

The early protists, especially the **protozoans**, are thought to have been the first complex organisms on Earth, evolving from the primitive cells about 1.5 billion years ago. The entire lineage of these single-celled organisms is highly debated. One reason is that multicellular organisms evolved many times—and some of these multicellular organisms are more closely related to their ancestral single-celled lineages than to other multicelled organisms. This explains why some members of the plant kingdom, such as multicellular algae, are sometimes considered to be multicellular protists. Some fungi are more closely related to the protist lineage than their own; and some single-celled protists are classified within the animal kingdom.

Some scientists speculate that the primitive protists were both plant- and animal-like, and had the capacity to obtain food by different mechanisms, including being able to photosynthesize food internally. It is also thought that early symbiotic relationships between prokaryotes (bacteria-like organisms) and protists led to the early eukaryotes' (and eventually the multicelled organisms') development of internal structures such as chloroplasts in plants, and mitochondria—structures found in both protists and many of the more advanced eukaryotes cells.

Protista Studies

The earliest discovery of **protists** (or **protoctista**) did not truly come about until 1677, when Anton van Leeuwenhoek saw them under his crude microscope. He published *Mysteries of Nature* in 1696, calling protists "animalculae," or small animals. It took a little over a century to determine the true existence of the organisms; in 1846, Karl von Siebold determined that protists were actually single-celled organisms. But he did make an incorrect assumption about these organisms, believing they were the basis for all other organisms in the world.

In the late nineteenth century, a burst of discovery concerning not only bacteria and viruses, but also their "cousins" the protists occurred. For example, in 1880, malaria—the first **disease** known to be caused by a protist—was discovered in the blood of a soldier. More abundant and common microorganisms were also discovered, such as the **amoebas** and paramecium.

By the early 1900s, protists were known to generate many diseases, not only in humans, but also in plants and other animals, most notably in cattle, fish, game, and poultry. But similar to bacteria, some protists such as protozoans were found to be beneficial—and sometimes essential to some animals. For example, ciliates are a necessary element of the stomach in cud-chewing animals, allowing for the digestion of the cellulose in the animals' diets. In the past few decades, humans also discovered that protozoans are useful in the treatment of sewage, where the tiny organisms help to remove bacteria during processing.

Today, more than 50,000 species of protozoa have been described, and many scientists believe that these organisms represent a peak of unicellular evolution. Recent DNA studies of protists show that the organisms are even more diverse than previously believed, with some scientists suggesting that Protista should be classified as a multiple kingdom. In fact, depending on how you view it, this kingdom can be divided into 20 to 50 phyla! *See also* CLASSIFICATION OF ORGANISMS, CYTOLOGY, EVOLUTION, EXTINCTION, FOSSILS, FUNGI, GENETICS, GEOLOGIC TIME SCALE, HUMANS, LIFE, and PLANTS.

Topic Terms

amoeba—The amoeba is probably the most well-known protozoan, as it has been readily seen under a microscope in numerous biology classes across the world. (Under the classification Protoctista, an amoeba is considered a fungus-like protoctista.) The microorganism moves by pushing out stubby projections called pseudopods (essentially "fake foot") from its body; the cytoplasm, or the liquid within the cell, flows into the "foot," enlarging it until all the cytoplasm has moved, and thus, the amoeba as a whole has moved. Their feeding habits are just as simple: As a food particle (such as a bacterium) is perceived, a pseudopod reaches out and encloses the food with a food vacuole. Digestive enzymes from the microorganism break down the food; what is not used is discharged by the vacuole and left behind—and the amoeba moves on. These organisms reproduce by mitosis and fission, and will grow this way as long as food and moisture are available.

diseases—Many diseases are caused by protists, including one of the most common and widespread of all human diseases—malaria. In this case, the protozoan *Plasmodium* causes the problem. The organism is transmitted by female *Anpheles* mosquitoes that feed on infected human blood after mating.

protoctista—Besides the classification of Protista, some scientists use a kingdom classification called Protoctista. In this classification, the true algae are included with the protists, along with the water molds, slime molds, and slime nets. These algae are primitive photoautotrops, thought to have evolved into modern flowering plants; in addition, these algae are thought to have generated 50 to 70 percent of the Earth's oxygen.

The following is only one classification of the Kingdom Protoctista (note: some classification systems consider several of these algae as plants):

Kingdom Protoctista

Subkingdom: Phycobionta
  Division: Xanthophyta (yellow-green algae)
  Division: Chrysophyta (golden-brown algae)
  Division: Dinophyta (dinoflagellates)
  Division: Bacillariophyta (diatoms)
  Division: Cryptophyta (cryptophytes)
  Division: Haptophyta (haptonema organisms)
  Division: Euglenophyta (euglenoids)
  Division: Chlorophyta
    Class: Chlorophyceae (green algae)
    Class: Charophyceae (stoneworts)
  Division: Phaeophyta (brown algae)
  Division: Rhodophyta (red algae)

---

protists—Actually, it is difficult to define the Kingdom Protista. It is not a true grouping of organisms such as animals or plants. This is because many protists may be more closely related to animals or plants than they are to other protists. Thus, the confusion, and often debates, as to the true meaning of a protists. In fact, the protists are usually defined on the absence of characteristics, such as the lack of complex development from embryos, or the lack of cell differentiation.

Take the variability of the protista: Some protista live in extreme conditions, and discovering and understanding these protista is difficult—not only because they are hard to reach, but also because it is hard to accurately duplicate the conditions in which they are found. Because of this, scientist know little about these creatures—except that they grow in extreme conditions in which there is sufficient energy to sustain them. For example, protists called psychrophiles need water as the solvent for life. This sets their lower limit for growth slightly below 32° F (0° C). In fact, the coloring of snow can be caused by a variety of photosynthetic eukaryotes such as *Chlamydomonas nivalis* and certain dinoflagellates—all well-known illustrations of cold-adaptation. Another example is the *Heteromita globosa*, an flagellate growing in the Antarctic in a physical environment characterized by highly variable moisture and temperatures, such as short-term freeze-thaw cycles and daily temperature fluctuations of up to around 40°F (20°C) (occasionally up to 80°F [40°C]). Bacteria, at least 24 species of protists, and some lichens and mosses can grow in these conditions.

protozoans—The protozoans are single-celled microorganisms that live mostly in water or liquids. They either drift, actively swim, or crawl along within their environment; some are static or live as parasites in animals. Nearly 50,000 species have been described, with many not yet discovered; the largest can reach about 0.08 inches (2 millimeters) long. They are extremely diverse—from the well-known, blob-like amoebas to organisms with strangely shaped features for catching prey, moving, or feeding. They generally reproduce by simply dividing into two or more cells, although some occasionally exhibit sexual reproduction, in which two cells fuse to form a larger cell, then divide into smaller cells. Some are photosynthetic and thus, plant-like; the majority are nonphotosynthetic, extracting nutrients by absorbing organic detritus or other microorganisms.

27. Reptiles

Introduction

According to the fossil record, reptiles were the first creatures on our planet that were completely land-dwelling. They were extremely important to the development of many species on Earth, as they were the first animals that did not have to depend so much on water—opening the way for their spread across land areas. Probably most important to natural history was the eventual evolution of the reptiles into mammals—which would, hundreds of millions of years later, evolve into humans.

Timeline

(note: mya=million years ago)

Date	Event

Prehistoric Events

Date	Event
~340 mya	During the Carboniferous period of the Paleozoic era, a group of amphibians evolves into a form that we now call a reptile.
~320 mya	*Hylonomus*, one of the earliest known reptiles, descends from the amphibians.
~270 mya	The reptiles, with their adaptations, start to thrive at the expense of the amphibians, developing into thousands of species; the main lines established are the anapsids, diapsids, synapsids, and euryapsids.
~250 mya	A great extinction event occurs, killing off almost 96 percent of all living species on Earth, including many of the reptiles; the surviving reptiles diversify and fill vacant ecological niches.
~230 mya	The first dinosaurs emerge from the thecodont reptiles; the first mammals evolve from the endothermic (warm-blooded) therapsid reptiles.
~100 mya	Many crocodiles become massive, including the *Deinosuchus*, a large terrestrial crocodile that reached 50 feet (15 meters) in length.
~205–65 mya	Dinosaurs, a line of the reptiles, become dominant on the land; the earliest known snakes appear, and true lizards evolve and diversify.
~65 mya	Another mass extinction occurs, leading to the disappearance of the dinosaurs and most other reptiles; some groups of reptiles make it past the extinction, eventually leading to modern turtles, tortoises, lizards, snakes, and crocodiles.

Modern Events and Discoveries

Date	Event
1824	English cleric and geologist William Buckland (1784–1856) becomes the first person to publish a paper describing and naming a prehistoric reptile that subsequently came to be known as a dinosaur; the *Megalosaurus*, or "big reptile."
1825	English physician and amateur paleontologist Gideon A. Mantell (1790–1852), extrapolating from some fossil teeth found by his wife in 1822 in Tilgate Forest, England, describes an "extinct reptile," the *Iguanodon* (iguana-tooth).
1841	English anatomist and paleontologist Sir Richard Owen (1804–1892) concludes that giant prehistoric reptiles were part of a new sub-order of animals that he called Dinosauria, or "terrible reptile."
mid-1800s	Longtime rivals Edward Drinker Cope (1840–1897), a U.S. paleontologist and herpetologist, and Othniel Charles Marsh (1831–1899), a U.S. professor of paleontology at Yale University, are instrumental in discovering many new types of reptiles, particularly dinosaurs.
1887	Harry Seeley organizes dinosaurs into two suborders, based on pelvic structure: the saurischian, or reptile-hipped, and the ornithischian, or bird-hipped, dinosaurs.
1917	The fossil remains of *Seymouria* (around 280 million years old), an organism showing both amphibian and reptilian characteristics, is discovered.

History of Reptiles

Early Reptiles

The amphibians dominated most of the Paleozoic era, flourishing in the warm and wet conditions. Sometime during the Carboniferous period, approximately 340 million years ago, an unknown group of amphibians began to evolve the characteristics that would eventually enable them to become fully land-dwelling. These characteristics included watertight skin with scales, and the ability to reproduce without needing a body of water present. With amphibians, the young had to be hatched in the water and went through their larval stage there, until they metamorphosed into adult form.

Of these characteristics, the ability of **reptiles** to reproduce without needing a body of water may have been the most important. The development of the cleidoic, or **amniote egg**, allowed fertilization to occur in the female reptile's body; the young, when hatched on land, were miniatures of their parents and able to fend for themselves. The development of this type of egg freed the reptiles completely from the water, allowing them to hunt, feed, and proliferate across the planet. It also allowed their young to develop in a stable environment until they were ready to hatch and cope with the world. As a result, reptiles diversified very rapidly, producing thousands of species in the relatively short period of about 40 million years.

One of the first reptiles that we have a fossil record of was *Hylonomus*, a small, lizard-like creature resembling an amphibian. But it also had a high skull, an astragalus bone in the ankle, and evidence of additional jaw muscles—all reptilian characteristics. There was one exception though: *Hylonomus* did not seem to have that essentially reptilian trait of an amniotic egg.

Around 270 million years ago, in the Permian period, the climate became hotter and drier. The reptiles, with their adaptations, thrived at the expense of the amphibians. The reptiles quickly spread, developing into thousands of species during the late Carboniferous period and throughout the Permian period. Four main lines of reptiles were established, distinguished from each other by the openings in their skulls (low-stress areas serving as attachment points for the jaw muscles): the anapsids, synapsids, euryapsids, and diapsids.

The anapsids were the earliest group to form and had no openings in the sides of their skulls (*Hylonomus* was an early anapsid); this line survived and evolved into today's **turtles** and tortoises. The synapsids, or "same hole," had a low skull opening; they became endothermic (warm-blooded) and were the ancestors of the mammals. The euryapsids, or "one hole" skull opening, returned to the water, evolving into the placodonts, ichthyosaurs, and plesiosaurs—all marine reptiles that went extinct when the dinosaurs disappeared. The diapsids, or "two skull openings," were more diverse, giving rise to many different species of reptiles. This line branched into the lepidosaurs, which evolved into today's lizards and snakes, and the archosaurs, or "ancient lizards." The archosaurs, in turn, evolved into the thecondonts—the common ancestors of the dinosaurs, the pterosaurs, and the **crocodiles**.

Around 250 million years ago, at the end of the Permian period, a great extinction event occurred, killing off almost 96 percent of all living species on Earth, including many of the reptiles. The reptiles that survived the mass extinction were able to take advantage of the changing climate and geography and began to diversify again.

In the Mesozoic era, it was generally drier and hotter than in the previous era. Reptiles became the dominant species on Earth during the this time, giving rise to a famous line of land-dwelling reptiles known as the dinosaurs, which flourished for about 150 million years. About 230 million years ago, during the early Triassic period, the first dinosaurs emerged from the thecodont reptiles, and the first mammals evolved from the endothermic (warm-blooded) therapsid reptiles. During the Jurassic period, 205 to 145 million years ago, dinosaurs became the dominant animal on the land. In addition, flying reptiles, such as pterosaurs, diversified, as did the dominant ocean reptiles, such as plesiosaurs. The earliest bird, also a reptile, made its appearance around 145 million years ago. During the Cretaceous period, 145 million to 65 million years ago, the dinosaurs still dominated on land, with other reptiles evolving into the early ancestors of turtles, lizards, and snakes. By 65 million years ago, the dinosaurs became extinct, along with close to half of the other ocean and terrestrial reptiles on Earth—leading to the rise of the mammals.

The reptiles that survived to modern times have somewhat obscure origins. The earliest turtles evolved during the Triassic period, but they probably could not withdraw into their shells like modern turtles. **Lizards** and **snakes** have poor fossil records, probably due to the animals' tendencies to live in dry uplands, far from the areas that are most likely to produce fossils. (Most animal bones survive if they are quickly buried with sediment, such as along riverbanks.) It is thought that lizards appeared in the late Triassic; the earliest known remains of snakes are found in late Cretaceous layers of rock in North America and Patagonia, South America.

Modern Reptiles and the Study of Reptiles (Herpetology)

Similar to amphibians, the study of reptiles throughout human history has not attracted as much attention as other animals. More than anything, most studies are tied up in the subject of dinosaurs, the most prolific reptiles in the Mesozoic and a topic of intense study since the mid-nineteenth century.

Today's reptiles, like many animals present at the end of the Cretaceous period, are almost like the survivors of a shipwreck. After the Cretaceous period, most reptile species were wiped out. About 6,000 reptile species exist today, fewer in numbers and much smaller in size than their ancestors—but great in diversity.

But before we dismiss the reptiles, we should realize two things. First, reptiles still exist today, having evolved over some 300 million years. The only animals that are alive today and classified as reptiles are the crocodiles, lizards and snakes, **turtles** and tortoises, and *Sphenodon*. All the others have become extinct over the millions of years of history. Some scientists also believe that birds are truly dinosaurs, and thus are considered unique reptiles.

Second, the mammals did not just appear out of thin air—it was a separate line of reptiles that evolved specific characteristics that eventually became today's mammals. On the tree of evolution, it was the reptiles' division into the four groups (anapsids, euryapsids, diapsids, and synapsids) that eventually led to most of the animals on the earth today—from birds and whales, to humans. Thus, mammals—today's dominant species—have the reptiles as their common ancestor. The true reptiles—crocodiles, snakes, lizards, and turtles—have not gotten as much attention over the years as the other descendants of the early reptiles. *See also* AMPHIBIANS, ANIMALS, BIRDS, CLASSIFICATION OF ORGANISMS, DINOSAURS, EVOLUTION, EXTINCTION, FOSSILS, GEOLOGIC TIME SCALE, and LIFE.

Topic Terms

amniote egg—This type of egg, known as cleidoc (closed), or amniotic, has a hard shell, with numerous tiny pores that allow air to enter and waste gases to leave, but keeps the inside from drying out as long as the surroundings are humid. Inside the shell are specific areas, lined with a membrane, that are necessary for the development of the young embryo. They in-

The common garter snake, *Thamnophis sirtalis*, is usually found in grasslands, marshes, woodlands, and even in suburban backyards. Two forms of this nonpoisonous snake are found in the eastern United States.

clude the amnion, a sac which contains the liquid that acts as a private pond for the developing embryos; the yolk, which provides food for the embryo; and a membrane, which contacts the air that has diffused into the shell, providing air to the embryo. The development of this type of egg millions of years ago freed the reptiles completely from the water.

crocodiles—One remarkable group of reptiles—also close relatives of the dinosaurs—are the crocodiles. They appear to have evolved from archosaurian ancestors during the late Triassic, but unlike most of their contemporaries, they survived through to the present. They are also remarkable because these moderate to large-sized semi-aquatic predators have remained relatively unchanged since the Triassic period.

Crocodiles—and alligators—are found only in certain areas of the world. The two types of modern crocodiles are found in tropical and subtropical environments. The gavialids, found in India, eat fish and have slender snouts. The crocodylids are found almost worldwide and consist of crocodiles and alligators. They have long bodies and powerful tails used for swimming or defense. Their limbs allow the animals to maneuver and steer in the water; on land, they use their limbs to walk with a slow gait, with their bellies held high off the ground. These animals eat a wide variety of food, including fish, large vertebrates, and carrion.

Although the habitats of alligators and crocodiles overlap occasionally, it is rare to see both together. The best way to tell the difference between the two animals is by checking the size and head: Crocodiles are slightly smaller and less bulky than an alligator. In addition, the crocodile has a larger, narrower snout, with a pair of enlarged teeth in the lower jaw that fit into a "notch" on each side of the snout. The alligator has a broader snout, and all the teeth in its upper jaw overlap with those in the lower jaw.

lizards—About 3,000 species of lizards are found around the world. Most have four legs, but others, such as the glass, legless, and worm lizards, have none. They are usually active during the day, but similar to most ectothermic (cold-blooded) animals, they often bask in the Sun for warmth. Their bodies range from slim to massive, from the small short-horned lizard (from 2 to 5 inches [5 to 13 centimeters] long) to the huge Gila monster (from 18 to 24 inches [46 to 61 centimeters] long). They usually have smooth or spiny scales, with the scales varying in color based on region, age, and sex. In some lizard species, the males are larger and brighter in color than the females. They usually hibernate in colder regions' soil in the winter; most lay eggs that hatch in the late summer or early fall. Their diet depends on the species: Many eat other animals, such as insects, spiders, and scorpions; the large Gila monster feeds on rodents and the eggs and young of ground-nesting birds.

reptiles—Modern reptiles include the alligators and crocodiles, turtles, lizards, and snakes. They all have several typical characteristics: They have a protective covering of scales or plates, five clawed toes on each foot (with exceptions, of course, such as snakes), and lungs instead of gills. Most lay eggs (although most poisonous snakes in the United States, except coral snakes, produce live young), and eat animals (except for a few species like the land tortoise).

The classification of modern reptiles is short, as there are not as many species of these creatures as other groups. The following lists two of the more commonly used classifications of the reptile class. (Note: The more ancient classes of the reptiles, including those of the dinosaur [Archosauria], do not have common names associated with the classification.):

Classification of Reptiles

Version 1: Class Reptilia

Order Testudines (turtles)
Order Sauria
Order Lepidosauria
Order Rhynchocephalia
Order Squamata (snakes and lizards)
Order Archosauria
Order Crocodylia (alligators and crocodiles)
Order Aves (birds)

Version 2: Class Reptilia

Subclass Anapsida
    Order Testudines (turtles)
        Suborder Cryptodira
        Suborder Pleurodira
Subclass Lepidosauria
    Order Rhynchocephalia (tuataras)
    Order Squamata
        Suborder Sauria (lizards)
        Suborder Serpentes (snakes)
        Suborder Amphisbaenia (amphisbaenians)
Subclass Archosauria
    Order Crocodylia (crocodiles)

---

snakes—Snakes are somewhat of a mystery—not to mention an attraction to some humans. Snakes have no legs (although the rubber boa and related species have spurs, or short remnants of legs); most species molt (shed their skin) several times a year; they have no eyelids; and they rely only on animals for food—unlike other reptile groups that usually include some plant-eating species. In addition, certain snakes cannot only injure, but have been known to kill humans. Thus, these animals inspire fear and fascination among humans.

The anaconda is the biggest snake in the world—even bigger than the Old World python. It is found in the tropical rivers of South America, mostly the Amazon and the Orinoco. They average about 16 feet (5 meters) long, with the longest recorded at over 33 feet (10 meters)—and they can weigh more than 550 pounds. Constriction is their favorite way to kill their prey, with their diet including large rodents, wild pigs, deer, birds, fish, and aquatic reptiles.

turtles—Turtles are most often found in or near water, especially resting on logs that are partially submerged. The turtle literally carries its home on its back; its home is actually a bony shell covered with hard shields called scutes. The upper part of the shell is called the carapace and the lower part, the plastron. Not all turtles have the same types of shells: Some of the softshell turtles that spend most of their time in the water have smaller shells. Unlike all other reptiles, turtles are all toothless (although they have a kind of "beak" that can cause a painful bite). They pop their head in and out of the shell, usually in response to danger; their diet consists of other small animals (such as mollusks for the pond turtles) as well as aquatic plants. North America has about 50 species of turtles—three of which belong to the tortoise family. You can tell the tortoises by their short legs, webless feet, and high-domed shells—and they only live on land.

28. Solar System

Introduction

The natural history of the solar system is obviously connected to Earth's natural history. The study of the planets (the composition and origins), satellites, Sun (its changes over time, and how it affects Earth), and other space objects (such as comets, asteroids, and stars) help scientists determine the composition, and often conditions, of the early Earth and solar system. The natural history of the solar system is intertwined with its planets and satellites and the universe at large. Most of our knowledge of the planets, satellites, galaxy, and universe is a reflection of our own planet. We base many of our interpretations of the atmospheres, chemistry, and physical features of the planets and satellites on the Earth and Moon—our only real examples of how the solar system developed and evolved over its natural history.

Timeline

(note: bya=billion years ago)

Date	Event

Prehistoric Events

~10–5 bya	A primitive solar nebula forms in this part of the Milky Way galaxy.
~4.55 bya	The solar system begins to form from the primitive solar nebula.
~4.0–3.8 bya	The Late Heavy Bombardment occurs, creating many of the impact basins viewed on our Moon and other satellites in the solar system.

Modern Events and Discoveries

~150 B.C.	Greek astronomer and geographer Hipparchus (c. 170 B.C.– c. 120 B.C.) works out the epicycle model of the solar system.
~140 A.D.	Egyptian astronomer Ptolemy (Claudius Ptolemaeus, c.100–c.170 A.D.) pushes the idea that Earth is the center of the universe, although he did not originate the idea; his resulting Ptolemaic system would influence astronomy for close to 2,000 years.
1543	Polish astronomer Nicolaus Copernicus (1473–1543) starts a revolution in astronomy by circulating the work *De revolutionibus orbium coelestium* ("On the revolutions of celestial bodies"); his idea is that Earth and other planets travel around the Sun; this concept went against the established idea that Earth was the center of the universe.
late 1500s	Danish astronomer Tycho Brahe (1546–1601) makes some of the most accurate positional measurements of stars and planets—all before the invention of the telescope.
1609	German astronomer Johannes Kepler (1571–1630), assistant to Tycho Brahe, formulates the principal laws governing the motions and orbits of planetary bodies.
1609	Italian astronomer and physicist Galileo Galilei (1564–1642) is the first to use a very crude telescope; he will use it, some say, as an astronomical instrument the next year; he discovers four of Jupiter's largest moons, Saturn's rings (although he identified them as "handles"), and the phases of Venus.
1682	British astronomer and physicist Edmond Halley (1656–1742) discovers the comet (and its 76-year period) that bears his name—Halley's Comet.
1694	Halley submits a paper to the Royal Society of England suggesting that Noah's Flood occurred as a comet passed unusually close to Earth, causing a tidal wave.
late 1700s	The Titius-Bode law is published by German astronomer Johann Elert Bode (1747–1826), a calculation originally determined by German astronomer Johann Daniel Titius (1729-1796); it mathematically determines the distances to the planets in astronomical units.
1783	French mathematician, astronomer, and physicist Marquis de Pierre-Simon Laplace (1749–1827) works out equations of motion of celestial bodies—including the planets of the solar system—using gravitational, electromagnetic, and other potentials; his theory of solar system development, called Laplace's nebular hypothesis, suggests that the planets formed from the same primitive mass of material.

Date	Event
early 1900s	U.S. astronomer Percival Lowell (1855–1916) proposes the idea of life on Mars, based on his observations at the Lowell Observatory in Flagstaff, Arizona, a facility he established.
1917	English mathematician, astronomer, and physicist Sir James Hopwood Jeans (1877–1946) first develops the theory that gravity from a passing star might have pulled material from the Sun, creating the planets and other members of the solar system.
1930	U.S. astronomer Clyde William Tombaugh (1906–1997) discovers the planet Pluto, the only planet discovered in the twentieth century.
1950	Dutch astronomer Jan Hendrik Oort (1900–1992) proposes the theory of the Oort Cloud, a collection of comets that surrounds the solar system far outside the orbit of Pluto.
1991	The first extrasolar planet is found by astronomers Alex Wolszczan and Dale Frail around pulsar PSR B1257+12 in the constellation Virgo.
1994	U.S. astronomer Carolyn Shoemaker (1929–), her husband, U.S. astrogeologist Eugene Shoemaker (1928–1997), and astronomer David Levy, discover Comet Shoemaker-Levy 9, the only known observed object to strike a planet—Jupiter.
1995	The first extrasolar planet found around a star similar to our own, 51 Pegasi, is discovered by astronomers Michel Mayor and Didier Queloz.

History of the Solar System

Early Solar System

The solar system is thought to have developed from a primitive solar nebula of dust and gases. Simply put, as the nebula slowly rotated, the center began to contract. As it rotated even faster, conserving its angular momentum, the nebula soon collapsed—eventually igniting in a nuclear reaction and forming the proto-sun (early sun-like star). The flattened disk of dust and gas continued to spin and became hot, especially closer to the proto-sun; within this area, only hardy substances with high melting temperatures—rock and iron—could condense. The results were planetesimals, asteroid-like objects that eventually smashed together (accreted), forming larger bodies, including the inner planets of Mercury, Venus, Earth, and Mars. Others either crashed into one another and broke apart, or were ejected from the solar system by collision or the gravitational effects of larger bodies.

In the cooler regions of the nebula, ice began to condense. This ice became part of the newly forming gaseous **planets**; with the extra mass, the outer planets attracted even more gases, especially hydrogen and helium, from the nebula. Jupiter and Saturn attracted the most gases, with Uranus and Neptune collecting far less.

One of the major events in the early solar system was the Late Heavy Bombardment. This bombardment happened about 4 to 3.8 billion years ago, after the planets had accreted and began to cool; large and small planetesimals (large, sometimes planet-sized, chunks of rock) began to bombard the **terrestrial planets** of the inner solar system. The same bombardment probably occurred in the **gaseous planets** of the outer solar system, but because of their physical make-up it is difficult to tell if they were affected at the same time.

Evidence of the bombardment is seen on many of the planets and **satellites** of the solar system. Two of the best examples are Mercury and our Moon. For example, based on rock samples from the Moon and analysis of the cratered landscape, scientists believe that during the first few hundred million years of the Moon's history, the bombardment caused outer layers of the lunar surface to completely melt several hundred miles deep. Earth itself shows little, if any, evidence of the Late Heavy Bombardment. Billions of years of erosion and deposition, continental movements, and volcanics have erased any evidence of these effects.

Other smaller leftovers from the formation of the solar system include **meteoroids**, small chunks of rock traveling around the system. Scientists have also theorized the existance of two other major leftovers from the formation. The first is the Kuiper belt objects, a collection of asteroid-comet-like objects that orbit the **Sun** just outside the orbits of Pluto and Neptune. Another collection of objects, most likely similar to com-

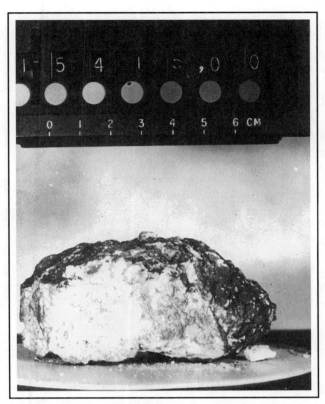

This rock is one of the oldest known on Earth—but it is from the Moon. The "genesis rock" is around 4 billion years old, a discovery made by using scientific dating tools. *(Photo courtesy of NASA)*

ets, surrounds the solar system about 100,000 astronomical units (an astronomical unit is 93 million miles) away from the Sun in an area called the **Oort cloud**. It is thought to be the source of long-term **comets**.

Studies of the Solar System

Although the planets Mercury, Venus, Mars, Jupiter, and Saturn were discovered in the ancient past as they were naked-eye objects, the other planets, being dimmer, took more time to find.

Around 150 B.C., Hipparchus worked out the epicycle model of the solar system, an idea that would exist in various forms and dominate astronomy for centuries. In this system, the other planets and the Sun move in their own circles in the sky (epicycle); all the planets and the Sun, in turn, revolve around Earth. Hipparchus also played an important role in the sciences, discovering the precession of the equinox (the natural wobble of Earth, like a spinning top), calculating the length of a year within 6.5 minutes, and devising the first known star map. Egyptian astronomer Claudius Ptolemy (Claudius Ptolemaeus) continued the idea of epicycles, and added one special factor—that Earth was the center of the universe (although he did not originate the idea). His model of the universe, with the planets revolving in epicycles, and then all circling Earth in large circles called deferents—would influence astronomy for close to 2,000 years.

True astronomical studies began around the sixteenth century, with the list of influential astronomers continuing to the present. Although many scientists contributed to the advancement of the field, it was truly Polish astronomer Nicolaus Copernicus who started the revolution in astronomy by circulating the work *De revolutionibus orbium coelestium* ("On the revolutions of celestial bodies"). His idea that Earth and other planets traveled around the Sun went against the established idea that Earth was the center of the universe and that everything revolved around a stationary Earth. Because of fear of reprisals from the church—who thought it was heresy to believe that Earth was not the center of all things—Copernicus showed his work only among the scientific community. As Copernicus neared death, he was finally convinced to publish the book. The Copernican system was quickly embraced as astronomical observations proved that the Sun was the center of the solar system.

Some of the most accurate positional measurements of stars and planets—all made before the advent of the telescope—were carried out by Danish astronomer Tycho Brahe, one of the bright spots in the history of astronomy. (The telescope was not used for astronomical purposes until around 1610, by Galileo Galilei.) Brahe also determined an alternative model of the solar system, combining the Aristotle model (who still maintained, like Hipparchus and Ptolemy, that Earth was at the center of the solar system, also known as the geocentric model) with the Copernican model. Brahe placed all the planets in orbit around the Sun, with a sphere of fixed stars centered on Earth. The model had many flaws and was soon forgotten due to the eventual popularity of the Copernican model.

German astronomer Johannes Kepler was one of the greatest astronomers of all time; an assistant to Tycho Brahe, he inherited (and expanded on) all of Brahe's planetary and stellar observations after the Danish astronomer's death. Because he believed in the Copernican model of the solar system, Kepler formulated the three principal laws governing the motions and orbits of planetary bodies. His work eventually helped to eliminate the epicycle models that had governed astronomy for close to 2,000 years, but only posthumously.

Astronomy entered a new era thanks to Italian astronomer and physicist Galileo Galilei, one of the world's most famous scientists; he was the first to introduce experimentation into science, thus laying the foundation for science as we practice it today. From his scientific observations, he deduced several basic laws (for example, in a vacuum, all bodies fall at the same speed, no matter what their weight or constitution), and he invented the first thermometer. He was the first to use a very crude telescope as an astronomical instrument (in 1609, although some say he did not use the instrument astronomically until the next year) and discovered four of Jupiter's largest moons (the Galilean satellites), Saturn's rings (although he identified them as "handles"), and the phases of Venus. His belief in the Copernican model caused problems between Galileo and the Catholic church. The church refused to accept that the Sun was at the center of the solar system and forced Galileo to retract his Copernican beliefs.

The Sun-centered idea was gradually accepted, especially by the astronomical community, and by the 1700s, scientists were searching the heavens for new planets. One impetus to this search was the **Titius-Bode law**—a calculation originally worked out by Johann Titius and published by Johann Bode in the late 1700s—that mathematically determines the distances to the planets along the ecliptic in astronomical units. Although Johann Titius wrote about the association earlier in a book, Bode was the first to bring it to the attention of the scientific community. (Bode originally saw it mentioned in a footnote in Titius's book; once called Bode's law, it is now called the Titius-Bode law.)

The law is yet to be truly explained, as nature is not often so

---

The Unusual in the Solar System

The solar system has always held a fascination for those of us on Earth. And some of the more recent occurrences within the system have made many of us realize that science fact can be just as strange as science fiction. Here are a few examples:

"big bombardments": Earth is bombarded with hundreds of tons of meteoric debris every day. Because most of the microparticles are too small to see, the evidence of the dust is not obvious. In addition, most of the material falls into the oceans, where it mixes with ocean water and eventually settles to the ocean floor. But scientists know that bigger objects are out there. And with the discovery of Shoemaker-Levy 9, now thought to be a fragmented comet, scientists have become more aware of the impact of space objects on other planets. That comet broke up into about 20 fragments, with each piece plunging into the atmosphere of Jupiter in July 1994. It was the first time humans had ever witnessed the impact of a space object on another planet—including Earth.

Kuiper belt: This is a theoretical collection of small bodies found beyond the orbits of Neptune and Pluto. The Kuiper belt, named after astronomer Gerald Kuiper, is thought to be remnants from the formation of the solar system. Several objects were found in 1992 that appear to be from the Kuiper belt. One of the possible telltale indications that the object 1992 QB1 was from the belt was that the orbit appeared to be nearly circular—and it's thought that the Kuiper belt objects have more circular orbits as they are farther from the gravitational effects of the Sun and planets. About over a dozen objects have been found since that time, with some bodies around 128 to 192 miles (200 to 300 kilometers) in diameter.

zodiacal light: Particles of interplanetary dust can be seen at dusk around the spring equinox and at dawn around the fall equinox. The zodiacal light is seen only in the Northern Hemisphere and is caused by meteoric particles that concentrate along the plane of the solar system's planets.

solar wind: The solar wind is also considered interplanetary gas and is composed of charged particles emitted by the Sun. The effects of the solar wind are visible when observing comets: A comet's tail of gas and dust is pushed away from the Sun by the solar wind as the comet swings around our star. The invisible particles affect Earth and other planets in several ways. For example, at times of high solar wind output, usually when the Sun becomes more active during its about 11-year solar cycle, the particles interact with Earth's upper atmosphere and magnetic poles, producing the aurora.

snowballs above the earth: Images taken by a trio of visible and ultraviolet cameras aboard the NASA Polar spacecraft have discovered a previously unknown class of objects. Apparently, a steady stream of cosmic snowballs—what some scientists have called mini-comets—bombard the Earth every day. Most of the objects disintegrate as they approach within 14,913 to 4,971 miles (24,000 to 8,000 kilometers) of Earth, depositing huge water vapor clouds in the upper atmosphere. The minisnowballs hit the atmosphere five to 30 times a minute, and some are reportedly about the size of a house. If the estimates are true, it would take about 10,000 to 20,000 years of such bombardment to raise the water all over the world about an inch. The data about the snowballs are still coming in. But many questions about these snowballs have made some scientists skeptical: Why have such large objects not been found by ground-based telescopes that watch the sky for such fast moving objects? Why has no one detected such strikes on the Moon, as such large objects would cause noticeable impacts? And most important, what is the affect—past, present, and future—of such a bombardment on Earth if it is truly taking place?

mathematically orderly. At the time that the law was developed, scientists did not know about the **asteroids**. Because the Titius-Bode law predicted that a planet should be between Mars and Jupiter, astronomers began a search for a planet in this region; thus, the asteroid belt was found—the first asteroid, Ceres, found by Giuseppe Piazzi in 1801. And even before the discovery of asteroids, the new planetary discoveries of Uranus and Neptune fell into step with the Titius-Bode law.

The search for another planet beyond Neptune became a major task for many astronomers, including U.S. astronomer Percival Lowell in the early 1900s. But it was U.S. astronomer Clyde William Tombaugh who finally discovered the planet Pluto in 1930—the only planet discovered in the twentieth century. He also searched for Planet X, a possible planet orbiting beyond Pluto. (The reason for the initial search were perturbations in the orbits of the outer planets; such disturbances in orbits are indicators that objects are exerting a gravitational pull on these planets.)

Tombaugh's search, and other astronomer's attempts since then, have been unsuccessful. But recently, asteroid and comet-sized bodies have been found in the far fringes of the solar system. They may not be considered a "planet" but may account for the perturbations. This group of objects is in the so-called Kuiper belt; and it took until August 1992 to find the first potential Kuiper belt object—a 23.5 magnitude object called 1992 QB1, found by David Jewitt at the University of Hawaii and Jane X. Luu of Stanford University.

Probably the most important discoveries concerning the planets have occurred in the latter part of the twentieth century, as spacecraft flew by or landed on the various planets and satellites in the solar system. Every planet and satellite, with the exception of Pluto and its moon Charon, has been visited; with these astronomical data, scientists have pieced together many new ideas about the overall solar system. For example, the Voyager spacecraft discovered **ring systems** around several planets besides Saturn, additional moons, and physical data, such as magnetic fields, strange geologic features, and atmospheric phenomena on certain planets and satellites.

More recent discoveries have turned from the inside to outside of our solar system: the search for **extrasolar planets**. The first extrasolar planet was found in 1991 by astronomers Alex Wolszczan and Dale Frail around pulsar PSR B1257+12 (a rapidly spinning neutron star) in the constellation Virgo. It is 1,300 light years from our solar system and has at least three Jupiter-sized planets. The first extrasolar planet found around a star similar to our own was 51 Pegasi, discovered in 1995 by astronomers Michel Mayor and Didier Queloz. It is in the constellation of Pegasus and is about 50 light years away from our solar system. Since then, over a half-dozen more extrasolar planets have been reported, although some are still in dispute. *See also* EARTH, MOON, and UNIVERSE.

Topic Terms

asteroid—An asteroid is a large rocky body—from about 600 miles (965 kilometers) wide to the size of a boulder. They are often called minor planets because of their large size (smaller chunks of rock are called meteoroids) and their tendency to orbit along the same plane, or ecliptic, as the major planets. Most of the asteroids revolve around the Sun in a tight band, called the asteroid belt, between the orbits of Mars and Jupi-

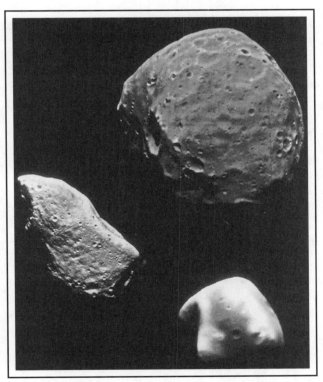

This collage represents, from lower left clockwise, the elongated asteroid Gaspra (the first asteroid imaged by a spacecraft flyby), and the moons of Mars—Phobos (the larger moon) and Deimos. *(Photo courtesy of Dr. Peter Thomas, Cornell University)*

ter. Italian astronomer Giuseppe Piazzi discovered the first asteroid, Ceres, in 1801.

The majority of these space objects are thought to have formed at the beginning of the solar system, about

4.55 billion years ago. Several theories have been advanced as to the origin of asteroids. Scientists once believed that the asteroids were leftovers from a shattered planet, but further analysis showed few, if any, sources—internal or external—would have had enough energy to crack a planet. A more accepted theory is that the asteroids formed at the same time as the other planets, but the pull of Jupiter did not allow huge chunks of rock called planetesimals to create a planet. Instead, the rocks settled between the orbits of Mars and Jupiter, colliding and creating the smaller asteroids we see today.

The majority of the asteroids stay within the asteroid belt. Over hundreds of thousands of years, because of the gravitational pull of the planets or other space objects, an asteroid may stray from the belt. Such asteroids that come close to Earth or cross Earth's orbit are called near-Earth asteroids.

It is known that in the past some near-Earth asteroids struck Earth, creating impact craters on the planet's surface; the larger impacts also seem to be associated with the extinction of a large number of species during the planet's history. The Meteor Crater in Arizona is an good example of an impact crater formed by an asteroid. To date, more than 6,000 asteroids have been found; of those, close to 300 are near-Earth asteroids. Scientists estimate there may be more than 10,000 asteroids in the solar system, and more than a 1,000 near-Earth asteroids.

comets—Comets are a collection of dust, gases, and ice that orbit the Sun. Once described as "dirty snowballs," many comets are now thought to be more like "mudballs," most carrying more dust than ice. In general, comets are composed of carbon dioxide, frozen water, methane, ammonia, and materials such as silicates and organic compounds.

Comets are relatively common in the solar system. Two comets that were readily visible to the naked eye added to our knowledge of comets. Comet Hyakutake in 1996 was merely a ball of ice, measuring only about 1 to 1.8 miles (2 to 3 kilometers) in diameter. Comet Hale-Bopp, about 10 times as big as Hyakutake, sped by the Sun in 1997 and was one of the brightest comets ever to be viewed.

Short-period comets, or those with orbits from a few to 200 years, are thought to have originated in the Kuiper belt, a fat disk of objects that exist beyond the orbits of Neptune and Pluto. Long-term comets, or those that travel into the solar system once in thousands of years (or may never return at all), are thought to originate in the Oort cloud, a theoretical cloud of comets that surrounds the solar system about 100,000 astronomical units from the Sun. This cloud was first proposed by Dutch astronomer Jan Oort.

eclipses—See MOON, Topic Terms.

ecliptic—The apparent path the Sun, Moon, and planets take in the sky. As seen from Earth, the ecliptic runs through the 12 major constellations of the zodiac: Taurus, Gemini, Cancer, Leo, Virgo, Libra, Scorpius, Sagittarius, Capricornus, Aquarius, Pisces, and Aries. Over hundreds of thousands of years, Earth's polar axis naturally shifts position in the sky because of gravitational tugs by the Moon and Sun. Because of this, the ecliptic will appear to move up or down in the sky. In reality, the ecliptic itself is not moving—Earth's axis shift will cause the ecliptic to be in different locations in the sky.

extrasolar planets—Planets around stars that are beyond our solar system are called extrasolar planets. Not all stars have solar systems; and it was only recently that scientists had conclusive physical proof of planets around other stars. Just over a half-dozen more extrasolar planets have been reported, although some are still in dispute.

Extrasolar Planet Discoveries		
Extrasolar Planet	Year	Distance from Central Star (astronomical units)
PSR B1257+12 b	1991	0.36
PSR B1257+12 c	1991	0.47
PSR B1257+12 a	1993	0.19
51 Pegasi b	1995	0.05
47 Ursae Majoris b	1995	2.1
Rho Cancri A b	1996	0.11
Tau Bootis A b	1996	0.05
Upsilon Andromedae b	1996	0.06
Rho Coronae Borealis b	1997	0.23

gaseous planets—Because of the thick blanket of gases that surround the outer planets of Jupiter, Saturn, Uranus, and Neptune, they are collectively called the gaseous planets. All gaseous planets have a dense atmosphere surrounding a small, dense core. It is thought that these planets gained their gases during the formation of the solar system. As the solar nebula disk's gas and dust revolved around the newborn Sun, ice began to condense in the cool outer reaches of the nebula. This ice became part of the newly forming gaseous planets; with the extra mass from the ice (and just being larger planets), the planets attracted even more gases, especially hydrogen and helium, from the nebula. Jupiter and Saturn attracted the most gases, with Uranus and Neptune collecting far less.

meteoroids—Meteoroids usually refer to the smaller chunks of rocks in space; they are also often used erroneously in reference to asteroids that strike Earth. Meteoroids that are seen streaking and leaving a short, seconds-long trail in the night sky are called meteors. (The common names include "falling star" or "shooting star.") A collection of meteors that radiate from a certain spot in the sky (radiant) at a particular time of the year is called a meteor shower. A meteor shower is produced as Earth's orbit passes through a collection of meteoroids, usually rocky remnants associated with the orbit of comets.

A meteor that is large enough to land on the surface of the planet is called a meteorite. Most meteorites are thought to have originated as chunks of rock broken off of the asteroids. Others, called the Shergotty-Nakhla-Chassignys (SNGs), are thought to have originated from the planet Mars after large asteroids or comets impacted the red planet's surface. Only about a dozen SNGs have been found on Earth. In 1997, an SNG found in Antarctica revealed possible evidence of life when it was analyzed. If scientists do prove that the small microbe-looking features are indeed evidence of life, it will be the first direct indication of life found on another planet besides Earth.

Oort cloud—Dutch astronomer Jan Oort proposed the theory of the Oort cloud in 1950—a collection of comets that surrounds the solar system far outside the orbit of Pluto, at a distance of about 10,000 light years. Because of the distance, the actual discovery of the cloud may be difficult. It is thought that long period comets come from the Oort cloud and that the Oort cloud originated as the leftovers from the formation of the early solar system.

planets—(See also individual planet within this definition.) The planets, in order from the Sun, are Mercury, Venus, Earth, Mars, Jupiter, Saturn, Uranus, Neptune, and Pluto. Here are some planetary highlights (Moon numbers are only approximate for most of the outer planets):

Mercury—Mercury is the first planet from the Sun; the average distance from Mercury to the Sun is 36 million miles (57.9 million kilometers). It is the most cratered planet in the solar system, its surface resembling our Moon. The heat of the Sun plus Mercury's low gravitational field caused the planet's original atmosphere to boil off into space, but the planet still has an extremely thin atmosphere composed of hydrogen and helium from the action of the Sun's solar wind on the surface of the planet. It is only about one-millionth-billionth the density of the earth's atmosphere.

Mercury's Characteristics	
Diameter (average—miles/kilometers)	3,031/4,878
Equatorial gravity (Earth = 1)	0.28
Year (revolution)	88 days
Day (rotation at equator)	58.64 days
Mass (Earth = 1)	0.0553
Moons	0

Venus—Venus is the second planet from the Sun; the average distance from Venus to the Sun is 67.2 million miles (108.1 million kilometers). Venus's atmosphere is extremely thick and is composed of 96 percent carbon dioxide, 3.5 percent nitrogen, and traces of sulfur dioxide, water vapor, argon, and carbon monoxide. The thick, dense atmosphere produces great pressures on the planet's surface—almost 90 times greater than at sea level on Earth's surface. In addition, the carbon dioxide creates a runaway greenhouse effect, with temperatures at the surface reaching close to 900° F (482° C). Venus also spins on its axis in a retrograde (from east to west, opposite of most planets) direction; scientists do not know why Venus spins in the opposite direction of the majority of bodies in the solar system.

Venus's Characteristics	
Diameter (average—miles/kilometers)	7,521/12,104
Equatorial gravity (Earth = 1)	0.88
Year (revolution)	224.7 days
Day (rotation at equator)	243 days (retrograde)
Mass (Earth = 1)	0.815
Moons	0

Mars—Mars is the fourth planet from the Sun; the average distance from Mars to the Sun is 141.5 million miles (227.9 million kilometers). The planet has a thin atmosphere, which has a reddish hue from the presence of iron. The atmosphere is composed mainly of carbon dioxide (95 percent), with 2.7 percent nitrogen, 1.6 percent argon, 1.3 percent oxygen, and 0.3 percent water vapor. Because carbon dioxide is an efficient radiator of heat when thin, temperatures are very cold—to as low as -207.4°F (-133°C) at the winter poles. It has two polar caps, composed either of water-ice or dry-ice (carbon dioxide).

Similar to Earth, Mars undergoes seasonal changes: In the spring, certain areas are less red, becoming almost green or gray, and the polar caps begin to shrink; in the Martian winter, the polar caps grow. Because Mars has no oceans, the surface responds quickly to changes in temperature; massive dust storms often appear in the summer months.

Mars's Characteristics	
Diameter (average—miles/kilometers)	4,217/6,787
Equatorial gravity (Earth = 1)	0.38
Year (revolution)	687 days
Day (rotation at equator)	24 hours, 37 minutes, 22.3 seconds
Mass (Earth = 1)	0.1074
Moons	2

Jupiter—Jupiter is the fifth planet from the Sun; the average distance from Jupiter to the Sun is 483.3 million miles (778.3 million kilometers). The huge gaseous planet accounts for the major proportion of the solar system's orbiting planetary mass—about 71 percent. Jupiter is essentially a immense ball of gas with a dense, but small, central core. In general, the atmosphere is composed mainly of hydrogen and ices of water, ammonia, and methane. The planet also emits more energy than it receives from the Sun, by about a factor of one, probably due to shrinkage of the planet's core. A thin ring system also surrounds the planet.

Jupiter's Characteristics	
Diameter (average—miles/kilometers)	88,730/142,800
Equatorial gravity (Earth = 1)	2.69
Year (revolution)	11.86 years
Day (rotation at equator)	9 hours, 50 minutes
Mass (Earth = 1)	317.83
Moons	~17 (more may be in the ring system)

Saturn—Saturn is the sixth planet from the Sun; the average distance from Saturn to the Sun is 886.2 million miles (1,427 million kilometers). In general, the atmosphere is composed of 93 percent hydrogen and 7 percent helium, which is why its density is only 0.69. (Water has a density of 1; Saturn would float if one could find a big enough body of water.) Saturn also emits more radiation than it receives from the Sun, putting out 1.76 times more energy than it receives. Saturn also has the most extensive ring system in the solar system; spacecraft images have revealed that the ring system is actually composed of thousands of rings, some found to contain small moonlets that were previously unknown.

Saturn's Characteristics	
Diameter (average—miles/kilometers)	74,900/120,540
Equatorial gravity (Earth = 1)	1.19
Year (revolution)	29.46 years
Day (rotation at equator)	10 hours, 39 minutes
Mass (Earth = 1)	95.18
Moons	~22 (more moons may be in the planet's giant ring system)

Uranus—Uranus is the seventh planet from the Sun; the average distance from Uranus to the Sun is 1,783.1 million miles (2,869.6 million kilometers). The atmosphere is composed mainly of hydrogen, helium, and methane. Uranus appears to have elements within its core that produce heat, as the planet emits more heat than it receives from the Sun by a factor of 0.1. Uranus's axis differs from the other planets: Like a rolling billiard ball, it is tilted so that the planet rotates almost in the plane of its own orbit. Uranus also has an extremely narrow ring system.

Uranus's Characteristics	
Diameter (average—miles/kilometers)	31,763/51,118
Equatorial gravity (Earth = 1)	0.79
Year (revolution)	84.01 years
Day (rotation at equator)	17 hours, 54 minutes
Mass (Earth = 1)	14.53
Moons	~17

Neptune—Neptune is the eighth planet from the Sun; the average distance from Neptune to the Sun is 2,794 million miles (4,496.6 million kilometers). The atmosphere is composed mainly of hydrogen, helium, and methane; white clouds of methane crystals have been detected in the upper atmosphere of the planet. Neptune emits more heat than it receives from the Sun, by a factor of 2.8. The planet also has its own ring system, consisting of about three rather dim rings, with some particles clumping into arc-like sections.

Neptune's Characteristics	
Diameter (average—miles/kilometers)	30,775/49,528
Equatorial gravity (Earth = 1)	1.12
Year (revolution)	168.79 years
Day (rotation at equator)	19 hours, 12 minutes
Mass (Earth = 1)	17.13
Moons	~8

Pluto—Pluto is the last planet from the Sun; the average distance from Pluto to the Sun is 3,666 million miles (5,900 million kilometers). (Note: In 1976, Pluto's eccentric orbit moved it within Neptune's orbit, making the blue planet Neptune the farthest known planet in the solar system; in 1999, Pluto moved to a more distant orbit—becoming once again the farthest planet in the solar system.)

Pluto's Characteristics	
Diameter (average—miles/kilometers)	1,430/2,300
Equatorial gravity (Earth = 1)	0.04
Year (revolution)	247.69 years
Day (rotation at equator)	6 days, 9 hours
Mass (Earth = 1)	0.002
Moons	1

ring system—All of the outer planets except Pluto have ring systems, accumulations of ice, rocks, and dust that orbit the planet in a specific belt. Saturn has the most extensive and noticeable ring system; Jupiter, Uranus, and Neptune have much dimmer rings. The origin of ring systems is still debated, but two general theories have been proposed: 1) They are debris left behind after a satellite was torn apart from the gravitational pull of the planet; or 2) they are material that did not coalesce into a satellite at the time the planet was forming.

satellite—A satellite is a small body that orbits a larger body. Natural satellites are also called moons. Only two of the planets in the solar system have one moon (Earth and Pluto); Mercury and Venus are the only planets without moons. Overall, more than 60 moons have been found around the planets of the solar system—with many more thought to exist.

The majority of the satellites move in approximately circular orbits, lining up with the plane of the equator of the major planet. According to astronomers, most satellites formed about the same time that the planets themselves formed. Other moons—usually the smaller ones—that follow more irregular paths (inclined orbits) may have been captured by the gravity of the planet, although it is not known how a moon could be captured and eventually go into orbit around a planet. In addition, certain moons have characteristics that are difficult to explain, including the retrograde (orbiting in the opposite direction) orbit of Neptune's moon, Triton.

Our own moon is an enigma. Unlike the other planets and satellites of the solar system, the Moon is very large compared to its mother planet. (Pluto and its moon Charon are the only exception—another very large moon compared to the mother planet's size.)

Sun—See UNIVERSE, Topic Terms.

terrestrial planets—The terrestrial planets are Mercury, Venus, Earth, Mars, and Pluto. These planets all have a thin or no atmosphere; they are composed mostly of rock (as opposed to the gaseous planets). It is thought that these planets gained their rock and iron masses during the formation of the solar system. As the solar nebula disk's gas and dust revolved around the newborn Sun, asteroid-like bodies called planetesimals began to form. (Around the same time, in the outer regions of the solar system, the gaseous planets were beginning to attract more gases.) The planetesimals collected together, or accreted, forming the four inner terrestrial planets of Mercury, Venus, Earth, and Mars. Pluto is the exception to terrestrial planet formation. It is thought that the small planet is a captured remnant of the Kuiper belt.

Titius-Bode law—This law mathematically determines the distances to the planets in astronomical units. The law is not yet truly explained, as nature is not often so mathematically orderly. The law falls apart around Neptune, but then predicts the mean distance of Pluto.

Titius-Bode Law Calculations		
Planet	Calculation*	Planets' Actual Mean Distance (astronomical units)
Mercury	(0+4)/10 = 0.4	0.387
Venus	(3+4)/10 = 0.7	0.723
Earth	(6+4)/10 = 1	1.000
Mars	(12+4)/10 = 1.6	1.524
Asteroids **	(24+4)/10 = 2.8	2.4–3.0
Jupiter	(48+4)/10 = 5.2	5.203
Saturn	(96+4)/10 = 10.0	9.539
Uranus	(192+4)/10 = 19.6	19.191
Neptune	—	30.071
Pluto	(384+4)/10 = 38.8	39.158

* The calculation is based on the formula: $a = (n+4)/10$, in which a is the astronomical units. The n is the number doubled for each planet; n starts with Mercury at 0, then Venus at 3, doubling from there.

** At the time this formula was developed, the existence of the asteroids was unknown.

29. Universe

Introduction

The universe includes all the stars, galaxies, planets, satellites, gases, nebulas, and other space objects. Although it may seem that the universe has nothing to do with the natural history of Earth, that assumption is incorrect. Our planet and all the organisms that live on Earth are made of starstuff—we owe our very existence to the universe. The actual stars and other distant objects can also affect us, from the Earth-Sun interaction to shock waves from exploding stars that can influence our Milky Way galaxy.

Timeline

(note: bya=billion years ago)

Date	Event
Prehistoric Events	
~20–15 bya	The universe first forms, although no one agrees on the actual date.
~10–5 bya	A solar nebula begins to form and spin in a small region of the galaxy we now call the Milky Way, eventually forming our solar system.
~4.55 bya	The planets and satellites of the solar system form; Earth begins to form a hardened crust.
Modern Events and Discoveries	
~1400 B.C.	It is thought the Chinese had a working astronomical calendar around this time or the century before.
~350 B.C.	Chinese astronomer Shih Shen prepares what is thought to be the earliest star catalog.
~350 B.C.	Greek philosopher and scientist Aristotle (384–322 B.C.) sets down numerous theories held as standards of scientific thought for years, such as Earth was the center of the universe, an idea that would dominate astronomy for more than 2,000 years.
~140 A.D.	Egyptian astronomer Ptolemy (Claudius Ptolemaeus, c.100–170 A.D.) pushes the idea that Earth was the center of the universe, although he did not originate the idea; his resulting Ptolemaic system would influence astronomy for close to 2,000 years.
1054	The Chinese and other cultures record a supernova in the area now called the Crab Nebula; it is visible for 22 months.
late 1500s	Danish astronomer Tycho Brahe (1546–1601) makes some of the most accurate positional measurements of stars and planets—all before the advent of the telescope.
1734	Swedish philosopher Emanuel Swedenborg (1688–1772) notes that stars form one vast collection of which the solar system is only one constituent.
1755	German philosopher Immanuel Kant (1724–1804) suggests that the small nebulous objects seen in telescopes are other Milky Way systems, or "island universes."
1915	German-born U.S. theoretical physicist Albert Einstein (1879–1955) introduces his general theory of relativity, revolutionizing the way we view the universe; his theories on quantum physics and relativity establish him as one of the world's greatest scientists.
~1917	English mathematician, astronomer, and physicist Sir James Hopwood Jeans (1877–1946) first proposes that matter is continuously created throughout the universe.
1918	U.S. astronomer Harlow Shapley (1885–1972) determines the true size of the Milky Way galaxy.
1923	U.S. astronomer and cosmologist Edwin Powell Hubble (1889–1953) derives the distances to a spiral nebula; his additional studies also reveal the true distances to deep space objects, and his Hubble constant is instrumental in determining the age of the universe, although this age is still debated to this day.

Date	Event
1934	Dutch-born U.S. astronomer Bart Jan Bok (1906–1983) discovers the Bok globules, or clumps of matter found in bright nebulas which are thought to be the incubator of young stars.
1942	U.S. astronomer Grote Reber (1911–) makes the first radio maps of the universe; he is also the first to build a radio telescope in 1937.
1965	Radio wave remnants of the big bang are found by U.S. astronomers Arno Penzias (1933–) and Robert Wilson (1936–).
1970s–present	English theoretical physicist and astronomer Stephen William Hawking (1942–) proposes many theories on black holes, the origin and evolution of the universe, and the nature of time-space; he is considered to be the greatest theoretical astronomer of the latter twentieth century.
1987	The first observable supernova since 1604 is seen from Earth, called 1987A.
1990	The satellite Cosmic Background Explorer (COBE) detects background radiation thought to be from the big bang.
1992	The Hubble Space Telescope detects what may be a black hole; many more are later inferred by other data from the telescope.
1998	The Hubble Space Telescope orbiting Earth sends back data on the farthest galaxies ever imaged—at about 12 billion light years away.

History of the Universe

Early Universe

The big bang theory states that our **universe** began as a single point, then expanded into what is known to exist today—taking about 15 to 20 billion years (although these numbers are highly debated). About a million years after the big bang, the universe was a featureless sea of hydrogen and helium gas. Eventually, the universe developed regions of higher and lower density; about 2 billion years after the big bang, embryonic **galaxies** appeared in the high density regions, forming clusters and superclusters of galaxies.

Scientists theorize that the big bang itself had several stages after its initial expansion. Here is only one scenario (many others have been proposed): At time zero, the single point started to expand (although some scientists refer to this as a fireball erupting); at less than 0.0001 seconds, elementary particles began strong interactions with each other; by about 0.0001 and 100 seconds, there was a rapid expansion and cooling of the area, with helium nuclei forming, while electrons, positrons, neutrinos, and photons reached thermal a stable temperature; by 1,000 to 1,000,000 years after the big bang, radiation separated from that matter, and deuterium and helium formed; after 1 million years, quasars and clusters of galaxies formed; and by now, galaxies and stars have fully formed—and are still forming.

Study of the Universe

The study of the universe has had a long and diverse history in terms of human understanding and discovery. But because of its ancient past, it is difficult to state when the earliest observations were made, or when the science of astronomy began. Most of the early studies of the universe were based on cultural beliefs and superstition. For example, most ancient peoples believed the celestial sphere (or the sky above us) rose and set, the sphere actually rotating as it carried the stars across the sky; others believed that the sphere was carried around on the backs of an animals—or even a giant human who carried the entire universe.

It is known that certain civilizations, including the Chinese in the early fourteenth or thirteenth centuries B.C.—and probably much earlier—had recognized and attempted to keep track and predict celestial phenomena, in particular, the movement of the Moon, **planets**, **Sun,** and **stars**. Some scholars claim that the Chinese had even determined the true length of an Earth year as 365.25 days as early as the twelfth century B.C. In 350 B.C., astronomer Shih Shen put together what was probably the first star catalog, which contained about 800 entries. (The Chinese also recorded a visible **supernova** in A.D. 1054.)

Another way of interpreting the sky by many cultures was to define the stars themselves into constellations, the apparent configuration of the stars. The Chinese, Egyptians, and Greeks all divided the sky into these features—with the modern astronomers still using these constellations to denote approximate locations in the sky. In fact, the 88 recognized modern constellations are of Greek origin, with names that are Latin translations of those given to them by the Greeks. And although we note today that the constellations rarely resemble actual people or objects, that was probably not the intent of the Greeks. They probably grouped the stars in honor of the mythological characters or animals—not because the stars actually resembled the character or animal.

While most astronomers believe that natural curiosity about the sky started the study of the universe (or field of astronomy), others believe one of the major impetuses was the practice of astrology, considered an ancient religion. The origin of astrology was probably in Babylonia; it was greatly developed and expanded by the ancient Greeks. For example, the Greek philosopher and scientist Ptolemy presented a systematic description of the subject in his book, *Tetrabiblos* in the second century A.D.

Thus, astronomy owes a great deal to astrology, because its practice necessitates a detailed knowledge of the movements of the celestial bodies. This is why many of the astronomers from antiquity to the time of Galileo devoted most of their time and energy to careful observations of the heavenly bodies—and also prepared horoscopes as part of their duties, including major figures such as sixteenth-century astronomers Johann Kepler and Tycho Brahe.

One scientist who had a major influence on astronomy (and other sciences) of his time, around 350 B.C., and beyond—was Aristotle, the teacher of Alexander the Great. Aristotle set down numerous theories that were held as standards of scientific thought for years. For example, his idea that Earth was the center of the universe dominated astronomy for almost 2,000 years.

By the fourth century A.D., the Greek golden era of astronomy was over, and the field (along with many other science studies) became mired in religious debates and superstition. By the seventh century, the Arab peoples revived astronomy; but in general, they elaborated on the Ptolemaic (Earth-centered) notion of the universe.

By the middle of the thirteenth century, knowledge of astronomy spread throughout Europe, as Greek manuscripts were translated into Latin—and with the Renaissance blossoming in the next two centuries, knowledge of the universe blossomed, too. One of the major scientists who changed astronomy forever was Nicolaus Copernicus, a Polish canon of ecclesiastic law and an astronomer—who pushed the revolutionary idea that the Sun was the center of the solar system and that Earth was not the center of the universe.

Perhaps the greatest turn of events in terms of understanding the universe came after the invention of the telescope in the early seventeenth century. Finally, observers were able to pick out some of the details that lay beyond our own solar system, including clusters, galaxies, and countless stars. As telescopes improved, interest in what was "out there" increased. Scientists such as Emanuel Swedenborg speculated that the solar system was only one constituent of the base collection of stars we view from Earth; Immanuel Kant believed that the small oval objects he viewed in his telescope were "island universes"; and scientists like William Herschel and his son John categorized more than 5,000 nebulous objects. With the advent of photography in the late 1800s, which allowed scientists to view and compare images better (film also has more light-gathering properties than the eye) from multiple nights, more and more galaxies, stars, and clusters were discovered, categorized, and speculated on—as the universe opened up even more to observers.

A plethora of discoveries concerning the universe were made in the twentieth century. For example, Edwin Powell Hubble was responsible for proposing the theory of Hubble's constant, which states that the rate of expansion of the universe is tied to the age of the universe. (So far, no one has agreed on the true value for the Hubble's constant. Thus, the stated age of the universe varies from about 9 billion to 20 billion years old, but the more accepted values range from 15 to 20 billion years.) And other studies were made, including Bart Jan Bok's 1934 discovery of the Bok globules, clumps of matter found in bright nebulas that are the incubator of young stars—seen in photographs of bright nebulaes as tiny dark blobs. Studies have even been conducted on **interstellar dust and gases**.

In addition, ideas about space-time, and the new quantum physics and the universe, also were advanced during this time. One of the major scientists involved in these discoveries was theoretical physicist Albert Einstein, who introduced his general theory of relativity in 1915, thus revolutionizing the way we view the universe. His theories on quantum physics and relativity established him as one of the world's greatest scientists. Along with Einstein in determining the true nature of the universe have been such science greats as Stephen Hawking (1942–) (cosmologist and

> **The Fate of the Universe**
>
> The overall fate of the universe may be connected to how much mass is actually located in it. In the past decade, scientists who study the stars in the universe have realized that, based on the size and distribution of matter in the universe, a great deal of mass was missing. In fact, it is thought that the universe contains between 20 and 200 times more mass than all the stars and galaxies we see, the invisible material called "dark matter." And it is this mass that will eventually determine the fate of the universe: if it exists, gravity will pull the universe together again; if not, the universe will keep expanding—or so some scientists believe.
>
> Where did all the missing mass go? Several theories have been proposed: First, the missing mass may be in dead stars called brown dwarfs. Most stars—at least those that don't go nova or supernova—eventually die, losing their light. These dark masses are called brown dwarfs; they may be out there, but because they send out no light, we cannot find them.
>
> Another theory is that the missing mass may be black holes: Black holes—stars toward the end of their lives that have so much gravity associated with them that no light can escape—may also be the "missing mass." In particular, the primordial black holes formed just after the big bang may contain this mass. Again, because we can't see black holes, it is unknown whether or not they are numerous in the universe and could account for the missing mass.
>
> Could it be that there really is no missing mass? Does the universe have clumps of matter throughout, hidden by dark nebulae or other space materials, so we cannot see it? So far, the only way for scientists to find these answers is to seek out the universe's invisible material indirectly, observing how certain seen objects seem to be affected gravitationally by what may be unseen matter.

black hole specialist), Sir Fred Hoyle (1915–) (proposer of the steady-state-model of the universe), and the planetary studies of Carl Edward Sagan (1934–1996).

More recent studies of space focus on the Sun—especially the seemingly 11-year (or 22-year) **solar cycle**, when the Sun's activity increases. Scientists are now trying to predict solar storms with the help of several satellites, including the Solar and Heliospheric Observatory (SOHO), a craft that uses ultraviolet cameras and other instruments to "stare" at the Sun and its corona 24 hours a day. This satellite and others support the Space Weather Program, coordinated by several U.S. agencies. The reasons for concern and need for prediction are obvious: Our society has become more reliant on electricity and electronic devices than ever before—from satellite-based cellular phones to global positioning systems that determine our location on the planet. Major solar storms not only could knock out power grids in many cities, but also could affect the numerous electronic devices on which we rely.

Other modern studies of our universe deal with objects in deep space and the limits to which the universe extends. Some of these objects, such as **black holes**, are still theoretical in their nature and mechanisms. Recent data has inferred that black holes actually exist—and may even be found in the heart of our own **Milky Way galaxy**. How far our universe extends is still a matter of conjecture; new equipment has given us a chance to find solid evidence of objects from the furthest reaches of space. For example, in 1998, the orbiting Hubble Space Telescope sent back images of the farthest galaxies known, about 12 billion light years away. *See also* EARTH, MOON, and SOLAR SYSTEM.

Topic Terms

black hole—A theoretical star at the end of its life cycle. (The star has burned up all its fuel and then implodes.) Because of its immense concentration of material within such a small volume, a black hole has a huge gravitational pull on nearby objects. Space-time also curves over upon itself at a black hole, and no energy, such as light, can escape. Although no one has ever, or can ever, visually see a black hole, scientists believe they can infer a black hole's existence by its gravitational effects on other visible matter surrounding it. Several scientists have recently claimed to have discovered black holes, including one at the center of our Milky Way galaxy—and one at the center of the galaxy Centaurus A. In the case of Centaurus A, it is believed that intense radio energy emitted by the galactic core is caused by a supermassive black hole. The minimum density of this black hole is immense—apparently 14 billion Suns per cubic light year.

galaxy—A galaxy is a collection of stars, dust, and gases seen as spiral, oval, or irregular shapes in the nighttime sky. Many number between 1 million and 1 billion stars—but they can also number several billion; only a few can be seen with the naked eye. Earth is a member of a solar system traveling through space in the Milky Way galaxy. Most galaxies are billions of years old; the first galaxies probably formed about 2 billion years after the big bang, as proto-galaxies.

interstellar dust and gas—Interstellar dust and gas particles are composed of many substances, including carbon, silicon and silicates of iron, magnesium, aluminum, and other molecules. Cosmochemists have so far discovered nearly 100 types of molecules in interstellar clouds. The presence of carbon-containing and water molecules found in these clouds is important: Such elements may indicate the possibility of life elsewhere in the universe.

Although it is not easy to distinguish the difference between interstellar and interplanetary dust and gas in our own solar system, it is logical to assume that some of the particles that rain down on Earth come from beyond our solar system. If a supernova occurred within about 130 light years from Earth (or if Earth traveled through a supernova remnant cloud), high-energy particles could react with the atoms in our atmosphere, producing high concentrations of certain isotopes (atoms of a given chemical element with different numbers of neutrons).

One study may indicate such interactions: Anomalous concentrations of certain elements (such as beryllium-10) in ice cores from Antarctica seem to indicate periods when the earth was irradiated by nearby supernovae. The two layers, dated at 35,000 and 60,000 years in age, still need to be studied to verify a supernova connection.

Milky Way galaxy—Our solar system resides in a spiral arm of the Milky Way galaxy. Observations by radio telescopes indicate that the galaxy is a flat, disk-like collection of stars, clusters, and dust and gases. It has several spiral arms surrounding a dense nucleus. The Sun is one of billions of stars in the galaxy and resides in the Orion spiral arm. Several scientists believe they have found a black hole at the center of the Milky Way galaxy.

planets—See SOLAR SYSTEM, Topic Terms.

solar cycle—Our Sun goes through several cycles, including an 11-year period called the solar (or sunspot) cycle. On the average, this cycle produces a time of greater solar activity— at the cycle's peak, the number of solar flares (streams of gas shooting upward), sunspots (dark spots on the Sun caused by intense magnetic fields), and solar wind increases.

The Sun's activity has an effect on the Earth—mainly through its output of energy in the form of light. But the Sun also produces powerful solar storms, usually in the form of gusts of solar wind or giant solar flares. As the particles from flares or the solar wind buffet the Earth's magnetic field, we see several effects. In most cases, we see auroras, colorful displays of light around the north (aurora borealis) and south (aurora australis) poles. If the intensity of a storm is great, the particles can raise havoc with radio communications, power grids and transmissions, and artificial satellites. In addition, at the peak of a solar cycle, such storms become more frequent, and even cause the earth's atmosphere to swell, causing drag on artificial satellites.

stars—The majority of stars are giant (compared to Earth) stellar furnaces that produce energy in the form of light and heat by the process of fusion (a nuclear reaction that converts hydrogen to helium). Not all stars produce energy—they are usually at the end of their lifecycle, such as brown dwarfs. Earth's star, the Sun, is an absolute necessity for life on this planet.

Closest Stars Compared

Star	Distance from Earth (light years*)	Apparent Magnitude **
Sun	93,000,000 miles avg. (149,637,000 kilometers)	-26.85
Moon (for comparison only)	238,866 (384,000 kilometers)	12.6
Proxima Centuri	4.3	+10.7 ***
Alpha Centuri A	4.3	0.0
Alpha Centuri B	4.3	+1.4
Barnard's Star	6.0	+9.5
Wolf 359	8.1	+13.5
Lal 21185	8.2	+7.5
Sirius A	8.7	-1.5
Sirius B	8.7	+8.5

* Except where noted.

** Apparent Magnitude—The brightness of a star based on an observer on Earth; a star's apparent magnitude depends on the type of star and its distance. Negative numbers indicate brighter starts; higher positive numbers indicate dimmer stars.

*** Closest star system to Earth.

Temperatures of Several Bright Stars Compared to Our Own Sun

Star	Approximate Temperature (degrees Kelvin)
10 Lacertae	>30,000–60,000
Rigel, Achernar, Hadar	10,000–30,000
Sirius, Vega, Altair, Acrux	7,500–10,000
Canopus, Polaris, Procyon	6,000–7,500
Sun, Capella	5,000–6,000
Aldebaran, Arcturus	3,500–5,000
Betelgeuse, Antares	<3,500

Sun—The Sun is a star at the center of our solar system. It comprises about 99.86 percent of the mass of the solar system. It is about 864,000 miles (1,390,180 kilometers) in diameter, or about 109 Earth diameters across.

It is thought that the Sun arrived at its present brightness about 800 million years ago, or about 200 million years before the end of the Precambrian era (or the beginning of the Cambrian period 600 million years ago). The Sun's brightness will not change for another 1.5 billion years. (Earth is about 4.55 billion years old). The Sun's fusion reaction that creates its energy has existed for about 6 billion years; it is thought that the Sun will burn for another 7 billion years.

The Sun is vital to life on Earth. Earth is located in what has been called the "life zone." This zone is where the energy from the Sun did not bake away the atmosphere, such as on Mercury or create an overreactive greenhouse effect in the atmosphere, such as on Venus. Overall, the energy from the Sun warms the atmosphere, keeping the planet just right for life, and it is that engine that keeps the atmospheric and ocean currents moving.

supernova—A huge release of energy from a massive star at the end of its life. From the Earth, a supernova is seen as a sudden brightening of a star. Such explosions are important for the formation of such elements as iron in the universe—in fact, many scientists believe the majority of heavy elements form from such shock waves throughout the universe. Some scientists believe that certain stars and solar systems are born as shock waves from supernovas travel through interstellar dust and gaseous nebulas—the waves causing the nebulas to collapse faster.

Two types of supernovas have been identified: Type 1 occurs in binary systems (two stars), in which one of the stars becomes a red giant at the end of its life, loosing its matter to its less massive companion and becomes a white dwarf. The other star evolves into a giant stage, shedding matter to the dwarf and triggering a giant supernovae explosion. Type 2 occurs in massive stars at the end of its regular life cycle. After the star depletes its hydrogen, it collapses and begins to burn helium. As it further collapses, other elements also burn, eventually creating a giant explosion.

universe—Several modern theories concerning the life and death of the universe have been advanced, including the following:

Open Universe Model—The theory that the universe expands in all directions, expanding forever, with the density becoming less over time.

Closed Universe Model—The theory that the universe is expanding, but at some point, gravity will take over, and the universe will collapse.

Steady-State Model—British astrophysicist Sir Fred Hoyle developed the steady-state hypothesis of the universe, which states, in general, that the universe will always be the same density and distribution forever.

30. Volcanoes

Introduction

One of the most awe-inspiring of natural phenomena, volcanoes are found in many areas of the world and have probably been present since the crust of the Earth began to cool. They have figured prominently in the destruction and collapse of ancient civilizations, have generated whole new islands or eliminated existing ones, and have continued to capture our attention—in the past and in recent years. For volcanoes are not some long-ago occurrence that has no relevance today; rather, they are an ongoing phenomenon that deeply affects the people who live in the areas of the world in which they exist.

Timeline

(note: bya=billion years ago; mya=million years ago)

Date	Event

Prehistoric Events

Date	Event
~4.55 bya	The Earth's crust forms.
~4.35 bya	The Earth's atmosphere and oceans are established as volcanoes spew gases and water vapor into the atmosphere; the Earth is also being almost continually bombarded by objects from space during this period.
~3.8 bya	The heavy bombardment of the planet dies down, allowing the planet to continue to cool down; life probably begins to form around this time even though erupting volcanoes are prolific.

Modern Events and Discoveries

Date	Event
~1600 B.C.	Santorini, or Thera, massively erupts, leading to the destruction of the Minoan civilization; this eruption may be the destruction of Atlantis to which Plato referred.
5th century B.C.	The oldest recorded description of an actual volcanic eruption is written by the poet Pindar; scientists believe that this description accurately describes the eruptions of Etna that occurred in 479 B.C.
~30–20 B.C.	Greek geographer Strabo (c.53 B.C.–c.39 B.C.) publishes his *Geography*, which includes accurate descriptions of the volcano Etna and recognizes Vesuvius as being volcanic in nature.
79 A.D.	Mt. Vesuvius, a small 4,200-foot (1,280-meter) volcanic mountain lying above the Bay of Naples in Italy, erupts, destroying the towns of Pompeii, Heraculaneum, and Stabiae; these towns are buried and forgotten until the seventeenth century.
1546	German geologist Georgius Agricola (Georg Bauer, 1494–1555) proposes that the Earth's subterranean heat is located at volcanic centers and is seen in volcanic eruptions; he speculates that the heat is derived from the combustion of coal, bitumen, or sulfur, ignited by intensely heated vapors.
1583	Monte Nuovo (New Mountain) is born near Pozzuoli, Campania, Italy, the first time that a volcano is scientifically studied from its inception through the various phases.
1631	Mt. Vesuvius erupts after lying dormant for 600 years, killing more than 4,000 people and ruining the coastal town of Portici; it is when Portici is being rebuilt that the previously unknown ruins of Heraculaneum, destroyed by the A.D. 79 eruption of Vesuvius, are discovered lying beneath.
1644	French philosopher and mathematician René Descartes (1596–1650), writing in his *Principia Philosophiae*, speculates that Earth was originally hot like the Sun and that the energy that drives volcanoes comes from heat associated with this original state.
1750s	French geologist Jean-Etienne Guettard (1715–1786) notices the similarity between black rocks used in construction in Auvergne and those found on Vesuvius and Etna.

Date	Event
1772	British volcanologist Sir William Hamilton's (1730–1803) letters to the president of the Royal Society in London concerning his observations about Mount Vesuvius are published; these letters constitute the first modern work of volcanology.
1784	U.S. diplomat, scientist, and inventor Benjamin Franklin (1706–1790) suggests that the severe Northern Hemisphere winter of 1783–84 was due to powerful volcanic eruptions in Iceland in the summer of 1783; he asserts that the Sun's warming rays were blocked because of the ash and other particles released into the air by the volcanoes.
1790	German geologist Abraham Werner (1749–1817) erroneously suggests that volcanoes are the result of combustion of underground coal and other inflammable substances.
1790s	Scottish geologist James Hall (1761–1832) performs the first laboratory experiments that show how molten rock could cool to have a crystalline structure; he calls the molten rock magma, a chemical term then in use for a pasty substance.
1700s	Electricity is studied as a source of volcanic energy.
1815	The greatest volcanic eruption in recorded history occurs on the island of Tambora, south of Borneo; this eruption disrupts the climate around the world for the next few years.
1883	On August 27, Krakatoa (Krakatau), near Sumatra and Java, explodes in a giant eruption that completely destroys the mountain, is heard 3,000 miles away, and generates a tsunami that kills 36,000 people.
1906	U.S. geologist Clarence Dutton (1841–1912) suggests that radioactivity is the cause of all the Earth's volcanic activity; we know now that much of the Earth's heat is from radioactive decay, but most volcanic activity is from the movement of crustal plates.
1963	U.S. geologist Harry Hammond Hess (1906–1969) delivers an address to the Geological Society of America, titled "Further Comments on the History of Ocean Basins"; in it, he describes the phenomena of sea-floor spreading due to the action of magma erupting from the earth's interior along mid-ocean ridges.
1963	The sudden appearance of a volcano in the sea off Iceland leads to the formation of a new island, where once nothing but water existed; the name of this island, and volcano, is Surtsey, after the giant Surtur. (In Norse mythology, Surtur will set fire to the earth at the Last Judgment.)
1980	Mt. St. Helens (Washington State, U.S.), long dormant, erupts; the fine ash dust is propelled into the upper atmosphere at heights of up to 13 miles (20 kilometers) and a flow of ash and hot gas flows swiftly down one side of the volcano, destroying everything in its path up to 16 miles (25 kilometers) away; 61 people die.
1985	The Ruiz volcano, Columbia, South America, erupts, melting the snow on its flanks; a wall of water and ash flows quickly down a canyon and out onto the town of Armero, resulting in the death of approximately 22,000 people.
1991	Mt. Pinatubo in the Philippines violently erupts; approximately 750 people die, one-third as a direct consequence of the eruption, the rest from mudslides or disease; a year afterward, the skies around the world see the effects of the eruption, especially with brilliant red sunsets, an indication of an influx of particles in the air.
1998	A new volcano is being monitored off the Hawaiian Islands chain—another volcanic eruption on the sea floor that will one day create a new island in the chain.

History of Volcanoes

Early Volcanoes and Volcanology

Volcanoes have been around since the Earth first formed its crust 4.55 billion years ago. Scientists believe volcanic activity over time has been responsible for several major events in the Earth's natural history. For instance, early volcanoes may have been instrumental in providing early atmospheric gases to the planet, including carbon dioxide and water vapor. This theory is based on the release of gases from modern volcanoes. And early volcanoes may also have contributed to periodic climate changes over the geologic history of the planet as debris from eruptions may have dimmed the Sun's light at various times.

The impact of volcanoes has affected not only wildlife and other organisms, but also various cultures living near such eruptive areas around the world. And because of this, most of the "early studies" of volcanoes were couched in terms of religion, myths, or superstition. Ancient peoples were extremely frightened of volcanoes, and with good reason—eruptions could destroy their land, crops, and villages, killing many people in the process. Without modern-day scientific knowledge, they could not find a rational reason for this devastating phenomena and therefore attributed the origins and activities of volcanoes to the actions of various gods and monsters.

Some Greek philosophers also attempted to explain volcanoes using more scientific explanations. Plato thought that the eruptions of volcanoes, with their accompanying earthquakes, were caused by the escaping of air from the "Pyriphlegethon," a river of fire that Greeks believed coursed through the inside of the earth. This explanation was not too far off the mark, if you compare this theory to the modern explanations of **magma** and gas. Aristotle proposed a "pneumatic" theory to explain volcanoes; he thought that the action of the waves breaking on the seashore compressed and forced air through caves deep into the Earth, where it came in contact with sulfur and bitumen, creating a fire. The result of this ignition—flames, smoke, **lava** fragments (scoria), and ashes—escaped through the openings in volcanoes. This theory, like many of Aristotle's, would dominate the thinking of scientists for more than 2,000 years.

The first written records of volcanoes were from the Mediterranean Sea area, where several ancient civilizations were centered. The oldest recorded description of an actual volcanic eruption was written by the poet Pindar in an ode describing the myth of Typon. In it, he describes the volcano of Etna (in Sicily) as a "pillar of the sky" from which fire shot with a loud roar. He described in detail the flow of lava down the flanks of the volcano and the crashing of white-hot rocks into the land and sea. Scientists believe that this accurately describes the eruptions of Etna that occurred in 479 B.C.

Around the time of Christ, Greek geographer Strabo published detailed descriptions of nature in his *Geography*, including volcanoes like Etna and Mons Vesbius (Vesuvius). He alone accurately recognized the volcanic nature of Vesuvius, though it was covered with vegetation and thought to be a typical mountain. Thus, the peoples around the mountain were surprised when in A.D. 79 Vesuvius erupted, destroying the towns of Pompeii, Heraculaneum, and Stabiae.

The fall of the Roman Empire interrupted this increasing scientific study of volcanoes, at least in the West; by the Middle Ages, volcanoes were thought to be gateways to hell or the prisons of the damned. Hekla, a volcano in Iceland, made noises that were thought to be the sounds of tormented souls. Mt. Etna, in Sicily, was thought to be the place of torment of Anne Boleyn, the cause of England's Henry VIII's secession from the Roman Catholic church when it wouldn't grant him a divorce from Anne.

With the coming of the Renaissance, explorations and scientific inquiries once again flourished. In 1583, Monte Nuovo (New Mountain) was born near Pozzuoli in Campania, Italy, and was studied from inception through its various phases. This was the first time a full scientific study had been done of the birth and evolution of a volcano. It also generated great debates between those who sought a religious explanation of the eruption and the scientists who cited Aristotle's theory.

The central issue that scientists and naturalists were concerned with during the next centuries was finding the source of the heat that fed the volcanoes, the fundamental question thought of by Aristotle. Indeed, his "combustion" theory—that air far underground combined with various substances to produce fire—was accepted by naturalists and scientists until near the end of the eighteenth century. Until then, the nature of the combustible substances was thoroughly debated, with coal and pyrites being added to Aristotle's original supposition of sulfur and bitumen.

Rene Descartes, writing in his *Principia Philosophiae* in 1644, speculated that at one time, the earth had been white-hot like the Sun. Cooling down, it began to separate into different layers. Volcanoes were localized areas where the residual energy from this long-ago heat had worked its way to the surface through the cooled crust—a theory later supported by geologists.

Modern Volcanoes and Volcanology

In the 1750s, Jean-Etienne Guettard of France noticed the similarity between black rocks used in construction in Auvergne and those found on Vesuvius and Etna. He traced the source of the rocks back to a quarry at Volvic and exploring the area, found the weathered and vegetated remains of ancient volcanic cones. He delivered a paper in 1752 to the French Academy of Science entitled "Memoir on Certain Mountains in France That Have Once Been Volcanoes," and spent the next decades studying and mapping the extinct volcanoes in that country.

In 1772, Sir William Hamilton's letters to the president of the Royal Society in London, concerning his observations about Mt. Vesuvius, were published. These letters constitute the first modern work of volcanology. Hamilton was the British envoy to Naples from 1764 to 1800; he studied Vesuvius's nine violent eruptions during this period and made more than 200 trips up the volcano. He also began to compile a historical list of the dates when the mountain had erupted, using the display of sacred images by the priests of Naples and the surrounding villages and towns as a guide.

Sir James Hall conducted a series of experiments in the 1790s that resolved the question of how molten rock could cool to form a crystalline structure. He took samples of a local basaltic stone and melted them in a forge. Calling the resulting fluid magma, after a Latin chemical term of the time for a pasty substance, he then cooled the molten rock under different conditions. If allowed to cool quickly, the magma formed a glass. If cooled over several hours, it formed a rock with a crystalline structure. Hall found that he could vary the size of the crystals formed by changing the time of cooling. He also experimented with the effects of enormous pressure, as well as temperature, on rock.

In 1815, the volcano on the island of Tambora, south of Borneo, erupted in the greatest explosion in recorded history. This eruption went unstudied until 1847, when a scientific expedition was able to reach the 3-mile wide crater. The force of this eruption blew approximately 20 cubic miles of debris into the sky, where it stayed in the upper atmosphere and was distributed around the world. This layer of debris prevented some of the Sun's radiation from penetrating through to the ground, thus cooling the planet. Strange and colorful sunsets were seen around the world, and the year 1816 became known as the "year without a summer," since record low temperatures and summer snow were reported in parts of Europe and the United States. Crop failures occurred in 1816 and 1817, leading to the outbreaks of famine, food riots, and typhus epidemics. But no one suspected a correlation between the eruption of Tambora and the weird weather.

By the time Krakatoa (Krakatau) erupted in 1883, however, this connection was better understood, and many scientific reports were published about the optical effects during the years 1883 to 1886. Krakatau, located between Java and Sumatra, finally exploded on August 27, 1883, in a blast that was heard almost 3,000 miles away, completely destroying the mountain and generating a tsunami that killed more than 36,000 people.

In 1902, Mt. Pelee in Martinique began a series of eruptions that cumulated on May 8, when a pyroclastic flow burst forth from the mountain and swept over the town of St. Pierre. This eruption produced no lava and little ash, just a boiling black cloud of superheated gas, estimated at a temperature between 2,370 and 3,270° F (1,300–1,800° C) that silently killed more than 30,000 inhabitants of the town. Only two people survived. Alfred Lacroix, of the French Academy of Sciences, who published a treatise on the 1902 eruptions of Pelee, termed this phenomenon a *nuee ardente*, or "glowing cloud." The more general, modern term is a pyroclastic (fire-broken) flow, which acts like a fluid in many ways.

In 1963, Harry Hess, a professor of geology at Princeton University, delivered an address to the Geological Society of America, titled "Further Comments on the History of Ocean Basins." In it, he described the phenomenon of sea-floor spreading due to the action of magma erupting from the Earth's interior along mid-ocean ridges. The new sea floor that was created slowly spread away from the ridges, and later sank back into the Earth's interior in the area of deep sea trenches. These findings gave mechanism and credence to Alfred Wegener's long discredited ideas of continental drift.

Despite all these studies, the vital question as to the source of the energy that manifests itself in volcanic activity had not been answered even though it has been studied and speculated on since Aristotle's time. In the last couple of centuries, numerous explanations have been offered to try to explain the source of the energy that drives volcanoes, ranging from Descartes's theory (1600s) of a white-hot origin of the planet; to the effect of electricity, a newly discovered phenomenon; and the action of friction and compression. In the 1800s, with new advances in chemistry, it was

> ### Volcanic Legends
>
> Different cultures have explained the eruptions of nearby volcanoes in hundreds of ways. Here are just a few of the legends associated with volcanoes around the world.
>
> The people of Chile believed a giant whale lived inside volcanoes; while those of India thought the monster was a giant mole or boar. The ancient Japanese felt that a horrible, giant spider lay inside a volcano. And the Indonesians thought that the world was held up by the snake, Hontobogo, whose movements shook the ground and caused fire to erupt from the mountains. To appease these monsters, ancient cultures made sacrifices to the volcanoes—sometimes of animals, sometimes of humans—throughout history. These explanations were handed down from generation to generation in stories of legend and myth—some of which still persist today.
>
> Other peoples throughout world have ascribed the origins and activities of volcanoes to various gods. Some Greeks thought that volcanic eruptions were the exhalations of the Titans, giants that had been buried beneath mountains by Zeus. In Peru, Indians tell of a bad spirit that had occupied the volcano of El Misti, which is located near the city of Arequipa. The Sun God punished this spirit by plunging its head into a flow of molten lava, then blocked the crater of the volcano with a cap of ice.
>
> The Polynesians thought that volcanoes were under the control of the demi-goddess Pele, who could appear in human form as a beautiful, young woman or an old, ugly woman. According to legend, after losing a bone-sled race down the side of a volcano to a young Hawaiian chief, she stamped on the ground, which split open, shooting out smoke and hot rocks. The chief ran for his life, pursued by Pele and showered with the rocks she threw at him.
>
> Reaching the shore, he escaped in a canoe. Even today there is a long row of spatter conelets down the side of the volcano which mark where Pele hurled the hot rocks at him.
>
> The people of the Russian Kamchatka peninsula, on the Pacific border of the Ring of Fire, recall in their legends how the volcanoes that populate this area of the world formed. Long ago, Kamchatka was flat, with neither mountains nor volcanoes. The young men there lived in peace, tending deer on the plains. But one day, the most beautiful young girl appeared among them, and each of them in turn fell in love with her. The men soon began to quarrel, because each thought she should marry him. An old shaman woman who lived in the area threatened the men to try to get them to stop fighting, but it did no good. Carrying out her threat, she turning the young men into mountains. Their love for the young girl was so strong and hot that it melted the hearts of the mountains, and their blood rushed out as hot lava. Their hearts could not be turned to stone by the shaman woman, because their love was stronger than her magic—and thus, the Kamchatka volcanoes came into being.
>
> In Africa, the volcanic peaks that surround Kilimanjaro, named Kibo and Mawenzi, are featured in the legends of the Chagga people who live nearby. It seems Kibo and Mawenzi were good neighbors, so when Mawenzi's fire went out, he went to Kibo to ask for some embers. Traveling back, he decided to play a trick on Kibo, dumped the embers, and went back for more, claiming they had gone out. He did this three times, until Kibo picked up his giant pestle and beat Mawenzi black and blue. According to the Chagga, this explains why the peak of Mawenzi is so battered and scarred.

thought that exothermic reactions between water and metals such as sodium, potassium, and calcium were responsible for the heat energy. And in the 1900s, the discovery of radioactivity led to speculations that this process was the source of volcanoes' origins. It was thought that uranium, thorium, and other radioactive elements generated enough heat to melt the rocks inside the Earth. Other scientists thought that natural nuclear reactions, similar to those that occur in nuclear weapons, were going on within the interior of the Earth, leading to the molten rock.

Where the energy comes from that turns rocks into magma is still not settled, with some volcanologists feeling that radioactivity, meteorite collisions, and compression are not sufficient to account for the enormous amounts of energy that power volcanoes. Some volcanologists feel that the hot origin of the planet is the best explanation for volcanoes. In other words, the latent heat left over from the formation of the planet may be the reason why volcanoes still crack the Earth's crust.

In the latter part of the twentieth century, violent volcanoes still erupt—and we still have no way of telling which volcanoes will blow. For example, on June 15, 1991, with little warning, Mt. Pinatubo in the Philippines violently erupted. Approximately 750 people died, one-third as a direct consequence of the eruption, the rest from mudslides or disease. The ash generated by this eruption closed Manila airport for several days, forced the permanent abandonment of the

U.S. Clark airbase, and buried almost 200,000 acres of farmland. In addition, ash was sent almost 49,000 feet (15,000 meters) into the stratosphere, where it affected global climate for several years.

Not all volcanoes are a menace because of ash, lava, and noxious gases—sometimes the eruptions result in other hazardous events, such as **lahars**. For instance, on September 29, 1996, an earthquake of magnitude 5 on the Richter Scale shook the Vatnajökull icecap in southeastern Iceland. By the morning of October 1, a deep basin was found at the surface of the glacier exactly where an eruption had occurred in 1938; as the day wore on, the size of the bowl increased, and three more bowls formed, indicating that the glacier was melting at its base—and that the underlying volcano was becoming more active. In addition, ice covering a volcanic crater some 15 kilometers to the south began to rise—the result of the meltwater flowing into the crater and lifting the ice cover. By October 2, the eruption began through the ice. The volcano had not only become a threat—but so did the melting ice from the eruption.

The prediction of the formation of a new volcano or the eruption of an existing one is, like earthquake prediction, an inexact science at present. Scientists currently use many indicators of potential volcanic activity, including the seismicity (amount and magnitude of tremors and earthquakes), the tumescence and tilt of the volcano, chemical emissions from inside, and changes in gravitational, magnetic, and electrical characteristics. Seismicity is important because earthquakes and tremors almost always precede volcanic eruptions; volcanologists are not sure why, but they speculate that these tremors occur because of the action of magma forcing its way to the surface, perhaps through numerous cracks in the crust. Unfortunately, the prediction of an eruption based on seismicity alone is not accurate; the eruption may occur a few hours, months, or even years, after the tremors begin.

Because of this inability to predict volcanic eruptions and their possible side effects, scientists are currently examining and discovering more ways to understand how a volcano might possibly erupt. They are monitoring the shapes of some volcanoes, as the volcano can change as the magma rises very close to the surface; at this point, an eruption may occur with minutes, or take days. Researchers are presently replacing the older method of surveying numerous fixed reference points on the surface of a volcano with the use of radio receivers and satellites of the global positioning system (GPS). The receivers continually deliver information about the changes in the shape and deformation of the volcano with a resolution of a few centimeters.

More recent studies also include relatively new techniques to take more specific measurements from a volcano. One method monitors the emission of sulfur dioxide from a volcano. In the case of Mt. Pinatubo, the rate of emission of this substance went from approximately 500 tons per day to approximately 5,000 tons per day in a period of a couple of weeks. But then it fell sharply to approximately 280 tons per day, leading scientists to speculate that a blockage had occurred—and therefore pressure would begin to build up underground. Within a week of this finding, the focus of the earthquakes moved to the summit of the volcano, which began to bulge, and the main eruption of Pinatubo began. Small changes in gravity also occur as a consequence of the presence and movement of magma and can be measured with a gravimeter. Also, currently in the investigation stage is the use of magnetic and resistive (electrical) changes as an indicator of magma movement.

One of the more amazing discoveries about volcanoes occurred in 1977, when the U.S. Navy submersible *Alvin* (now at the Woods Hole Oceanographic Institution in Massachusetts) found volcanic **vents** on the ocean floor off the Galapagos Islands. Less than a few decades ago, no one really believed that any organisms could live without light around deep, hot, ocean vents. The *Alvin* carried two people to depths of about 13,000 feet (3,962 meters). At thermal vents around a volcanically active region, they found a previously unknown community of organisms, clustered around the hot springs and vents along the deep-sea bed—organisms never seen before.

Today, we realize volcanoes produce many different phenomena during their eruptions, and different volcanoes eject differing amounts of these phenomena, of which lava is only one part. Some **volcano types**, such as those underwater and in Hawaii, produce mostly lava, while rift volcanoes on land, such as in Iceland and East Africa, may only produce 60 percent of their ejecta as lava. The land volcanoes associated with subduction zones may only produce 10 percent lava. In general, the main ejecta of most land volcanoes is **tephra**, a collective term for all of the material that comes out of volcanoes that is not lava. *See also* EARTH, EARTHQUAKES, LIFE, and MOUNTAINS.

Topic Terms

lahars—Lahars are mud flows associated with volcanic activity. They can take the form of the breaching of a volcanic crater lake, the melting of snow on the volcanic peak, or the interaction between the fallen ash and large amounts of rainfall.

lava—Scientists have identified two basic types of lava, or ground-flowing magma, found around the world, called pahoehoe and aa. Their names come from Hawaii, where lava flows are a way of life.

Pahoehoe (pa-hoy-hoy)—Pahoehoe is a smooth kind of lava flow that tends to form a ropy surface; it comes from higher temperature and lower-volume eruptions, and has low viscosity, allowing it to flow easily and form a skin. The rate of flow of this type of lava tends to be a slow creeping motion, moving along at approximately a yard (or meter) per minute. But, under the right conditions, e.g., a steep slope and higher rates of emission, it can move at speeds up to 14 miles per hour. The thickness of the flow of pahoehoe is typically around 1 foot (30 centimeters).

Aa—Aa lava is sharp, rough, and coarse; the name is very expressive of the way you feel if you should fall on it. This lava normally occurs under the opposite conditions that produce pahoehoe and has very different characteristics. Aa tends to flow in surges, with the front advancing slowly, building up its height as it moves along at a rate of a few yards per hour. Then, suddenly, it will quickly surge forward, returning to its original thickness while covering ground at the rate of 100 yards in a few minutes. The normal thickness of an aa flow is between 6.5 and 16.5 feet (2 and 5 meters) thick, and these flows tend to be large, extending more than 100 yards wide, fed by numerous smaller streams of lava.

Other features that form from these lavas include the following:

Pele's hair—Pele's hair is a very viscous lava that has been forced through a very small opening and modified by the wind.

pillow lava—Pillow lava is a form of pahoehoe lava that is normally created underwater, at ocean rifts and ridges. The lava that flows out of these openings does not solidify instantly; the characteristic skin that forms on the surface of the pahoehoe acts as an insulating layer, allowing a sack-like structure of lava to form. This sack inflates as the lava accumulates inside, then a tear forms and the lava flows out, forming another sack. This process continues, and the accumulated sacks resemble piles of pillows, hence the name. The ocean floor has a great deal of this type of lava, and it can only be seen on land where the sea floor has been uplifted.

pumice—Pumice is an acidic lava that has solidified with a large number of gas bubbles. Because

This photo shows "splatter" activity in a fissure on the Kilauea, Hawaii, eastern rift. This volcano is among the most extensively studied in the world. *(Photo courtesy of U.S. Geologic Survey, Hawaii Volcano Observatory)*

of the multitude of gas bubbles and holes within the cooled rock, it is one of the only types of rock known to float.

magma—Scientists now know that the chemical composition of the rocks in a volcano and in the magma upwelling from beneath are crucial to the way it behaves. In fact, they believe that the silica content of magma, combined with the water content, determines the eruptive characteristics of a specific volcano.

Silica (silicon dioxide or SiO_2) is the most abundant mineral on Earth and is the formula for quartz and glass. Magma with a high silica content is more viscous and thus thicker and less free-flowing, while lower-silica magma is thinner and flows more freely. Basalt has approximately 55 percent silica and is the runniest lava, while rhyolite has approximately 73 percent and is the most viscous. In between, both in silica content and viscosity, are andesite and dacite. When high-silica magma cools quickly, it forms a volcanic glass without a crystalline structure known as obsidian; this was used by ancient civilizations for tools and decorations. Because basalt is the least viscous of lavas, it can flow, cover, and thicken over large areas, such as the 88,000-square-mile Columbia River plateau in the United States' northwest, or the 200,000-square-mile Deccan Traps area of northwest India.

The interaction of the silica content of the magma and the amount of water present, superheated to a high pressure gas by the high temperature, has been suggested as the cause of the different types of volcanic eruptions. When a low amount of water is present, and a low amount of silica, the resulting eruption is quiet, with mostly runny lava. But if the water content is high, the resulting steam will spew out the runny lava in fire fountains. The combination of low water content and high silica content results in the release of a viscous lava that will gradually build up a large lava dome. When high water content is present in conjunction with high silica content, the chance of geyser-like explosions is greater.

tephra—Tephra is the general name given to the all the material that is ejected from a volcano that is not lava. These pyroclastic materials (a pyroclast is material ejected during the explosive eruption of a volcano in the form of fragments) come in all sizes and shapes, from very tiny, fine ash, to large house-sized blocks or bombs. Volcanologists have a classification system for the tephra that comes out of a volcano. In general, anything smaller than approximately a tenth of an inch (2 millimeters) is called ash; anything between 1 inch and 25 inches (2 and 64 centimeters) is called lapilli, or "little stones." Anything larger than is either a block (a solid, ejected rock) or a bomb (still molten inside). Bombs, being fluidic, are shaped by their passage through the air and may explode and gush on contact; they have been known to travel more than 3 miles (5 kilometers), with initial velocities above 1,000 miles (1,609 kilometers) per hour.

vents—Vents are areas where volcanic activity reaches a crack in the ocean floor. These vents create what are known as "smokers," high chimneys that release sulfur-rich hot water into the ocean. Worms up to 10 feet (33 meters) long and 4 inches (10 centimeters) in diameter have been discovered around the smokers, as have clams more than 10 inches (25 centimeters) long, strange crabs, sea anemones, and some fish. Most marine organisms survive because of the Sun, feeding on plants that need sunlight for photosynthesis, or feeding on the animals that eat the plants. Along the vents, no photosynthetic plants can survive; the animals make their own food using the sulfur and bacteria from the vents.

The existence of these creatures has added a new dimension to how organisms survive—and brings up two major questions concerning our natural history: Could some life have begun on Earth in the oceans, along the vents? And if organisms live without the need for plants or sunlight, could other such life have gained a foothold in the possible deep oceans of other satellites—such as Jupiter's large moon, Europa—in our solar system?

volcano—The word "volcano" comes from the Mediterranean area; it was derived from the island of Vulcano, located in the Tyrrhenian Sea off the northern tip of Sicily, where an active volcano was located in ancient times. This island was thought to be the entrance to the nether regions that were the domain of Vulcan, also known to the Greeks as Hephaestus, where, on his forge, he fashioned armor for the gods and thunderbolts for Jupiter.

Overall, three designations are assigned to volcanoes: active (eruptions present), extinct (no longer a chance of eruption), and dormant (presently not erupting). However, even this classification suffers, since it appears that on the average, one "extinct" volcano erupts every five years!

Volcanic features develop when hot, liquid rock called magma pushes up at vents, or openings, in the surface from deep within the Earth's interior. The mechanism of formation of magma is still under debate by volcanologists. Although volcanoes do cause widespread destruction, they are our only means of studying and understanding the very deep recesses of our planet. For example, the magma of the Hawaiian volcano Kilauea originates from tens of miles down or deeper, while our drilling cannot at present go more than 6 miles down.

Volcanoes occur at three general regions on Earth, and the theory of plate tectonics neatly explains two of them. Most magma reaches the surface along major weakened areas in the crust called plate boundaries. These plate boundaries represent areas in which the earth's many crustal plates are moving—either sliding past, subducting, or pushing up the crust where they come in contact with each other. Along these boundaries, there is an abundance of volcanoes and earthquakes.

The rift, or ridge, volcanoes occur mainly at oceanic ridges, at the boundary where plates are pulling apart. Here, magma flows out of the cracks, creating new sea floor. Examples of these areas include the Mid-Atlantic Ridge and the East Pacific Rise; the number of oceanic volcanoes associated with these rifts is unknown, but is thought to be quite large. More rare, but much easier to study, are the rifts found on land—mostly on Iceland (the only above-water part of the Mid-Atlantic Ridge) and the Great Rift Valley in East Africa. These two areas have approximately 160 potentially active volcanoes combined.

The other area in which plate tectonics best explains volcanoes is in subduction zones, where one plate is slipping under the other. As the subducting plate sinks into the mantle, it generates magma (in mechanisms that are not fully understood); this hot rock rises up to form volcanoes. These types of volcanoes are found on continents, such as the Andes of South America, or in the ocean, such as the islands of Japan. In fact, most of the Pacific Ocean rim is ringed with volcanoes because of this subducting action of the plates, leading to a large amount of volcanic activity in the so-called Ring of Fire (also referred to as the "Circle of Fire"). For example, the western United States has 69 subduction-related volcanoes, Alaska and the Aleutian Islands have 68, Kamchatka has 65, Chile 75, and Japan 77.

Other types of volcanoes are not formed at plate boundaries, but elsewhere, such as the Hawaiian chain of islands found in the middle of the Pacific plate. To account for these types, J. Tuzo Wilson, one of the pioneers of plate tectonics, theorized the presence of a "hot spot," in which a stationary upwelling of hot magma seeps through weaker spots in the continental plates from the mantle below. As the plate slowly moves past this plume-shaped mass of molten material, a magma chamber continues to provide molten rock for continuing eruptions—thus creating a chain of volcanic islands over millions of years. The Hawaiian hot spot has generated approximately 200 volcanoes over 75 million years—and in fact, another island is building up under the ocean surface even now, but it will probably take thousands of years to reach the surface. Other hot spots may be responsible for the Azores, the Galapagos and Society Islands, and the geysers and other activities at Yellowstone National Park in Wyoming.

Some volcanic mountain chains form far from a plate boundary, but are still associated with the moving crust. For instance, the Cascade Mountains, running from British Columbia's Frazer River to Lassen Peak in northeastern California, are a chain of volcanic mountains formed about a million years ago. The Cascades sit on an ancient plateau that was pushed up and buckled from north to south by the movement of the Pacific, Juan de Fuca, and North American crustal plates. Millions of years ago, volcanic eruptions occurred along the area of the Cascades. About a million years ago, more and larger volcanoes erupted along the plateau, including Mounts Hood, Shasta, and Rainier. Today, many of the volcanoes in the Cascades are dormant, but others are still active—the result of the continuing movement of crustal plates more than 200 miles (322 kilometers) away.

Volcanoes are one of the most destructive natural disasters on Earth. The volcanic hazards to humans depend on the type of volcano and eruption. Hazards include asphyxiation from volcanic ash, dust, and/or gases; a combination of a thick layer of ash and water (usually from torrential rains) that causes a mudslide; molten lava flowing out and covering everything in its path; the explosive concussion of the eruption that knocks down everything in its path for miles around the volcano; roof collapse from ash, dust, and lava "bombs" (hot rock spit out by the volcano); tsunamis from volcanically precipitated seismic waves or huge landslides washing into nearby oceans; and firestorms started by falling molten rock.

Major and Notable Volcanic Eruptions *			
Eruption	*Year*	*Casualties*	*Major Causes of Casualties*
Mt. Pinatubo, Philippines	1991	350	roof collapse; disease; mudslides from excessive rains after the eruption
Nevado del Ruiz, Columbia	1985	25,000	mudflows
Mt. St. Helens, Washington, USA	1980	61	asphyxiation from ash
Mt. Pelee, Martinique	1902	30,000	pyroclastic ashflows
Krakatau, Indonesia	1883	36,000	tsunami
Tambora, Indonesia	1815	92,000	starvation
Unzen, Japan	1792	15,000	volcano collapse; mudslide and resulting tsunami
Lakagigar, Iceland (Mount Skaptar)	1784	9,800	starvation; poisonous gases kill livestock and crops
Mt. Etna, Sicily	1669	20,000	ash and dust
Mt. Etna, Sicily	1683	60,000	ash and dust
Kelut, Indonesia	1586	10,000	unknown

* Modified from the United States Geological Survey/Cascades Volcano Observatory and *The Universal Almanac* 1997 by John W. Wright (ed.), New York: Andrews & McMeel, 1996

Famous Volcanoes			
Name and Local	*Elevation (ASL*)*	*Type*	*Comments*
Cotopaxi, Ecuador	19,347 ft/5,897 m	composite	one of the tallest active volcanoes
Etna, Sicily, Italy	10,902 ft/3,323 m	composite	one of the most active in Europe, with records of more than 200 eruptions
Kilauea, Hawaii, USA	4,078 ft/1,243 m	shield	Kilauea is an active volcano
Krakatau, Indonesia	2,667 ft/813 m	composite	the great eruption of 1883 was heard more than 2,900 miles (4,700 kilometers) away

Famous Volcanoes (cont'd.)

Name and Local	Elevation (ASL*)	Type	Comments
Mauna Loa, Hawaii, USA	13,677 ft/4,169 m	shield	the world's largest volcano; it is more than 9,600 cubic miles (40,000 cubic kilometers) in volume
Mt. Fuji	12,388 ft/3,776 m	composite	the cone is nearly symmetrical
Mt. Katmai, Alaska, USA	6,715 ft/2,047 m	composite	the Valley of Ten Thousand Smokes formed here in 1912 from an ash flow
Mt. Pelee, Martinique, West Indies	4,583 ft/1,397 m	composite	the 1902 eruption eliminated the city of St. Pierre in minutes
Mt. St. Helens, Washington, USA	8,366 ft/2,550 m	composite	one of the most memorable eruptions in recent record, in 1980
Paricutin, Mexico	8,990 ft/2,740 m	cinder cone	the volcano formed from a small crack in a farmer's field in 1943, and grew to its present height
Stromboli, Italy	3,038 ft/926 m	composite	eruptions have been continuous for at least 2,000 years
Surtsey, Iceland	586 ft/173 m	cinder cone and lava flow	the island appeared above the ocean water in 1963, and currently has an area of about 1.1 square miles (2.8 square kilometers)
Tamora, Indonesia	9,354 ft/2851 m	composite	the 1815 eruption released the most amount of material of any volcano ever
Vesuvius, Italy	4,203 ft/1,281 m	composite	the A.D. 79 eruption was recorded as destroying many cities, including Pompeii and Stabiae

* Above Sea Level (in ft [feet] and m [meters])

Appendix A
Careers in Natural History

Introduction

Our knowledge of the natural history of the Earth is tied to the efforts of scientists—past and present—who have examined the intricacies of our universe and planet. From paleontologists who examine ancient bones to determine the earth's previous life to the astronomers who examine features and compositions of other planets to determine the Earth's past, scientists were, and continue to be, the people who fill in the gaps of knowledge in natural history. Presented here are just a few of the careers in the expansive field of the natural sciences.

Types of Scientists

Scientist name	A person who studies ...
Astronomy:	
archaeoastronomer	ancient astronomy (also called astroarchaeologist).
astrochemist	the composition of other bodies in space, such as asteroids.
astrophysicist	the chemical and physical processes in the universe.
cosmochemist	the chemistry of objects in the universe, including molecules found in interstellar clouds.
cosmologist	the structure, dynamics, and development of the universe.
observational astronomer	the universe using a telescope.
planetologist	the physical or chemical features of the planets in the solar system.
solar astronomer	the Sun.
steller astronomer	stars.
theoretical astronomer	physics and mathematics to determine the nature of the universe.
Biology:	
anatomist	the structure of living organisms.
bacteriologist	bacteria.
biochemist	the chemical processes and substances that occur in living organisms.
biophysicist	living things using the tools and techniques in physics.
botanist	plants.
comparative anatomist	the similarities and differences in the body structures of animals.
cryobiologist	extremely low temperatures and how they affect living organisms.
crytologist	the structure, composition, and function of cells.
ecologist	the relationships between living organisms and their surrounding environment.
embryologist	the formation and development of plants and animals, from fertilization to the time they become independent organisms.
entomology	insects.
ethology	certain animal behavior under natural conditions.
evolutionary biologist	the evidence, physical or otherwise, that supports the theory of evolution.
exobiologist	life elsewhere in the universe.
geneticist	heredity.
gnotobioticist	organisms raised in environments free of germs or those that only contain certain specific germs.

Scientist name	A person who studies ... (cont'd.)
histologist	tissues from an organism.
ichthyologist	fish.
immunologist	the body's defenses from disease and foreign substances that may try to enter the body.
lepidopterist	insects in the order Lepidoptera, which includes butterflies and moths.
limnologist	freshwater bodies and the organisms that live within the water.
marine biologist	life in the oceans.
medicine (doctor)	the art of treating and healing humans.
microbiologist	microscopic organisms.
molecular biologist	the molecular processes that occur in cells of organisms.
mycologist	fungus.
naturalist	the natural world, including people involved in zoology or botany.
neurobiologist	the nervous system of animals.
oologoist	bird eggs.
ornithologist	birds.
parasitologist	the lifecycle and processes of parasites.
pathologist	the changes in the body that can cause disease, or the changes that are caused by disease.
pharmacist	the distribution of medicines for health.
physiologist	the functions of living organisms.
sociobiologist	the biological bases for the social behavior between animals and/or humans.
taxonomist	the scientific classification of organisms, also referred to as systematics.
virologist	viruses and the types of disease from viruses.
zoologist	animals.

Chemistry:

analytical chemist	the detailed analysis of chemical substances and their individual components.
biochemist	the compounds and chemical reactions in living systems.
chemical kineticist	the chemical reactions and the factors affecting the rates of reaction.
electrochemist	the relationships between flow of electricity and chemical changes.
environmental chemist	the chemical changes in the natural environment and how they affect humans and other species.
geochemist	the chemical analysis of rocks and minerals.
inorganic chemist	the compounds generally not containing carbon-to-carbon bonds and usually derived from minerals.
nuclear chemist	the chemical changes in the atomic nucleus.
organic chemist	the compounds from living organisms, as well as hydrocarbons and their derivations.
photochemist	the reactions caused by light or ultraviolet radiation on chemical elements.
physical chemist	the applications of physical principles to describe chemical processes.

Scientist name	A person who studies ... (cont'd.)
polymer chemist	plastics and other similar molecules.
qualitative chemist	the types of compounds and elements that make up chemical substances.
quantitative chemist	the amounts of the various chemicals that make up substances.
quantum chemist	the distribution of electrons and the interpretation of the chemical behavior of molecules in terms of the electron structures.
radiochemist	the production and identification of radioactive isotopes of chemical elements.
solid-state chemist	the composition of solids and changes that occur within and between solids.
stereochemist	the arrangement of atoms and the properties of the arrangement.
stoichiometrist	the mathematical relationships between components of chemical reactions and formulas.
synthetic chemist	how to combine chemical elements and compounds to duplicate naturally occurring substances; or the study of such combinations to produce compounds that do not occur naturally.
thermochemist	the interrelationship of heat and chemical reactions.

Geology/Earth Science:

anthropologist	the origin and development of human cultures; includes cultures, interactions, origins, evolution, distributions, and social forms; also the study of how humans have evolved certain physical characteristics over time.
archaeologist	cultural development by researching objects and things that peoples of various cultures have made and used over time; the field deals with finding, preserving, and interpreting the physical remains left by our ancestors, both modern and primitive.
climatologist	weather trends to determine general patterns, in hopes of eventually predicting weather patterns more accurately.
economic geologist	metals, coal, and other natural materials useful to industry.
environmental geologist	the applications of the geologic principles to surrounding environmental problems.
geochemist	the substances in the earth and the chemical changes that the planet goes through.
geochronologist	the geologic time scale.
geophysicist	many Earth conditions, including electromagnetic properties, the interior of the planet, and features of earthquakes.
glacial geologist	glaciers and how they alter the surface of the Earth.
hydrologist	the distribution and movements of water above and below the surface of the Earth.
ichnologist	a field of paleontology that deals with the study of footprints preserved in rock.
marine geologist	the geology of the oceans.
meteorologist	the Earth's atmosphere and the weather produced within the atmosphere.
minerologist	minerals.
oceanographer	oceans.
paleoanthropologist	ancient, fossil humans (a subset of anthropology).
paleoecologist	the relationships between ancient animals and plants, and how they interacted with their surroundings.

Scientist name	A person who studies ... (cont'd.)
paleontologist	prehistoric life.
petrologist	all forms of rock: igneous, sedimentary, and metamorphic.
sedimentologist	sediment and how it is deposited on the Earth's surface.
stratigrapher	the layers of rock in the Earth's crust.
structural geologist	the shapes and amount of rock deep in the Earth, and the causes for the changes in rock layers.
volcanologist	the activity, formation, and features of volcanoes.

Physics:

Scientist name	A person who studies ...
acoustic engineer	the property of sound.
aerodynamicist	the forces acting on an object as it flies through the air or other fluid.
atomic physicist	the structure, properties, and behavior of atoms.
biophysicist	living organisms and life processes with physics tools and techniques.
cryogenicist	extremely low temperatures.
electrodynamics	the relationship between electric and magnetic forces.
fluid physicist	the behavior and movement of liquid and gases (sometimes called fluid dynamics).
health physicist	the protection of people who work with or near radiation.
mathematical physicist	the mathematical systems that represent physical phenomena.
molecular physicist	the structure, properties, and behavior of molecules.
nuclear physicist	the structure and properties of atomic nuclei; also the study of nuclear reactions and their applications.
optical physicist	the nature and behavior of light.
particle physicist	the behavior and properties of elementary particles, also called a high-energy physicist.
plasma physicist	highly ionized gases.
quantum physicist	quantum theory that deals with the interaction between matter and electromagnetic radiation.
solid-state physicist	the physical properties of solid materials; also called a condensed matter physicist.
thermodynamicist	heat and other forms of energy; it also includes the conversion of energy from one form to another.

Appendix B
Natural History Sites on the World Wide Web

(Please note: We have attempted to give you the most current Internet addresses possible, but they do change frequently. We regret any inconvenience created by a changed WWW site address.)

General

Academy of Natural Sciences of Philadelphia (Philadelphia, PA)
http://www.acnatsci.org/

American Museum of Natural History (New York, NY)
http://www.amnh.org/

Carnegie Museum of Natural History (Pittsburgh, PA)
http://www.clpgh.org/cmnh/index2.html

Denver Museum of Natural History (Denver, CO)
http://www.dmnh.org/

Field Museum of Natural History (Chicago, IL)
http://www.fmnh.org/

Idaho Museum of Natural History
http://isu.edu/departments/museum/index.html

National Museum of Natural History, Smithsonian Institution (Washington, DC)
http://www.mnh.si.edu/

Natural History Book Store (NHBS) Mailorder Bookstore
http://www.nhbs.com/

Natural History Museum (London, England)
http://www.nhm.ac.uk/

Natural History Museums and Collections
http://www.lib.washington.edu/sla/natmus.html

Oxford University Museum of Natural History (Oxford, England)
http://www.ashmol.ox.ac.uk/oum/

Provincial Museum of Alberta (Edmonton, AB)
http://www.pma.edmonton.ab.ca/

Royal Ontario Museum (Toronto, ON)
http://www.rom.on.ca/

San Diego Natural History Museum (San Diego, CA)
http://www.sdnhm.org/

Swedish Museum of Natural History (Stockholm, Sweden)
http://www.nrm.se/

University of Colorado Museum (Boulder, CO)
http://www.colorado.edu/CUMUSEUM/

Web Sites by Topic

Amphibians

The Electronic Zoo—NetVet—Amphibians
http://netvet.wustl.edu/amphib.htm

Introduction to the Amphibia
http://www.ucmp.berkeley.edu/vertebrates/tetrapods/amphibintro.html

North American Amphibian Monitoring Program (NAAMP)
http://www.im.nbs.gov/amphibs.html

Society for the Study of Amphibians and Reptiles
http://falcon.cc.ukans.edu/~gpisani/SSAR.html

USGS Frog and Amphibian Research
http://www.usgs.gov/frogs.html

Animals

American Museum of Natural History—Expedition: Endangered!
http://www.amnh.org/Exhibition/Expedition/Endangered/index.html

Discovery Online -1998 Animal Olympics
http://discovery.com/area/nature/animalolympics/animalolympics.html

The Electronic Zoo—Animal Resources
http://netvet.wustl.edu/ssi.htm

Mammal Species of the World
http://nmnhgoph.si.edu/msw/

Mammal Species of the World
http://nmnhwww.si.edu/gopher-menus/MammalSpeciesoftheWorld.html

Mammalia
http://www.oit.itd.umich.edu/bio108/Chordata/Mammalia.shtml

Protected Marine Species
http://www.rtis.com/nat/user/elsberry/marspec.html

The University of Michigan Museum of Zoology's Animal Diversity Web
http://www.oit.itd.umich.edu/bio108/

Virtual Zoo: Mammals
http://mirrors.org.sg/vz/mammals.html

Zoology Resource Guide
http://www.york.biosis.org/zrdocs/zoolinfo/gp_index.htm

Arthropods

Arthropods of the El Paso Region
http://www.utep.edu/~epbionet/cheklist/arthropo/arthlist.htm

The Birmingham Zoo: Arthropods
http://www.birminghamzoo.com/ao/arthrop.htm

Department of Entomology, Virginia Tech: Characteristics of Arthropods
http://www.ento.vt.edu/Courses/Undergraduate/IHS/distance/html_files/char.html

University of Michigan Museum of Zoology—Phylum Arthropoda
http://www.oit.itd.umich.edu/bio108/Arthropoda.shtml
University of Wisconsin-Madison, Geology Museum: Phylum Arthropoda
http://www.geology.wisc.edu/~museum/arthroinfo.html

Bacteria and Viruses

Bayer Corporation©: Germs (Bacteria and Viruses)
http://www.bayerpharma-na.com/hottopics/hc0109.asp
Division of Biological Sciences, The University of Texas at Austin: Viruses
http://ccwf.cc.utexas.edu/~bogler/ecology/viruses.html
University of California Museum of Paleontology: Introduction to the Bacteria
http://www.ucmp.berkeley.edu/bacteria/bacteria.html
University of California Museum of Paleontology: Introduction to the Viruses
http://www.ucmp.berkeley.edu/alllife/virus.html
The University of Kansas, Department of Molecular Biosciences: Bugs in the News!
http://falcon.cc.ukans.edu/~jbrown/bugs.html

Birds

Cornell Laboratory of Ornithology
http://birds.cornell.edu/
Kaytee Discovery Zone: Bird Anatomy
http://www.kaytee.com/discovery/anatomy/bones.html
The Mining Company: Bird Anatomy Information
http://birding.miningco.com/msub20-anatomy.htm
National Audubon Society
http://www.audubon.org/
Smithsonian Migratory Bird Center
http://www.si.edu/natzoo/zooview/smbc/smbchome.htm

Classification

The Basics of Taxonomy
http://town.morrison.co.us/dinosaur/taxinfo.html
Introductory Glossary of Cladistic Terms
http://www.science.uts.edu.au/~davidm/glossary.html
Journey into Phylogenetic Systematics
http://www.ucmp.berkeley.edu/clad/clad4.html
On Taxonomy and Cladistics
http://town.morrison.co.us/dinosaur/tax-clad.html
Why Do Biologists Need Cladistics?
http://www.ucmp.berkeley.edu/clad/clad5.html

Atmosphere/Climate & Weather

The National Center for Atmospheric Research
http://www.ncar.ucar.edu/
National Oceanic and Atmospheric Administration
http://www.noaa.gov/
National Weather Service
http://www.nws.noaa.gov/
Royal Meteorological Society
http://itu.rdg.ac.uk/rms/rms.html
World Meteorological Organization
http://www.wmo.ch/

Cytology

Cells
http://www.botany.uwc.ac.za/SCI_ED/std8/cells/index.htm
Cells II: Cellular Organization
http://gened.emc.maricopa.edu/bio/bio181/BIOBK/BioBookCELL2.html
Cells: General
http://miramesa.sdcs.k12.ca.us/Organelles/home.html
The Dictionary of Cell Biology
http://www.mblab.gla.ac.uk/~julian/Dict.html
Picture It: Cells
http://www.clarityconnect.com/webpages/cramer/PictureIt/cells.htm

Dinosaurs

Dinamation International
http://www.dinamation.org/
Dino Russ's Lair
http://128.174.172.76:/isgsroot/dinos/dinos_home.html
Dinosauria On-Line
http://www.dinosauria.com/
Dinosaurs: Facts and Fiction
http://pubs.usgs.gov/gip/dinosaurs/
Dinosaurs: Facts and Fiction
http://pubs.usgs.gov/gip/dinosaurs/
Museum of the Rockies
http://museum.montana.edu/

Earth

Cutaway View of the Earth's Interior
http://www.hawastsoc.org/solar/cap/earth/earthint.htm
Earth Science Systems Programs (formerly Mission To Planet Earth [MTPE])
http://www.hq.nasa.gov/office/mtpe/
Earthwatch Institute International
http://www.earthwatch.org/
An Introduction to Plate Tectonics
http://www.hartrao.ac.za/geodesy/Plate.html
The Virtual Geosciences Professor
http://www.uh.edu/~jbutler/anon/anonfield.html

Earthquakes

Earthquakes
http://pubs.usgs.gov/gip/earthq1/
National Earthquake Information Center
http://wwwneic.cr.usgs.gov/
The San Andreas Fault
http://pubs.usgs.gov/gip/earthq3/contents.html

Evolution

Enter Evolution: Theory and History
http://www.ucmp.berkeley.edu/history/evolution.html

Evolution and Natural Selection
  http://www.sprl.umich.edu/GCL/paper_to_html/selection.html
The Origin of Species
  http://www.literature.org/Works/Charles-Darwin/origin/
Scientific American: In Focus. Evolution Evolving
  http://www.sciam.com/0997issue/0997infocus.html
The Talk.Origins Archive: Biology and Evolutionary Theory
  http://www.talkorigins.org/origins/faqs-evolution.html

Extinction

Dinosaur Extinction
  http://www.enchantedlearning.com/subjects/dinosaurs/extinction/index.html
Extinction
  http://tyrrell.magtech.ab.ca/tour/extinctn.html
Extinction—What Is It ?
  http://palaeo.gly.bris.ac.uk/communication/Belton/Extn.html
A Mathematical Model for Mass Extinction
  http://www.lassp.cornell.edu/newmme/science/extinction.html
On the Road to Extinction
  http://tqjunior.advanced.org/4073/

Fish

Animal Diversity Web: Actinopterygii (bony fish, osteichthyes, ray finned fish, spiny rayed fish)
  http://www.oit.itd.umich.edu/bio108/Chordata/Actinopterygii.shtml
Animal Diversity Web: Chondrichthyes (cartilaginous fishes: sharks, rays, chimaeras)
  http://www.oit.itd.umich.edu/bio108/Chordata/Chondrichthyes.shtml
Colorado State University
  Department of Fishery and Wildlife Biology—Ichthyology
  http://www.cnr.colostate.edu/~brett/fw300/300home.html
Ichthyology Resources
  http://muse.bio.cornell.edu/cgi-bin/hl?fish
NMFS—Northeast Fisheries Science Center: Fish FAQ
  http://gopher.wh.whoi.edu/homepage/faq.html

Fossils

The Field Museum of Chicago: Life over Time
  http://www.fmnh.org/exhibits/web_exhibits.htm
Fossils, Rocks, and Time
  http://pubs.usgs.gov/gip/fossils/contents.html
St. Louis Science Center: Fossils
  http://www.slsc.org/docs/mod3/mod3_2/mod3_22/ep2800ag.htm
University of California Museum of Paleontology: Learning from the Fossil Record
  http://www.ucmp.berkeley.edu/fosrec/Learning.html
University of California Museum of Paleontology: Paleontology: The Window to Science Education
  http://www.ucmp.berkeley.edu/fosrec/Stucky.html

Fungus

British Mycological Society
  http://www.ulst.ac.uk/faculty/science/bms/
Medical Mycology Research Center
  The University of Texas Medical Branch at Galveston, Texas, USA
  http://fungus.utmb.edu/f-atlas/medmyc.htm
Mycological Resources on the Internet: Collections
  http://muse.bio.cornell.edu/~fungi/fcollect.html
Mycological Resources on the Internet: Resources for Teaching
  http://muse.bio.cornell.edu/~fungi/fteach.html
University of Minnesota: Mycological Aspects of Indoor Environmental Quality
  http://134.84.147.72/fungus/myco.html

Genetics

Access Excellence: Graphics Gallery
  http://www.gene.com/ae/AB/GG/
A Gene Map of the Human Genome
  http://www.ncbi.nlm.nih.gov/SCIENCE96/
Mendelian Genetics
  http://esg-www.mit.edu:8001/esgbio/mg/mgdir.html
MendelWeb
  http://www.netspace.org/MendelWeb/
Picture It: Genetics
  http://www.clarityconnect.com/webpages/cramer/PictureIt/gene.htm

Geologic Time Scale

Geologic Time
  http://pubs.usgs.gov/gip/geotime/contents.html
Welcome to the UCMP Web Lift for Geologic Time
  http://www.ucmp.berkeley.edu/help/timeformold.html

Humans

Hominid Species
  http://earth.ics.uci.edu/faqs/homs/species.html
Human Evolution
  http://bioserve.latrobe.edu.au/vcebiol/cat3/u4aos2p5.html
More Skullduggery: Know Your Skulls!
  http://www.geocities.com/Athens/Acropolis/5579/skullduggery.html
Prominent Hominid Fossils
  http://earth.ics.uci.edu/faqs/homs/specimen.html
Timeline
  http://www.geocities.com/Athens/Acropolis/5579/timeline.html

Life

Astrobiology and the Origins of Life
  http://www.gene.com/ae/bioforum/bf02/awramik/bf02a1.html

The Beginnings of Life on Earth
  http://www.sigmaxi.org/amsci/articles/95articles/CdeDuve.html
The Origins and Early Evolution of Life
  http://www.chemistry.ucsc.edu/Projects/origin/home.html
Origins of Life
  http://nitro.biosci.arizona.edu/courses/EEB105/lectures/Origins_Of_Life/origins.html
Origins of Life May Be Darker Than We Think
  http://wupa.wustl.edu/nai/feature/1996/Feb96-Life.html

Moon

Everything You Ever Wanted to Know about the Moon
  http://www.tsgc.utexas.edu/everything/moon/
Exploring the Moon
  http://cass.jsc.nasa.gov/moon.html
Lunar Exploration
  http://nssdc.gsfc.nasa.gov/planetary/lunar/apollo_25th.html
The Moon
  http://www.seds.org/billa/tnp/luna.html
Moon Fact Sheet
  http://nssdc.gsfc.nasa.gov/planetary/factsheet/moonfact.html

Mountains

Assembly and Breakup of Pangea
  http://www.geo.ukans.edu/faculty/wrvs/105_S97/notes/breakup_of_pangea.html
Deformation, Mountain Building, and The Evolution of Continents
  http://bama.ua.edu/~blank005/geology/deformation.html
Mountain Erosion
  http://asterion.rockefeller.edu/marcelo/Mountain/mountain.html
Regional Climatic Influence on Tectonics: Compressional Orogenies
  http://earth.agu.org/revgeophys/sleep00/node2.html
Tectonics of the Appalachians
  http://www.geo.wvu.edu/~geol351/spring98/rviso/webpage2/apptect.htm

Oceans

International Ocean Institute
  http://is.dal.ca/~ioihfx/
National Oceanic and Atmospheric Administration
  http://www.noaa.gov/
The Oceanography Society
  http://www.tos.org/
Scripps Institute of Oceanography
  http://www-sio.ucsd.edu/
Woods Hole Oceanographic Institution
  http://www.whoi.edu/

Plants

Botanical Society of America
  http://www.botany.org/
Diversity of Life Web Index (Kingdom: Plantae [Plants])
  http://www.geocities.com/RainForest/6243/diversity4.html#Plant
Internet Directory for Botany: Vascular Plant Families
  http://www.helsinki.fi/kmus/botvasc.html
Tree of Life: Green Plants
  http://phylogeny.arizona.edu/tree/eukaryotes/green_plants/green_plants.html
University of California Museum of Paleontology: Introduction to the Plantae
  http://www.ucmp.berkeley.edu/plants/plantae.html

Protista

BIOSIS© Internet Resource Guide for Zoology: Protista
  http://www.york.biosis.org/zrdocs/zoolinfo/grp_prot.htm
Diversity of Life Web Index (Kingdom: Protista or Protoctista)
  http://www.geocities.com/RainForest/6243/diversity2.html#Protist
Natural Perspective: The Protoctist Kingdom (Protoctistae)
  http://www.perspective.com/nature/protoctista-index.html
University of California Museum of Paleontology, Eukaryota: Systematics
  http://www.ucmp.berkeley.edu/alllife/eukaryotasy.html
The University of Texas at Austin; Division of Biological Sciences: Protista
  http://ccwf.cc.utexas.edu/~bogler/ecology/protista.html

Reptiles

Crocodilians: Natural History & Conservation
  http://www.flmnh.ufl.edu/natsci/herpetology/brittoncrocs/cnhc.html
The EMBL Reptile Database
  http://www.embl-heidelberg.de/~uetz/LivingReptiles.html
Herp-edia
  http://members.aol.com/FRILLED/
Reptile Page
  http://tqjunior.advanced.org/4073/reptile/megrep.html
Sea Turtle Conservation Program
  http://www.co.broward.fl.us/bri00600.htm

Solar System

The Nine Planets: A Multimedia Tour of the Solar System
  http://www.seds.org/billa/tnp/
The Solar System
  http://www.geocities.com/CapeCanaveral/Lab/1000/solar_system.html
Solar System Links
  http://www.as.wvu.edu/~planet/lnk_solr.htm

Theories of the Solar System: Links
   http://macml-mac.lut.ac.uk/HistAstron.html
Views of the Solar System
   http://bang.lanl.gov/solarsys/eng/homepage.htm

Universe

Constructing the Cosmos
   http://www.sciam.com/explorations/012797cosmos/012797horgan.html
Cosmology and What Happened Before the Big Bang!
   http://www2.ari.net/home/odenwald/cosmol.html
Dark Matter, Cosmology, and Large Scale Structure of the Universe
   http://www.astro.queensu.ca/~dursi/dm-tutorial/dm0.html
Introduction to Cosmology
   http://map.gsfc.nasa.gov/html/web_site.html
Our Hierarchical Universe
   http://chico.ncsa.uiuc.edu/Cyberia/Cosmos/HierarchUni.html

Volcanoes

Cascades Volcano Observatory
   http://vulcan.wr.usgs.gov/home.html
Hawaiian Volcano Observatory
   http://hvo.wr.usgs.gov/
The Volcano Information Center
   http://www.geol.ucsb.edu/~fisher/
Volcanoes
   http://pubs.usgs.gov/gip/volc/cover2.html
Volcanoes of the United States
   http://pubs.usgs.gov/gip/volcus/titlepage.html

Glossary

absolute temperature—The temperature at which no energy loss is possible and where all molecular motion stops. Absolute zero is equal to –459.67° F (-273.16° C)—a temperature that has not yet been reached in a laboratory. The Kelvin temperature scale favored by physicists starts at absolute zero.

aerobic—Aerobic is often in reference to bacteria; it means that the organism thrives in an oxygen-filled environment.

albedo—The reflectance of an object. Usually light colored objects have a higher albedo (reflectance) than darker objects.

amphibious—An organism that lives on land and in the water.

anaerobic—Anaerobic is often in reference to bacteria; it means that the organism thrives in an environment of low or no oxygen.

apogee—The farthest distance of a space body from the central body it is orbiting. The earth's apogee occurs when it is farthest from the Sun, usually around July; the Moon's apogee occurs when it is farthest from the earth in its orbit, which occurs once a lunar month.

astrology—Often thought of as a religion in ancient times, it is the study of heavenly bodies and their influence on everyday life. Originally, space bodies were regarded as dieties, thus they had an influence on earthly and human affairs. The configuration of the Sun, Moon, and planets in the sky at the time and place of one's birth was believed to be a significant indicator of a person's character and destiny.

astronomical unit—Astronomical units are one of the ways astronomers use to measure distances in the universe, such as distances to stars and galaxies. One astronomical unit equals about 93,000,000 miles [149,673,000 kilometers], the average distance between the earth and the Sun; therefore, the earth is 1 astronomical unit from the Sun.

carnivorous—Animals (birds) that are exclusively meat-eating. They may hunt for prey, either other birds or animals, or scavenge dead animals.

chemistry—Chemistry is the science that studies the structure and properties of substances based on sound experiments and observations. It was derived from the Arabic for "gold cooking."

condensation—The process that occurs when a gas becomes a liquid; the temperature at which this occurs is known as the condensation point. Clouds are a common example of condensation: Warm air rises and is then cooled in the upper atmosphere. This process causes the water vapor (a gas) to condense, forming masses of very tiny water droplets, also known as clouds.

detritus—Fragmented dead material, usually mentioned in association with plants on the forest floor or on soil.

ectothermic—Gathering heat externally to regulate the body temperature; for example, using the rays of the Sun. Animals with this capability are termed "cold-blooded"; reptiles are one example.

ethology—The study of animal behavior.

endothermic—Generating heat internally to regulate the body temperature; for example, through internal chemical reactions. Animals with this capability are termed "warm-blooded." All mammals are warm-blooded.

erosion—Erosion is the wearing away of an object. In the case of natural history, it is usually in reference to a physical feature, such as the erosion of a mountain top, or in reference to rock, as the erosion of an outcrop of sandstone.

evaporation—The process whereby a liquid becomes a gas. As the temperature of a liquid increases, the molecules move more rapidly. Some of these speeded-up molecules break free from the surface of the liquid and escape as a vapor, or gas. A common example of this is the evaporation of liquid water into water vapor by the heating of the Sun's rays.

fauna—A name for animals is fauna. An additional definition for fauna is the entire animal life of a given region, habitat, or fossils found in a rock layer.

fission—A process whereby an atom with large mass is split, as a consequence of a collision, into atoms of smaller masses, accompanied by a release of energy. This process is found in nuclear power plants and weapons. For example, a heavy atom, whose nucleus collides with a neutron, will absorb that neutron. The resulting unstable atom splits into lighter atoms, or those with a smaller mass; the total mass of these lighter atoms is less than that of the original atom, and the extra mass is converted into energy. This relationship, first discovered by Albert Einstein, shows that very little mass is needed to create large quantities of energy; this is the reason why nuclear reactions are so powerful—and often thought to be dangerous. Natural fission is not present on the earth's surface because the rocks are virtually "too old" to produce the fission process.

formation—A rock formation is a layer of rock of a certain age. It is generally determined by specific fossils within the rock; or the dating of the rock, allowing scientists to determine the layer's age.

fossil record—The history of life on Earth, as compiled and interpreted from the fossils that have been recovered to date.

frequency—The number of identical displacements at a given point in a specified time interval as a wave passes through matter. For example, if, in 10 seconds, 100 wave crests pass a specific point, then the frequency is 100 cycles of waves per every 10 seconds, or 10 (wave) cycles per second. The term "cycles per second" is symbolized by Hz, or Hertz, named after Heinrich Hertz, the German physicist.

fusion—The process in which lighter-mass atoms join together to form heavier ones, with accompanying releases of energy. This process needs great amounts of heat to occur and is therefore termed a thermonuclear reaction. For example, when two lighter-mass atoms, in the presence of heat, combine to form a single, heavier-mass atom, extra mass is left over. The excess mass is converted to energy; this energy is what stars, including our Sun, radiate. Without this energy, life could not exist as we know it on Earth; indirectly fusion is the engine that powers our world, including its climate, weather, and life-forms.

genome—A genome is the complete genetic structure of a species.

genotype—A genotype is the genetic constitution of an organism.

hereditary—Hereditary is having a genetic basis; or transmitted from one generation to the next.

heredity—The mechanism of transmitting certain specific characteristics or traits from the parent to the offspring.

hertz—The number of cycles per second. It is also used as a unit of measurement for frequency.

hominid—This term refers to members of the family of humans, Hominidae, which consists of all species on our side of the last common ancestor of humans and living apes.

humanoid—This term refers to any animals that are human-like in appearance.

hybridize—The process of cross-mating between two or more different species.

ice—One phase of water, in its solid state. Some scientists consider ice, with its orderly structure, to be the most abundant mineral on Earth.

ice cap—An ice cap is considered, literally, a permanent thick ice and snow layer on the polar regions of a planet. On Earth, the polar ice caps are found at the North and South Poles. At the North Pole, the ice cap covers an region to around the Arctic Circle; at the South Pole, the ice caps covers the region of the continent of Antarctica. Ice caps often change with the seasons, shrinking in the respective summer and expanding in the respective winter. The term is also used, although less frequently, to describe smaller ice sheets, such as those on the smaller islands in the Arctic; and more erroneously, ice that caps a mountaintop. Mars also has ice caps at its poles.

insectivores—Birds that feed mainly on insects. These include warblers, flycatchers, wrens, and swallows.

invertebrates—Animals without a backbone.

ions—Ions are protons and other positively charged atoms.

libration—The Moon's rotational period is about the same as its period of rotation, thus, only one side of the lunar surface faces the earth. But the lunar orbit is not circular, and the velocity of the Moon varies. The result is an apparent "wobble" in the orbit, called libration, and allows a small sliver of the other side of the Moon to be observed from the earth. Overall, because of libration, about 59 percent of the Moon can be seen from the earth.

light year—The amount of time it takes for light to travel in one year; it is equal to about 5.88 trillion miles (9.46 trillion kilometers). It is used mostly in astronomy to measure distances to deep-space objects, such as stars.

lunar day—The lunar day is unique in the inner solar system: The Moon's day is almost as long as the time it takes to orbit the earth. Because of this, we constantly see the same side of the Moon facing toward us.

metamorphosis—Organisms metamorphose or change in form and structure; it is also referred to as transformation. Numerous species undergo metamorphosis, including amphibians (a tadpole to a frog) and butterflies (a caterpillar to a butterfly).

nebula—A collection of dust and gas; many result in stars.

nocturnal—Nocturnal usually refers to animals—and mostly to animals that forage and hunt for food at night, such as bats or raccoons.

nova—A nova is the sudden brightening of a star. In most cases, the burst can be variable; in a few hours, a nova can brighten from 10,000 to a million times its normal light. It then sinks back the original brightness in a few months. A nova can occur when a star suddenly throws off a shell of matter. Other novas, usually variable ones, flare from 10 days to tens of years. They are usually in binary systems, in which a white dwarf draws material from its companion. The material is burned in a violent nuclear reaction on the dwarf star, creating the periodic outburst.

omnivorous—An omnivorous animal eats both meat and vegetation.

ontogeny—The course of growth and development of an individual organism to maturity.

oviposition—The process of laying or depositing eggs.

paleontology—The field of science that deals with the study of fossils.

pelagic—Birds that are normally found in the ocean setting, far from land. They include such birds as shearwaters, petrels and fulmars.

penumbra—Both the earth and Moon cast long conical (cone-shaped) shadows out into space. The penumbra is the outside, lighter part of a body's shadow, and is usually only in association with a lunar eclipse. If the earth is between the Sun and the Moon, and is farther away from the Moon, or if the Moon passes above the earth's umbral shadow, the earth's penumbral shadow will cause a penumbral lunar eclipse.

perigee—The closest approach of a space body to the central body it is orbiting. The earth's perigee, or when it is closest to the Sun in its orbit, occurs around January; the Moon's perigee occurs when the satellite's orbit is closest to the earth, once every lunar month.

photosynthesis—The process plants go through to convert sunlight into food for the plant.

phylogeny—The ancestry or evolutionary history of an organism.

pressure—Pressure is the measure of the force, or weight, exerted by the air on everything it touches. Cool, dry air has greater weight and is referred to as a high pressure system. Warmer, moist air is lighter and is referred to as a low pressure system. The pressure of the atmosphere is measured by a barometer and can be expressed in inches of mercury or hectopascals (formerly known as millibars).

radiation—Radiation is the transmittal of energy by electromagnetic waves. The energy is dependent on the source: For example, the Sun radiates relatively short-wave radiation, including light; the earth radiates long-wave radiation in the infrared range, usually from the nuclear reactions in the interior of the planet.

revolution—Revolution is the movement of an object around a central body. All planets in our solar system revolve around the Sun in elliptical (oval) orbits. On a smaller scale, satellites revolve around the parent planet, most in elliptical orbits, but some in near-circular orbits. The amount of time it takes for a planet to revolve around the Sun is called its year. For example, the earth takes 365.256 days (one Earth-year) to revolve around the Sun. Pluto takes about 247.69 Earth-years to revolve around the Sun; thus one Pluto-year is equal to 247.69 Earth-years.

rotation—Rotation is the spinning of an object around a central axis. All planets and satellites in the solar system rotate on their axis. For example, the earth rotates on its axis every 23 hours, 56 minutes, and 4.1 seconds; Mars rotates in 24 hours, 37 minutes, and 22.3 seconds. Venus rotates every 243 days, but it spins in the opposite direction from the other planets of the solar system.

sediment—Sediment is the result of rock erosion. The sediment can be anything from fine particles of mud to sand, depending on the original rock being eroded.

sound—Sound begins as a vibratory shock to air or other media and travels through the medium in waves. Sound is the alternate compression and expansion of these waves as they travel at different speeds through various media such as water or air; sound waves cannot travel in a vacuum. (In contrast, light, an electromagnetic wave, can move through the vacuum of space.) Sound has been important to natural history, especially in the survival of animals. Without sound, species would not be able to communicate (especially for territorial and mating rituals) or hear dangerous activity. Each species has a certain range of hearing. The ability to hear varies in the animal kingdom; in general, the frequencies of sonic waves fall within 15 to 20,000 hertz, although some animals can hear higher frequencies.

spawning—The process of laying eggs. The term is used in reference to most of the aquatic, egg-laying animals, such as frogs, toads, and fish.

species—A group of organisms capable of breeding and producing fertile offspring.

speed of light—The speed of light has been measured at 186,291 miles per second (299,792.458 kilometers per second). This speed is important, because, based on our current understanding, it is the upper limit of speed attainable in our universe. Nothing, not even light, can travel at infinite speeds. This means that the light we see from distant stars and galaxies is from the distant past; the farther away these objects are, the longer it took for the light to reach us. Similarly, any radio signals we send out into space will take years to reach even the nearest star. This also puts a limit on the ways humans can travel to the stars—in other words, at current speeds of spaceships, it will take generations to reach any other worlds.

temperature—Temperature is the measure of the average kinetic energy, or motion, of the molecules of a system. Higher kinetic energy molecules are warmer and thus have higher temperatures. The air temperature reflects the warmth of the atmosphere, which is influenced primarily by radiation from the Sun as the earth rotates on its axis. Two temperature scales are commonly used when referring to the weather: Celsius and Fahrenheit. Both use the freezing point of ice and the boiling point of water as reference points; each assigns different numbers to these points, however.

thermal radiation—The transfer of heat by electromagnetic waves, as opposed to heat transferred by convection (heat transferred as the medium moves, such as gases in the atmosphere) or conduction.

topography—Topography is the natural up and down relationship of the landscape with respect to a standard starting point, such as sea level.

transformation—(See metamorphosis.)

tsunami—Tsunami in the oceans are most often caused by the effect of earthquakes on the ocean floor, or from earthquakes precipitated by volcanic activity. This moving body of water is thought to originate as water is displaced in one of these types of seismic events.

umbra—Both the earth and Moon cast long conical (cone-shaped) shadows out into space. The umbra is the inside, darkest part of a body's shadow. When the Moon is between the Sun and the earth, and is close enough to the earth, the Moon's umbral shadow will cause a total solar eclipse (and partial eclipses in the areas just outside of the total eclipse zone); if the Moon is farther from the earth, the Moon's umbral shadow will cause an annular solar eclipse. If the earth is between the Sun and the Moon, and is close enough to the Moon, the earth's umbral shadow will cause a total lunar eclipse.

vertebrates—Vertebrates are animals with backbones.

References

General Information on Natural History

Ackerman, Diane, Ann H. Zwinger, and David Rains Wallace. *The Curious Naturalist: Guide to Understanding and Exploring Nature*. Washington, DC: National Geographic Society, 1998.

Allaby, Michael, ed. *Illustrated Dictionary of Science*. New York: Facts on File, 1995.

Allstetter, William, exec. ed., *Science and Technology Almanac 1999*. Phoenix, AZ: The Oryx Press, 1999.

Angela, Piero, Alberto Angela, Gabriele Tonne (contributor), and Valter Fogato. *The Extraordinary Story of Life on Earth*. Albany, NY: Prometheus Books, 1996.

Asimov, Isaac. *Asimov's Chronology of Science & Discovery*. New York: HarperCollins, 1994.

Barnes-Svarney, Patricia, ed. *The New York Public Library Science Desk Reference*. New York: Macmillan, 1995.

Beebe, William, ed. *The Book of Naturalists: An Anthology of the Best Natural History*. Princeton, NJ: Princeton University Press, 1989.

Bynum, W.F., E.J. Browne, and Roy Porter, eds. *Dictionary of the History of Science*. Princeton, NJ: Princeton University Press, 1981.

Calder, Nigel. *Timescale*. New York: Viking Press, 1983.

Carnegie Library of Pittsburgh. *The Handy Science Answer Book*. Detroit: Visible Ink Press, 1994.

Durrell, Gerald. *A Practical Guide for the Amateur Naturalist*. New York: Knopf, 1982.

Garber, Steven D. *Biology: A Self-Teaching Guide*. New York: John Wiley & Sons, 1989.

Gould, Stephen Jay. *The Book of Life: An Illustrated History of the Evolution of Life on Earth*. New York: W.W. Norton & Company, 1993.

Hellemans, Alexander and Bryan Bunch. *The Timetables of Science*. New York: Touchstone Book, 1991.

Lincoln, R. J. and G. A. Boxshall. *The Cambridge Illustrated Dictionary of Natural History*. New York: Cambridge University Press, 1990.

Matthews, P., ed. *Guinness Book of Records*. New York: Bantum Books, 1994.

Ochoa, George and Melinda Corey. *The Timeline Book of Science*. New York: Ballantine Books, 1995.

Rees, Robin, ed. *The Way Nature Works*. New York: Macmillan, 1992.

Simonis, Doris, ed. *Scientists, Mathematicians, and Inventors* (Lives & Legacies: An Encyclopedia of People Who Changed the World). Phoenix, AZ: The Oryx Press, 1999.

Wilson, Edward O. and Laura Simonds Southworth (illus). *In Search of Nature*. Washington, DC: Island Press, 1997.

Topic References

Alexopoulos, C. J., C. W. Mims, and M. Blackwell. *Introductory Mycology*. New York: John Wiley and Sons, 1996.

Banister, Keith and Andrew Campbell. *Encyclopedia of Aquatic Life*. New York: Facts on File, 1985.

Bova, Ben. *Welcome to Moonbase*. New York: Ballantine Books, 1987.

Burroughs, William J., Bob Crowder, Ted Robertson, Eleanor Vallier-Talbot, and Richard Whitaker. *Weather*. New York: Time-Life Books, 1996.

Carroll, G.C., and D.T. Wicklow. *The Fungal Community: Its Organization and Role in the Ecosystem*. New York: Marcel Deker, Inc., 1992.

Cavendish, Marshall. *World Wildlife Habitats*. vol. 1, 2, & 3. New York: Marshall Cavendish, 1992.

Cogger, Harold G., Richard G. Zweifel, and David Kirshner. *Encyclopedia of Reptiles and Amphibians*. New York: Academic Press, 1998.

Darwin, Charles and Greg Suriano, eds. *The Origin of Species*. New York: Grammercy, 1998.

Decker, Robert and Barbara Decker. *Volcanoes*. New York: W.H. Freeman and Co., 1998.

Dennett, Daniel Clement. *Darwin's Dangerous Idea: Evolution and the Meanings of Life*. New York: Touchstone Books, 1996.

Dix, N.J. and J.W. Webster. *Fungal Ecology*. London: Capman and Hall, 1995.

Dow, J.A.T., ed., and J. M. Lackie. *The Dictionary of Cell Biology*. New York: Academic Press, 1995.

Farlow, James O. and M. K. Brett-Surman, eds. *The Complete Dinosaur*. Indianapolis: Indiana University Press, 1997.

Fisher, Richard V., Grant Heiken, Jeffrey B. Hulen, and Renate Hulen. *Volcanoes: Crucibles of Change*. Princeton, NJ: Princeton University Press, 1997.

Fortey, Richard. *Life: A Natural History of the First Four Billion Years of Life on Earth*. New York: Knopf, 1998.

Friedman, B. Ellen and Ellen R. Friedman. *Bacteria*. New York: Creative Education, 1998.

Gallant, Roy A. and Christopher J. Schuberth. *Earth: The Making of a Planet*. New York: Marshall Cavendish, 1998.

Ganeri, Anita and Luciano Corbella (photo.) *The Oceans Atlas*. London: DK Publishing, 1994.

George, Michael. *The Moon*. New York: Creative Education, 1992.

Gould, Edwin, Gregory McKay, David Kirshner (illus.), and George McKay. *Encyclopedia of Mammals*. New York: Academic Press, 1998.

Graedel, Thomas E. and Paul J. Crutzen. *Atmosphere, Climate, and Change*. New York: W H Freeman & Co, 1997.

Greeley, Ronald and Raymond Batson. *The NASA Atlas of the Solar System*. New York: Cambridge University Press, 1996.

Hamburg, Michael. *Astronomy Made Simple*. New York: Doubleday, 1993.

Harland, W.B. and Richard L. Arstrong. *Geologic Timescale* (Cambridge Earth Science Series). New York: Cambridge University Press, 1990.

Harris, Stephen L. *Agents of Chaos: Earthquakes, Volcanoes, and Other Natural Disasters*. Misoula, MT: Mountain Press, 1990.

Jones, Brian and Stephen Edberg, eds. *The Practical Astronomer*. New York: Simon & Schuster, 1990.

Jones, Steve, Robert Martin, and David Pilbeam. *The Cambridge Encyclopedia of Human Evolution*. New York: Cambridge University Press, 1992.

Kaufmann, William J. III. *Universe*. New York: W H Freeman & Co, 1993.

Klug, William S. and Michael R. Cummings. *Essentials of Genetics*. New York: Prentice Hall Press, 1998.

Lamb, Simon and David Sington. *Earth Story*. Princeton, NJ: Princeton University Press, 1998.

Lyons, Walter A. *The Handy Weather Answer Book*. Detroit: Visible Ink Press, 1996.

Morris, Neil. *Mountains (The Wonders of Our World)*. New York: Crabtree Pub, 1995.

Oldstone, Michael B.A. *Viruses, Plagues, and History*. New York: Oxford University Press, 1998.

Parker, Steven and Raymond L. Bernor, eds. *The Practical Paleontologist*. New York: Simon & Schuster, 1990.

Peacock, Graham Terry Hudson, and Jenny Hughes (illus.). *Exploring Habitats*. Austin, TX: Raintree/Steck Vaughn, 1993.

Ridley, Mark. *Evolution*. Boston: Blackwell Science, Inc., 1996.

Ricciuti, Edward R. and Vincent Marteka. *Amphibians (Our Living World)*. Woodbridge, CT: Blackbirch Marketing, 1994.

——— and William Simpson (illus.). *Birds (Our Living World)*. Woodbridge, CT: Blackbirch Marketing, 1994.

Ritchie, David. *The Encyclopedia of Earthquakes and Volcanoes*. New York: Facts on File, 1994.

Ronan, Colin A. *The Natural History of the Universe*. New York: Macmillan, 1991.

Silverstein, Alvin, Virginia B. Silverstein, and Robert A. Silverstein. *Monerans and Protists*. New York: Twenty First Century Books, 1996.

Svarney, Thomas E. and Patricia Barnes-Svarney. *The Handy Dinosaur Answer Book*. Detroit: Visible Ink Press, 1999.

Tattersall, Ian. *Becoming Human: Evolution and Human Uniqueness*. New York: Harcourt Brace, 1998.

Thompson, Ida and Carol Nehring. *National Audubon Society Field Guide to North American Fossils*. New York: Knopf, 1982.

Tompkins, Peter and Christopher O. Bird. *The Secret Life of Plants*. New York: HarperCollins, 1989.

For a good Web source of books, CDs, videos, etc. on natural history, contact the NHBS Mailorder Bookstore on the World Wide Web: <http://www.nhbs.com>.

Index

by Kay Banning

Aa (lava), 202
Absolute temperature, 217
Absolute time, 128, 130, 131
Abyssal plains, 162, 163
Acoustic engineers, 210
Acquired immunodeficiency syndrome (AIDS), 32, 33, 34, 35
Actinomycetes, 34
Adaptive radiation, 93, 96
Adenine, 125
Adirondack Mountains, 152
Aegean Sea, 107
Aegyptopithecus, 136
Aerobic, 217
Aerodynamicists, 210
Afar Depression, 136
Africa
    amphibians and, 4, 5, 6
    as continent, 79
    continental movement and, 77
    deserts and, 80, 81
    ethnic groups of, 135
    forests and, 80
    fossils and, 134, 143
    habitats and, 79
    human evolution and, 134, 135, 136, 138
    impact craters and, 104
    Neanderthals and, 138
    volcanoes and, 200, 201, 203
Agaricus campestris, 118
Agassiz, Louis, 77
Age, 131
Agnatha, 16, 109
Agriculture
    arthropods and, 18, 19
    droughts and, 54
    fungi and, 117, 119, 120
    genetic engineering and, 124
    insects and, 20
    plants and, 169
    volcanoes and, 199
    weather and, 52
AIDS (acquired immunodeficiency syndrome), 32, 33, 34, 35
Air masses, 52, 53, 54, 55, 56
Alaska, 79, 113, 204
Albedo, 24, 217
Albertosaurus, 68
Aldrovandi, Ulisse, 10, 45
Aleutian Islands, 163, 204
Alexander the Great, 192

Algae
    amphibians and, 4
    blue-green algae, 11
    brown algae, 162, 171
    cytology and, 60
    evolution of, 141, 142
    green algae, 2, 119, 170, 171
    hibernation and, 14
    as land organism, 9
    as nonflowering plant, 171
    protists and, 173, 174–75
ALH84001, 144
Alleles, 125
Alligators, 4, 6, 178, 179
Allitrichidae, 139
Alternative energy sources, 113
Alvarez, Luis, 69, 94, 101
Alvarez, Walter, 69, 94, 101
Alvin, 142, 158, 201
Amazon Rainforest, 50, 80, 118
Amazon River, 4, 159, 179
American bullfrog, 4
American cockroaches, 19
American Ornithologists' Union, 46
Amino acids
    cytology and, 61
    description of, 124
    discovery of, 123
    life's origins and, 142
    proteins and, 62, 126
Ammonia, 142, 186
Ammonites, 31
Ammonium, 119, 169–70
Amniote egg, 177, 178
Amoebas, 14, 174, 175
Amphibians
    description of, 3–4
    estuaries and, 161
    fish and, 106
    history of, 2–3
    introduction to, 1
    as land vertebrates, 9
    Mesozoic era and, 129
    reptiles and, 1, 3, 4, 9, 66, 177
    timeline of, 1
    topic terms and, 3–6
    tropical forests and, 80
    as vertebrates, 16
    Web sites for, 210
Amphibious, 217
Anabarites-Protohertizina layer, 10
Anaconda, 179

Anaerobic, 217
Analytical chemists, 208
Anapsids, 66, 177, 178
Anatomically Modern Humans, 138
Anatomists, 207
Anatomy, 42
Anaximenes, 85
Ancestors, 46–47, 134
Andes Mountains, 159
Andrews, Roy Chapman, 68
Angiosperms, 105, 129, 169, 171, 172
Animals
    bacteria and, 34
    boreal forests and, 79–80
    Cambrian Explosion and, 142, 168
    caves and, 78
    classification of organisms and, 7, 9, 10, 11, 12, 45–46
    coastline and, 158
    cytology and, 58, 60, 61–62, 63
    Darwin and, 10
    description of, 11
    deserts and, 80
    dinosaurs and, 72
    disease study and, 32
    endangered and threatened species and, 103
    estuaries and, 161
    evolution and, 11, 91
    extinction and, 101
    fossils and, 10, 96, 111, 112, 114, 128
    fungi and, 119, 120
    habitat and, 79
    history of, 9–11
    humans and, 134
    introduction to, 7
    marine animals, 157, 161
    marine plants and, 162
    Mesozoic era and, 129
    oceans and, 9, 13, 22, 155, 157, 159
    ozone layer and, 22–23, 27
    Paleozoic era and, 129
    pollination and, 172
    Precambrian era and, 129
    precipitation and, 55
    protists and, 174, 175
    reptiles and, 66
    Species 2000 Programme and, 47
    temperate forests and, 80
    timeline of, 7–8
    topic terms and, 11, 13–16
    tropical forests and, 80
    Web sites for, 210
Ankylosaurid dinosaurs, 67
Annular solar eclipses, 148
Anorthite, 146
Anorthosites, 146
Anorthositic gabbro, 148
Anpheles mosquitoes, 174

Antarctica
    climate and weather and, 50
    as continent, 79
    deserts and, 80, 81
    erosion and, 151
    fish and, 107
    fungi and, 118, 119
    glaciers and, 81
    humans and, 79
    Mars and, 144
    meteoroids and, 187
    ozone hole and, 27
    ozone layer and, 23
    protists and, 175
    supernovae irradiation and, 194
Anteaters, 80
Anthracite coal, 113
Anthrax, 32
Anthropogene period, 129
Anthropologists, 209
Anthropology, 96, 130, 134, 138
Antibiotics, 33, 116, 118
Anticyclones, 52
Antiviral agents, 32
Antrhopoidea, 139
Ants, 19, 20
Anura, 4, 5
Apatite, 115
Apennines, 112
Apes, 134, 136
Apoda, 4
Apogee, 217
Apollo manned missions, 147, 149
Apomorphic, 48
Appleton, Edward, 23, 25
Appleton layer, 23, 25
Aquifers, 28
Arachnids, 18–19
Archaeoastronomers, 207
Archaeologists, 209
Archaeopteryx lithographica, 37, 38, 42, 43
Archaic Homo sapiens, 138
Archelaus, 85
Archosaurs
    birds and, 37, 42
    crocodiles and, 178
    description of, 69
    dinosaurs and, 66, 72
    reptiles, 177, 179
Arctic, 14, 23, 81
Arctic Ocean, 161
Arduino, Giovanni, 76
Argentina, 42–43, 81, 162
Aristotle
    astronomy and, 192
    classification of organisms and, 9, 45
    earthquakes and, 85

fish and, 107
fossils and, 111
solar system model and, 183
volcanoes and, 198, 199
Arizona, 19, 114, 186
Armillaria bulbosa, 118
Armillaria mellae, 117
Armillaria ostoyae, 118
Artedi, Petrus, 107
Arthropods
    Burgess shale layer and, 13
    description of, 19–20
    as first land animals, 9, 142
    fossils and, 9, 10
    fungi and, 117, 119
    history of, 18
    introduction to, 17
    as invertebrates, 13
    timeline of, 17
    topic terms and, 18–20
    Web sites for, 211–12
Asaro, Frank, 69, 94, 101
Ascomycetes, 117, 118, 119
Asia
    climate and weather and, 56
    as continent, 79
    deserts and, 81
    ethnic groups of, 135
    forests and, 80
    habitats and, 79
    human evolution and, 134, 138
    human fossils and, 137
    mountains and, 152
Assyrians, 169
Asteroid belt, 185, 186
Asteroids, 60, 67, 72, 101, 185–86, 187
Asthenosphere, 82
Astrochemists, 207
Astrology, 192, 217
Astronomical unit, 217
Astronomy, 9, 76, 183–85, 191–93, 207
Astrophysicists, 207
Atacama Desert, 81
Atlantic giant squid, 14
Atlantic Ocean
    abyssal plains and, 162
    coastline and, 159, 161
    currents and, 160, 161
    hurricanes and, 56
    oceanography and, 158
    tides and, 164
Atmosphere
    climate and weather and, 23, 53
    description of, 24
    extinction and, 67
    history of, 22–24
    introduction to, 21

of Jupiter, 188
life's origins and, 142, 143
of Mars, 187
of Mercury, 27, 187, 195
of Moon, 146
of Neptune, 188
oceans and, 155, 157
ozone layer and, 22, 23–24, 141
of planets, 185, 186, 187, 188, 189
Precambrian era and, 129
of Saturn, 188
snowballs above Earth and, 184
solar cycle and, 194
Sun and, 195
supernova theory and, 105
timeline of, 21–22
topic terms and, 24–28
of Uranus, 188
of Venus, 21, 27, 187, 195
volcanoes and, 198, 199
water and, 162
water vapor and, 27, 76
Web sites for, 212
Atmospheric layers, 23, 24–26
Atomic physicists, 210
Audubon, John James, 37
Auroras, 194
Australia
    amphibians and, 5
    climate and weather and, 54, 55, 56
    coastline of, 158
    as continent, 79
    currents and, 160
    deserts and, 81
    dinosaurs and, 69, 72
    human evolution and, 138
    human fossils and, 134
    plants and, 170
    stromatolites and, 143
Australopithecus, 136–37
Australopithecus afarensis, 112, 136
Australopithecus africanus, 134, 136, 137
Australopithecus boisei, 136
Australopithecus robustus, 136
Avalanches, 165
Avery, Oswald Theodore, 123
Azores, 204

Babylonia, 192
Bacillary dysentery, 20
Background rate of extinction, 101, 102
Bacteria
    algae and, 171
    Antarctica and, 175
    caves and, 78, 79
    cytology and, 60, 61, 62
    description of, 34

Bacteria *(continued)*
    as first land organisms, 9
    fossil fuels and, 113
    fossils and, 31, 143
    genes and, 123
    genetic engineering and, 124
    history of, 31–33
    infections and, 118
    introduction to, 29
    life's origins and, 142, 143
    Lyme disease and, 19
    Mars and, 144
    Mondera and, 11
    protists and, 173, 174
    size of, 141
    Species 2000 Programme and, 47
    temperate forests and, 80
    timeline of, 29–31
    topic terms and, 34–35
    tropical forests and, 80
    variety of species of, 11
    Web sites for, 212
Bacteriologists, 207
Bacteriology, 32
Bailly crater, 146
Bakker, Robert T., 68–69
Bald cypress, 170
Banana frog, 4
Barberton Greenstone Belt, 143
Barking treefrog, 6
Barnacles, 20
Barometer, 52
Barrel, Joseph, 129
Basalt, 148, 202
Bases, 125
Basidiocarp, 118
Basidiomycetes, 118, 119–20
Basidium, 120
Bates, Henry Walter, 18, 93, 96
Batesian mimicry, 93, 96
Bats, 14, 41, 80
Bauhin, Gaspard, 45
Beaks, 39, 40, 43
Bears, 14, 79–80
Becquerel, Antoine Henri, 128
Beetles, 19, 20
Belgium, 68
Belon, Pierre, 107
Benguela Current, 161
Benioff, Hugo, 86
Big bang theory, 97, 191, 193
Big bombardments, 184
Bills, 39, 40, 43
Binary notation, 45
Binomial classification, 45
Biochemists, 207, 208
Biology, xi, 76, 93, 207–08

Biophysicists, 207, 210
Biosphere, 21
Biotrophs, 119
Bird anatomy, 37–41
Bird behavior, 38, 41
Bird flight, 38, 41
Birds
    Cenozoic era and, 129
    classification of organisms and, 38–39, 41–42, 46
    description of, 38
    dinosaurs and, 36, 37, 38, 42–43, 69, 178
    estuaries and, 161
    evolution and, 9, 38, 42–43
    history of, 37–38
    introduction to, 36
    Jurassic period and, 177
    magnetism and, 26–27, 38, 41
    pollination and, 171, 172
    Species 2000 Programme and, 47
    species of, 98
    temperate forests and, 80
    timeline of, 36–37
    topic terms and, 38–42
    tropical forests and, 80
    as vertebrates, 16
    Web sites for, 212
Bison, 80
Biston betularia, 94
Bituminous coal, 113
Bivalves, 10
Bjerknes, Jacob Aal Bonnevie, 52
Bjerknes, Vilhelm, 52
Black, Davidson, 136
Black holes, 193, 194
Bladderworts, 170
Blastula, 11
Blind cave beetles, 79
Blue-green algae, 11
Bode, Johann, 184
Body waves, 90
Bok, Bart Jan, 192
Bok globules, 192
Boleyn, Anne, 198
Boltwood, Bertram, 128
Bony fish, 16, 107, 108
Bora winds, 57
Boreal forests, 79–80
Borrelia burgdorferi, 33
Botanists, 207
Botany, 45, 48, 169
Boylston, Zabdiel, 31
Brachiopods, 10
Brahe, Tycho, 183, 192
Brazil, 160
Bristlecone pine, 170
Broca, Pierre-Paul, 94
Brown, Robert, 60

Brown algae, 162, 171
Brown bats, 79
Brown dwarfs, 193, 194
Bryan, William Jennings, 96
Bubbles, 64
Bubonic plague, 33
Buckland, William, 67, 68
Buffon, Comte de Georges-Louis Leclerc, 10, 45–46
Bufonidae, 5
Burbank, Luther, 170
Burbank potato, 170
Burgess shale layer, 13
Burke, Ann, 42
Butler law, 95–96
Butterflies, 19, 20, 47
Butterwort, 170
Buys Ballot, Christoph Hendrik Diederik, 52
Buys Ballot law, 52

C-12, 114
C-13, 114
C-14, 114
Cacti, 47
Caecilians, 2, 3, 4
Calcite, 115
California, 77, 81, 87, 89, 90
California Current, 161
California, Gulf of, 158
Calls, of birds, 41
Cambrian Explosion
    animals and, 9, 168
    description of, 13
    evolution and, 13, 92
    fossils and, 130
    life and, 142
    Paleozoic era and, 9, 129, 142, 157
    plants and, 168
    theories on, 10–11
Cambrian period, 9, 10–11, 13, 130, 142
Camerarius, Rudolph, 45
Campanius, John, 52
Canada
    birds and, 43
    Cambrian Explosion and, 13
    dinosaurs and, 68
    early life and, 143
    forests and, 79
    fungi and, 118
    geologic time scale and, 129
    impact craters and, 104
Canaries Current, 160, 161
Canary Islands, 161
Cancer, 35
Canterbury Swarm, 147
Capture theory, 146
Capuchin monkeys, 15
Carbon
    amino acids and, 124
    DNA and, 125
    fossil fuels and, 113
    interstellar dust and gas and, 194
    life's origins and, 142, 143
Carbon dating techniques, 114, 130, 134
Carbon dioxide
    atmosphere and, 22, 23, 27
    bacteria and, 34
    birds and, 39, 40
    comets and, 186
    fossil fuels and, 113
    greenhouse effect and, 26
    life and, 141
    photosynthesis and, 171
    plants and, 166, 168, 169
    Venus and, 27
    volcanoes and, 198
Carbon monoxide, 27
Carbon-14, 114
Carbonate rocks, 22, 27
Carboniferous period
    amphibians and, 2, 3, 177
    fossil fuels and, 113
    plants and, 171
    reptiles and, 66, 177
Careers, in natural history, 207–10
Caribbean Sea, 161, 164
Carlsbad Caverns, 78
Carnivorous, 217
Carpenter, William, 157
Cascade Mountains, 204
Caspian Sea, 77
Catastrophism
    description of, 96–97
    Earth's development and, 77
    evolution and, 94–95
    extinction and, 94, 96, 100, 105, 112
Caterpillars, 19, 20
Caucasian ethnic groups, 135
Cave crickets, 79
Cave pools, 79
Caves, 78–79, 136, 198
Cebidae, 139
Cell theory, 60
Cells. *See also* Cytology
    animal and plant cells, 63
    atmosphere and, 23, 25–26
    description of, 61–62
Cellular division, 61, 62
Cellular phones, 193
Cenozoic era, 37, 108, 129, 153–54
Centaurus A, 193
Centipedes, 17
Central America, 80
Cephalopods, 161
Ceratocysis ulmi, 117

Ceratopsian dinosaurs, 67
Cercopithecidae, 139
Ceres, 185
Cesalpino, Andrea, 45
CFCs (chlorofluorocarbons), 24, 26, 27
Challenger Deep, 158
Chandler, Seth Carlo, 77
Chandler wobble, 77
Charon, 185, 189
Chemical kineticists, 208
Chemistry, 199–200, 208–09, 217
Chemosynthesis, 142
Chestnut blight, 117
Chicken pox, 34, 35
Chicxulub crater, 94, 101
Chile, 81, 200, 204
Chimaeras, 108
Chimpanzees, 124, 134
China
    astronomy and, 191–92
    bacteria and viruses and, 33
    birds and, 42
    climate and weather and, 55
    dinosaurs and, 67
    earthquakes and, 84, 85, 86–87, 88
    extinction and, 102
    fossils and, 42, 112, 134, 136, 137
    mountains and, 151
Chitin, 119, 120
Chlamydomonas nivalis, 175
Chlorofluorocarbons (CFCs), 24, 26, 27
Chlorophyll, 169, 170, 171
Chloroplasts, 169, 170–71, 174
Chondrichthyes, 16, 108, 109
Chordates
    arthropods and, 19
    description of, 13
    dinosaurs and, 72
    evolution and, 92
    fish and, 108
    humans and, 134
    vertebrates and, 16
Chromosomes, 13, 61, 62, 123, 124–25
Chytridiomycota, 117
Ciliates, 174
Cladistics, 46–48
Cladograms, 46–47
Classical genetics, 123, 125
Classification of organisms
    amphibians and, 4
    animals and, 7, 9, 10, 11, 12, 45–46
    arthropods and, 20
    birds and, 38–39, 41–42, 46
    dinosaur classification and, 68, 72
    evolution and, 46, 92
    fish and, 107
    fungi and, 46, 117
    history of, 45–46
    introduction to, 44
    mammals and, 15–16
    plants and, 45, 46, 169
    protictista and, 175
    protists and, 46, 173, 174
    reptiles and, 179
    species and, 98
    systems of, 45, 48
    timeline of, 44–45
    topic terms and, 46–49
    Web sites for, 212
Clay, 64
Climate theory, 104
Climate and weather
    atmosphere and, 23, 53
    boreal forests and, 79
    Cambrian Explosion and, 11
    continental movements and, 73
    currents and, 159, 160, 161
    evolution and, 97
    extinction and, 72, 99, 101, 104, 105
    extraterrestrial impacts and, 102, 105
    fossilization and, 115
    history of, 52–53
    introduction to, 50
    life's origins and, 143
    Mesozoic era and, 69
    mountains and, 150, 151, 152
    oceans and, 155
    oil spills and, 113
    oxygen and, 23
    reptiles and, 177
    timeline of, 50–51
    topic terms and, 53–57
    tropical forests and, 80
    troposphere and, 24
    volcanoes and, 198
    Web sites for, 212
Climatologists, 209
Cloning, 124
Closed Universe Model, 195
Cloudina, 10
Clusters, 192
Coal products, 113
Coastal deserts, 81
Coastal margins, 157, 162–63
Coastal plains, 159
Coastlines, 56, 77, 79, 157, 158–59, 165
Coccidioides immitis, 118
Cockroaches, 18, 19
CODATA (Committee on Data for Science and Technology), 47
Codon, 124
Coelacanth, 108
Coelurosaurian dinosaurs, 42, 43
Cold front, 54

Comet Hale-Bopp, 186
Comet Hyakutake, 196
Comets, 60, 67, 182–83, 184, 186, 187
Committee on Data for Science and Technology (CODATA), 47
Common Access System, 47
Comparative anatomists, 207
Comparative anatomy, 10
Computers, 53
Concealed stratification, law of, 76
Condensation, 56, 217
Confuciusornis, 42
Conifers, 169, 172
Conjugation, 33
Connecticut River valley, 112
Conodont, 16
Constellations, 185, 191–92
Continental drift, 199
Continental plates, 54, 77, 159
Continental rises, 162, 163, 165
Continental shelf, 159, 162, 165
Continental slope, 162–63, 165
Continents. *See also* Plate tectonics
    crust and, 82
    description of, 79
    earth's formation and, 76
    movement of, 73, 76, 77, 86, 104, 162, 165, 182
    wetlands and, 159
Conveyor belt theory, 78
Cope, Edward Drinker, 68, 112
Copepods, 20
Copernican model, 183, 184
Copernicus, Nicolaus, 183, 192
Coprolites, 114
Coral reefs, 158
Corals, 47
Core
    iridium anomaly and, 94
    as layer of Earth, 75, 82
    of Moon, 148
Coriolis, Gustave-Gaspard, 23
Coriolis effect, 23, 26, 28, 56, 159
Cormorants, 37
Cormoro Islands, 108
Correns, Karl Franz Joseph, 123, 125
Cosmologists, 207
Covergent evolution, 96
Crabs, 20
Craters. *See also* Extraterrestrial impacts
    Earth and, 186
    Moon and, 146, 147
Crayfish, 20
Creationism, 92–93, 95, 97, 134
Cretaceous period
    birds and, 37
    dinosaurs and, 65, 67, 68, 69, 70–71, 72, 100, 177
    extinction and, 105
    plants and, 129, 169
    reptiles and, 177–78
Cretaceous-Tertiary boundary, 69, 94, 100, 101
Crick, Francis, 61, 123, 125
Crickets, 19, 20
Cro-Magnon, 138
Crocodiles, 66, 69, 177, 178–79
Crocodilians, 37, 42, 66
Crocodylids, 178–79
Cross-cutting relationships, 77
Cross-pollination, 172
Crossopterygians, 2
Crust
    continental plates and, 77
    earthquakes and, 83, 84, 86, 87
    faults and, 78
    fossils and, 110
    isostasy and, 152
    as layer of Earth, 75, 82
    of Moon, 148
    Moon's formation and, 147
    mountains and, 151, 152, 153
    oceans and, 158
    plate tectonics and, 154
    Precambrian era and, 129
    theories of, 76
    vents and, 142
    volcanoes and, 200, 203
Crustaceans
    as arthropods, 17
    cave pools and, 79
    description of, 20
    humans and, 19
    oceans and, 157
    Species 2000 Programme and, 47
Cryobiologists, 207
Cryogenicists, 210
Crytologists, 207
Cuba, 3, 4
Cumulonimbus clouds, 55
Curculionid beetles, 47
Currents, 11, 73, 157, 158, 159–61
Cuvier, Georges, 10, 94, 100, 112
Cycads, 169
Cyclones, 56
Cytology
    history of, 60–61
    introduction to, 58
    timeline of, 58–59
    topic terms and, 61–64
    Web sites for, 212
Cytosine, 125

D-layer, of ionosphere, 25
Dalton, John, 52
Daniell, John Frederic, 23
Dark matter, 193

Darrow, Clarence, 96
Dart, Raymond Arthur, 134, 136
Darwin, Charles Robert
   birds and, 37–38, 42
   evolution and, 10, 18, 42, 93–94, 95, 97, 98, 112, 134
   extinction and, 100–101
   geologic time scale and, 128
Darwin, George H., 146
Dating techniques, 112, 114, 128–30, 134, 135–36
Dawkins, Richard, 94
De Vries, Hugo Marie, 123, 125
Death Valley, California, 163
Deccan Traps, 101, 105, 202
Deciduous trees, 80
Decomposition, 117, 118–19
Deep-ocean currents, 159, 160
Deep-water waves, 165
Deer, 80
Deferents, 183
Deimos, 27
Deinonychus, 68
Deinonychus antirrhopus, 42
Deoxyribonucleic acid. See DNA
Descartes, René, 10, 198–99
Desertification, 79
Deserts, 50, 79, 80–81, 104
Detritus, 4, 217
Deuterium, 191
Devonian period, 2, 100, 108, 117
Dew point, 56
Dhytridiomycetes, 119
Diapsids, 66, 177, 178
Diatoms, 162, 171
Dinoflagellates, 171, 175
Dinosaur bones, 67, 71–72
Dinosaur classification, 68
Dinosaur extinction, 67, 72
Dinosaur fossil locations, 68, 72
Dinosauria, 67, 69, 72
Dinosaurs
   amphibians compared to, 3
   birds and, 36, 37, 38, 42–43, 69, 178
   description of, 69–71
   dominance of, 100
   extinction and, 31, 43, 67, 69, 72, 100, 101, 104, 129, 133, 177
   flowering plants and, 169
   fossils and, 65, 67–69, 72, 110, 112, 115
   history of, 66–69
   introduction to, 65
   Mesozoic era and, 65, 66, 67, 68, 69, 70–71, 129, 133, 177, 178
   timeline of, 65–66
   topic terms and, 69–72
   Web sites for, 212
Dionaea muscipula, 170
Dip-slip faults, 87

Diphtheria, 32, 34
Diphyodonty, 15
Diptera, 47
Disease
   bacteria and viruses and, 19, 31–33, 34, 35
   endangered and threatened species and, 103
   extinction and, 99, 101, 104
   fungi and, 116, 117
   protists and, 174
   volcanoes and, 200
Disease theory, 104
Disky seaside sparrow, 102
Diurnal tides, 164
Divergence, 96
DNA
   chromosomes and, 124
   cytology and, 60, 61
   description of, 125
   evolution and, 92, 135, 138
   genetics and, 61, 123
   protists and, 174
   RNA and, 126
   viruses and, 35
Dollo, Louis, 69
Dolly, 124
Dominican Republic, 159
Doppler radar, 53
Dormice, 14
Douglas fir, 170
Dragonflies, 13, 18
Droughts, 2, 53, 54
Dubois, Eugene, 112, 136, 137
Dung beetles, 79
Dust devils, 55
Dutch Elm disease, 117
Dutrochet, Henri, 60

E. coli, 35
E-layer, 25
Earth
   age of, xi, 100, 128
   epicycle model of the solar system and, 183
   evolution and, 91
   formation of, 73, 75–76, 97
   history of, 75–78
   introduction to, 73
   layers of, 78, 81–82
   life beginning on, 9
   Milky Way galaxy and, 193
   near-Earth asteroids and, 186
   reorientation of, 11
   solar cycle and, 194
   solar system and, 182
   solar wind and, 184
   Sun and, 195
   supernovae irradiation and, 194
   as terrestrial planet, 189

    timeline of, 73–75
    topic terms and, 78–82
    universe and, 190
    variety of species on, 11
    Web sites for, 212
Earth science, 76–78, 209–10
Earthquake measuring devices, 85, 88–89
Earthquakes
    deep-water waves and, 165
    description of, 87–88
    history of, 84–87
    introduction to, 83
    isostasy and, 152
    oceans and, 158
    plate boundaries and, 203
    rock layers and, 76
    San Andreas fault and, 77
    timeline of, 83–84
    topic terms and, 87–90
    volcanoes and, 201
    Web sites for, 212
East Pacific Rise, 203
Eastern newts, 5
Echinoderms, 161
Eclipses, 145, 147, 148
Ecliptic, 186
Ecologists, 207
Economic geologists, 209
Ecosystems, 101, 102, 142
Ectothermic
    amphibians as, 3
    definition of, 217
    fish and, 108
    lizards and, 179
    reptiles and, 68
Ecuador, 88
Ediacaran fauna, 9, 10
Ediacaran fossils, 10, 13
Ediacaran period, 10
Egypt, 76, 87, 169, 192
Ehrlich, Paul, 61
Einstein, Albert, 192
El Niño, 54, 160
Elastic rebound, 89–90
Eldredge, Niles, 96, 97
Electricity, 199
Electrochemists, 208
Electrodynamicists, 210
Electron microscopy, 61
Electron spin resonance, 114, 135
Eleuth frog, 3, 4
Eleutherodactylus, 3
Elk, 80
Embryologists, 207
Emu, 42
Enantiornithes, 37
Endangered Species Act of 1973, 101, 103

Endangered Species Protection Act of 1992, 103
Endangered and threatened species, 99, 101, 103
Endocarditis, 32
Endoplastmic reticulum, 13
Endothermic
    definition of, 217
    dinosaurs and, 38, 68, 69
    reptiles and, 177
Endothia canker, 117
Engineering, 76
England
    climate and weather and, 52, 55
    currents and, 160
    dinosaurs and, 67, 68
    evolution and, 94
    fungi and, 118
    Moon craters and, 147
    tides and, 157
Entomologists, 207
Entomology, 18
Environment. *See also* Habitats
    adaptive radiation and, 96
    amphibians and, 3, 6
    bacteria and, 34
    ethnological paleontology and, 68
    evolution and, 93
    extinction and, 101
    fish and, 107, 108
    fungi and, 117
    genetic engineering and, 124
    habitats and, 79
    life's origins and, 143
    natural selection and, 97
Environmental chemists, 208
Environmental geologists, 209
Eon, 131
Eoraptor, 67
Epicenters, of earthquakes, 86, 87, 88, 89
Epicycle model of the solar system, 183
Epoch, 131
Equinox precession, 183
Era, 131
Eratosthenes of Cyrene, 76
Erosion
    coastlines and, 158–59
    definition of, 217
    fossils and, 72, 91, 115
    geology and, 78
    Late Heavy Bombardment and, 182
    mountains and, 151–52
    oceans and, 155, 158
    precipitation and, 55
    rock layers and, 76
    seawater and, 164
    turbidity currents and, 165
    winds and, 57
Estivation, 3

Estuaries, 157, 161, 164, 165
Ethics, 124
Ethiopia, 112, 136
Ethnological paleontology, 68
Ethologists, 207
Ethology, 217
Etna, Mount, 198
Eucalyptus regnans, 170
Eukaryotic cells, 9, 13, 60, 129, 141
Eukaryotic organisms, 7, 11, 117, 171, 174
Europa, 144
Europe
    astronomy and, 192
    bacteria and viruses and, 33
    catastrophism and, 100
    caves of, 79
    climate and weather and, 52
    as continent, 79
    dinosaurs and, 72
    forests and, 79, 80
    fungi and, 117
    human evolution and, 134, 138
    ice sheets and, 77
    mountains and, 151
    oceanography and, 158
    volcanoes and, 199
European Alps, 151
Euryapsids, 66, 177, 178
Eutheria, 15, 16
Evaporation, 2, 28, 217
Everest, Mount, 152, 163
Evergeen forests, 80
Everglades, 159
Evolution
    amphibians and, 1, 2, 9
    animals and, 11, 91
    bacteria and viruses and, 33
    birds and, 9, 38, 42–43
    Cambrian Explosion and, 13, 92
    cladistics and, 47–48
    classification of organisms and, 46, 92
    cytology and, 60
    Darwin and, 10, 18, 42, 93–94, 95, 97, 98, 112, 134
    description of, 97
    dinosaurs and, 66
    extinction and, 11
    fish and, 106, 107, 108
    fossils and, 13, 91, 92, 93, 94, 110, 112, 114
    fungi and, 117
    genetics and, 121, 124, 135, 138
    geologic time scale and, 127, 129
    history of, 92–96
    of humans, 9, 91, 93, 96, 134–39
    insects and, 18
    introduction to, 91
    invertebrates and, 13, 92
    mammals and, 9, 15
    meiosis and, 62
    Mesozoic era and, 129
    mountains and, 151
    multicellular life and, 92, 174
    plants and, 91, 171
    protists and, 174
    reptiles and, 176, 177
    timeline of, 91–92
    topic terms and, 96–98
    Web sites for, 212–13
Evolutionary biologists, 207
Evolutionary biology, 97
Exobiologists, 207
Exoskeletons, 18, 19, 20
Extinction
    amphibians and, 3
    ancestors and, 47
    birds and, 37
    Burgess shale layer and, 13
    catastrophism and, 94, 96, 100, 105, 112
    classification of organisms and, 44
    conodonts and, 16
    creationism and, 97
    dinosaurs and, 31, 43, 67, 69, 72, 100, 101, 104, 129, 133, 177
    evolution and, 11
    extraterrestrial impacts and, 67, 69, 94, 101, 102, 103, 104, 105, 186
    fish and, 108
    fossils and, 13, 100
    history of, 100–01, 103
    introduction to, 99
    Mesozoic era and, 129
    Multituberculata and, 15
    Paleozoic era and, 100, 129
    Permian period and, 66, 177
    problems of, 102
    reptiles and, 177, 178
    timeline of, 99
    topic terms and, 103–05
    viruses and, 31
    Web sites for, 213
Extinction theories, 101, 103–05
Extrasolar planets, 185, 186
Extraterrestrial impacts
    big bombardments and, 184
    cells and, 60
    Earth's formation and, 76
    extinction and, 67, 69, 94, 101, 102, 103, 104, 105, 186
    impact craters and, 104, 105
    Late Heavy Bombardment and, 142, 146, 182
    life's origins and, 142–43, 144
    Moon and, 146, 147
    oceans and, 157
    puntuated equilibrium and, 96
    solar system and, 182
Eye cataracts, 28

F layer, 25
F-1 layer, 25
F-2 layer, 25
Fabricius, Johann Christian, 18
Fauna, 217. *See also* Animals
Feathers, 40
Feduccia, Alan, 42
Feldspar, 146
Ferguson, William, 170
Fermentation, 116, 117
Ferns, 172
Ferrel, William, 23
Ferrel cells, 23, 26
51 Pegasi, 185
Filomarino, Ascanio, 85
Finches, 38
Fireflies, 18
First International Zoological Congress, 46
Fish
   amphibians and, 2
   Cambrian period and, 142
   Cenozoic era and, 129
   currents and, 160
   description of, 108–09
   estuaries and, 161
   evolution and, 106, 107
   extinction and, 108
   history of, 107–08
   introduction to, 106
   as marine animals, 161
   Species 2000 Programme and, 47
   timeline of, 106
   topic terms and, 108–09
   Web sites for, 213
Fish and Wildlife Service, 103
Fishery industries, 160
Fission, 217–18
Fission theory, 146
Five-Kingdom Classification, 48
Fjords, 158
Flatworms, 79
Fleming, Alexander, 118
Flemming, Walther, 60–61, 123
Flies, 18
Florida
   amphibians and, 5
   continental shelf and, 162
   currents and, 160
   extinction and, 102
   sea level and, 163
   stromatolites and, 143
   wetlands and, 159
Florida Straits, 161
Flowering dicot plants, 172
Flowering monocot plants, 172
Flowering plants, 11, 105, 169, 171, 172
Flowers, 169, 171

Fluid physicists, 210
Focus, of earthquakes, 86, 87, 90
Food chain
   animals and, 15
   arthropods and, 19
   crustaceans and, 20
   fish and, 106, 107
   frogs and, 6
   ozone hole and, 28
   plants and, 171
Food webs, 78
Forests, 79–80
Formanifera, 3
Formation, 218
Fossil fuels, 26, 113
Fossil record, 218
Fossilization, 112, 113, 114–15
Fossils
   amphibians and, 2, 3, 4, 5
   animals and, 10, 96, 111, 112, 114, 128
   arachnids and, 18
   arthropods and, 9, 10
   bacteria and, 31, 143
   birds and, 37, 38, 42–43
   carbon dating and, 130
   creationism and, 95
   Darwin and, 10
   dating techniques and, 135–36
   dinosaurs and, 65, 67–69, 72, 110, 112, 115
   earth science and, 76
   evolution and, 13, 91, 92, 93, 94, 110, 112, 114
   extinction and, 13, 100
   fungi and, 117
   geologic time scale and, 128, 130
   history of, 111–12, 114
   humans and, 96, 111, 112, 115, 134, 136–38
   introduction to, 110
   Leonardo da Vinci and, 76
   life and, 141
   living fossils, 108
   Mars and, 144
   mountains and, 151
   plants and, 111, 112, 114, 128, 168, 169, 170
   puntuated equilibrium and, 96, 97–98
   reptiles and, 177–79
   Species 2000 Programme and, 47
   stromatolites and, 9
   timeline of, 110–11
   topic terms and, 114–15
   Web sites for, 213
Four-Kingdom classification, 48
Frail, Dale, 185
France, 33, 137, 138, 199
Franklin, Benjamin, 23, 52
Franklin, Rosalind, 123
Frequency, 218
Fries, Elias Magnus, 117

Frisch, Karl von, 18
Frogs
    as amphibians, 3, 4
    cloning and, 124
    description of, 4–5
    environment and, 6
    hibernation and, 14
    Temnospondyls and, 2
    toads compared to, 6
Fronts, 52, 53, 54
Fruit, 172
Fruit flies, 13, 61, 123
Fujita and Pearson Tornado Scale, 55
Fundy, Bay of, 165
Fungal spores, 118
Fungi
    caves and, 78
    classification of organisms and, 46, 117
    description of, 119–20
    diseases and, 32
    eukaryotic cells and, 13
    fossil fuels and, 113
    fossils and, 117
    history of, 117–18
    introduction to, 116
    Paleozoic era and, 129
    protists and, 173, 174
    Species 2000 Programme and, 47
    temperate forests and, 80
    timeline of, 116
    topic terms and, 118–20
    tropical forests and, 80
    variety of species of, 11
    Web sites for, 213
Fusion, 194, 195, 218

Galapagos Islands, 93, 142, 158, 201, 204
Galapagos Rift, 142
Galaxies, 191, 192, 193
Galilei, Galileo, 183–84, 192
Galileo spacecraft, 144
Galton, Francis, 52
Galton, Peter, 69
Gangrene, 34
Gas deposits, 113
Gaseous planets, 182, 186
Gastrula, 11
Gavialids, 178
GEF (Global Environment Facility), 47
General circulation models (GCMs), 53
Genes
    cells and, 62
    chromosomes and, 61, 124, 125
    description of, 125
    eukaryotic cells and, 13
    genetics and, 123
    plants and, 170

Genetic engineering, 33, 61, 123–24, 125
Genetic variation, 62
Geneticists, 207
Genetics
    adaptive radiation and, 96
    animals and, 11
    bacteria and viruses and, 32–33
    cellular division and, 62
    cytology and, 61
    dinosaurs and, 69
    DNA and, 61, 123
    evolution and, 121, 124, 135, 138
    hibernation and, 14
    history of, 123–24
    introduction to, 121
    natural selection and, 97
    plants and, 170
    timeline of, 121–22
    topic terms and, 124–26
    vaccination and, 33
    viruses and, 35
    Web sites for, 213
Genome, 218
Genotype, 218
Geocentric model, 183
Geochemists, 209
Geochronologists, 209
Geochronology, 127, 129
Geographic cycle, 151
Geography
    climate and weather and, 50
    forests and, 79
    human evolution and, 134
    Pliny the Elder and, 9
    reptiles and, 177
Geologic time scale
    description of, 130
    history of, 128–30
    introduction to, 127
    mountains and, 152
    timeline of, 127–28
    topic terms and, 130–31
    volcanoes and, 198
    Web sites for, 213
Geology
    careers in, 209–10
    Earth's composition and, 73
    habitats and, 79
    Leonardo da Vinci and, 76
    study of, 77–78
Geomorphology, 77
Geophysicists, 209
Geostationary Operational Environmental Satellites (GOES), 53
Germ theory of disease, 32
Germany, 60, 93, 136, 137, 138
Gervase of Canterbury, 147

Gesner, Konrad von, 9
Giant boars, 80
Giant cubmosses, 171
Giant sequoias, 170
Giant toad, 5
Gila monsters, 179
Giordano Bruno crater, 147
Glacial geologists, 209
Glacial ice, 151, 158, 162
Glass lizards, 179
Global Environment Facility (GEF), 47
Global positioning system (GPS), 152, 193, 201
Global warming, 34, 163
Globule theory, 60
Gneissic rock, 151
Gnotobioticsts, 207
Gobi Desert, 81
GOES (Geostationary Operational Environmental Satellites), 53
Golden-mantled ground squirrels, 14
Golgi bodies, 13
Goliath beetles, 20
Goliath frog, 4
Gondwanaland, 76
Goosen, Henrik, 108
Gorillas, 124, 134
Gould, Stephen Jay, 96, 97
GPS (global positioning system), 152, 192, 201
Gradualism
  evolution and, 94, 97
  extinction and, 100–101, 104, 105
  punctuated equilibrium and, 96
Grasshoppers, 19, 20
Gravity
  black holes and, 193
  layers of Earth and, 81
  Mercury and, 187
  mountains and, 151
  satellites and, 189
  sea level and, 163
  solar system and, 182
  tides and, 164
  universe and, 193
  volcanoes and, 201
Gravity waves, 165
Great Barrier Reef, 158
Great gray owl, 80
Great Lakes, 102
Great Rift Valley, 203
Great Smoky Mountains National Park, 5
Greater horseshoe-nosed bat, 45
Greece
  astronomy and, 192
  earthquakes and, 87, 88
  human fossils and, 137
  mountains and, 151
  volcanoes and, 198, 200

Green algae, 2, 119, 170, 171
Greenhouse effect, 23, 26, 27, 187, 195
Greenland
  amphibians and, 2
  climate and weather and, 52
  currents and, 160
  erosion and, 151
  geologic time scale and, 129
  glaciers and, 81
  oldest rock and, 143
Grizzly bears, 14
Groundwater, 28, 78, 162
Guanine, 125
Guano, 79
Guatemala, 88
Guettard, Jean-Etienne, 199
Gulf Stream, 159, 160, 161
Gurden, John B., 124
Gutenberg discontinuity, 82
Gymnophiona, 4
Gymnosperms, 129, 168, 169, 172
Gyres, 160, 161

Habitats. *See also* Environment
  continental movements and, 73
  crocodylids and, 179
  description of, 79–81
  Earth's crust and, 78
  endangered and threatened species and, 103
  extinction and, 101, 102
  fungi and, 119
Hadley, George, 23
Hadley cells, 23, 26, 28
Hagfish, 16, 109
Haiti, 159
Hall, James, 199
Hamilton, William, 199
Hawaii, 152, 158, 201, 204
Hawking, Stephen, 192–93
Health physicists, 210
Heaviside-Kennelly layer, 25
Hedgehogs, 14
Heidelberg Man, 136
Heilmann, Gerhard, 42
Hekla volcano, 198
Helium, 186, 191, 194
Hellas Planitia, 104
Heng, Zhang, 85
Henry VIII (king of England), 198
Heraculaneum, 198
Hereditary, 218
Heredity, 123, 218. *See also* Genetics
Herodotus, 111
Herpes virus, 35
Herpetology, 178
Herschel, John, 144, 192
Herschel, William, 192

Hertz, 218
Hesperornithiformes, 37
Hess, Harry, 78, 86, 199
Heteromita globosa, 175
Heterotrophs, 11
Hibernation, 3, 14, 179
Himalayas, 154
Hip structure, of dinosaurs, 67, 68, 69, 72
Hipparchus, 183
Histologists, 208
HIV (human immunodeficiency virus), 32, 34, 35
H.M.S. *Beagle,* 10, 37, 93
H.M.S. *Challenger,* 157
H.M.S. *Lightning,* 157
H.M.S. *Porcupine,* 157
Holmes, Arthur, 129
Holmes, Oliver Wendell, 31
Homeothermic, 68
Hominid precursors, 136
Hominids, 130, 133, 134, 136–38, 218
Homo erectus, 112, 134–35, 136, 137, 138
Homo ergaster, 137, 138
Homo habilis, 137
Homo heidelbergensis, 138
Homo sapiens, 110, 134, 135, 137, 138
Homo sapiens neanderthalensis, 93, 138
Homo sapiens sapiens
    ancestors and, 47, 134
    evolution and, 9, 133, 134, 135, 137, 138
Honeybees, 18
Hong Kong, 87
Hood, Mount, 204
Hooke, Robert, 60
Horner, John R., 69
Horoscopes, 192
Horsehoe crabs, 13
Hoyle, Fred, 193, 195
Hubble, Edwin Powell, 192
Hubble Space Telescope, 193
Hubble's constant, 192
Human evolution, 9, 91, 93, 96, 134–39
Human genome, 124
Human immunodeficiency virus (HIV), 32, 34, 35
Human theory, 104
Humanoid, 218
Humans
    amphibians and, 1
    ancestors and, 47
    animal disease studies and, 32
    animals and, 9
    aquifers and, 28
    arthropods and, 17, 19
    bacteria and, 34
    carbon dating and, 130
    chordates and, 13
    climate and weather and, 53
    earthquakes and, 84
    earth's origins and, 76
    evolution of, 9, 91, 93, 96, 134–39
    extinction and, 99, 102, 103
    fish and, 107
    fossils and, 96, 111, 112, 115, 134, 136–38
    frogs as prey of, 4
    fungi and, 116, 117–18, 119, 120
    genetic engineering and, 125
    genetics and, 124
    greenhouse effect and, 26
    habitats and, 79
    history of, 133–35
    insects and, 20
    introduction to, 132
    as mammals, 14
    mountains and, 150, 151
    oceans and, 157
    ozone and, 27
    ozone hole and, 28
    as placental mammals, 15
    protozoans and, 174
    reptiles and, 176
    salt and, 164
    snakes and, 179
    taxonomy of, 48
    timeline of, 132–33
    toads as prey of, 6
    topic terms and, 135–39
    volcanoes and, 204
    Web sites for, 213
Humboldt, Alexander von, 93
Hummingbirds, 42
Humoral theory of disease, 31
Hungary, 137
Hurricanes, 56, 159, 160. *See also* Tropical cyclones
Hutton, James, 76–77
Huxley, Julian, 96
Huxley, Thomas Henry, 42, 94
Hybridize, 218
Hydro-power, 113
Hydrocarbons, 113, 144
Hydrogen
    amino acids and, 124
    atmosphere and, 27
    big bang theory and, 191
    DNA and, 125
    fossil fuels and, 113
    fusion and, 194
    gaseous planets and, 186
    hydrogen power, 113
    killer cloud theory and, 105
    life's origins and, 142
    water vapor and, 56
Hydrologists, 209
Hydrothermal vents, 142
Hylaeosaurus, 67
Hylobatidae, 139
Hylonomus, 177

Ice, 218
Ice Ages
  Agassiz and, 77
  boreal forests and, 79
  continental shelf and, 162
  continental slopes and, 163
  human theory and, 104
  mammoths and, 100
  mountains and, 152
  Neanderthals and, 138
  polar regions and, 81
  sea levels and, 163
Ice cap, 218
Ice sheets, 28, 77, 81, 151, 162
Icebergs, 151–52, 161
Iceland, 198, 201, 203
Ichneumon wasps, 47
Ichnologists, 209
Ichthyologists, 208
Ichthyology, 107–08
Ichthyornithiformes, 37
Ichthyosaurs, 31, 177
Ichthyostega, 2
Igneous intrusion, 77
Igneous rocks, 146, 153
Iguanas, 67, 80
Iguanodon, 67, 68, 69
IGY (International Geophysical Year), 27
Immunologists, 208
Impact theory, 103, 104
Impennes, 42
Imperfects (Fungi Imperfecti), 120
India
  climate and weather and, 52
  crocodiles and, 178
  earthquakes and, 88
  forests and, 80
  magma and, 202
  mountains and, 152
  volcanoes and, 101, 105, 200
Indian Ocean, 56, 108, 154, 162, 164
Indonesia, 108, 134, 136, 137, 200
Influenza, 33, 34, 35
Inner core, 82
Inorganic chemists, 208
Insect mimicry theory, 93
Insectivores, 218
Insects
  as arthropods, 17, 18
  Cenozoic era and, 129
  description of, 20
  evolution and, 18
  flight and, 41
  flowering plants and, 169
  humans and, 19
  metamorphoses and, 19
  Paleozoic era and, 129
  plants and, 170
  pollination and, 169, 171, 172
  reduced metabolic rate and, 14
  reptiles and, 66
  species of, 98
Insulin, 124
Interior deserts, 81
Interior of the Moon, 147, 148
International Commission on Zoological Nomenclature, 46
International Geophysical Year (IGY), 27
International Union of Biological Sciences (IUBS), 47
International Union of Microbiological Societies (IUMS), 47
Interstellar dust and gas, 192, 194, 195
Introns, 170
Invertebrates
  arthropods and, 18
  Cambrian Explosion and, 142
  Cambrian period and, 11
  definition of, 218
  description of, 13–14
  Ediacaran fauna and, 9, 10
  evolution and, 13, 92
  oceans and, 161
Ionosphere, 23, 24–25
Ions, 218
Iran, 87, 88
Ireland, 117, 157
Iridium anomaly, 69, 94, 101
Isoseismal maps, 86
Isostasy, 152
Isostatic uplift, 152
Isotopes, 114, 131, 194
Isotopic dating techniques, 114
Israel, 138
Italy
  climate of, 160
  Cretaceous-Tertiary boundary in, 94, 101
  dinosaurs and, 69
  earthquakes and, 85, 86, 87
  mountains and, 151
IUBS (International Union of Biological Sciences), 47
IUMS (International Union of Microbiological Societies), 47

Jaguars, 80
Janssen, Zacharias, 60
Japan
  earthquakes and, 86, 87, 88
  mountains and, 151
  sea level and, 163
  volcanoes and, 200, 203, 204
  winds and, 53
Japan, Sea of, 164
Japanese giant salamander, 5

Java Man, 136, 137
Jefferson, Thomas, 23, 52
Jeffries, Gwyn, 157
Jenner, Edward, 31, 34
Jet streams, 53, 54
Jewitt, David, 185
Johanson, Donald, 112, 136
Juan de Fuca plate, 204
Jupiter
    asteroids and, 185–86
    atmosphere of, 21, 27
    description of, 188
    discovery of, 183
    extraterrestrial impact and, 147, 184
    as gaseous planet, 186
    moons of, 144, 184
    ring system of, 188–89
    solar system and, 182
    Titius-Bode law and, 185
Jurassic period
    birds and, 36, 37
    dinosaurs and, 65, 67, 70, 177
    extinction and, 100
    geologic time scale and, 129
    plants and, 169
    reptiles and, 177

K/T boundary, 69, 72, 100, 101, 105
Kalahari Desert, 81
Kamchatka, 200, 204
Kamen, Martin, 130
Kangaroos, 15
Kant, Immanuel, 192
Karst topography, 78
Kauai o'o, 102
Kenya, 136, 137
Kepler, Johannes, 183, 192
Keratin, 15, 40, 43
Kerogen, 113
Kilauea, 203
Killer bees, 18
Killer cloud theory, 105
King's Holly, 170
Kiwi, 39, 42
Klebs, Edwin, 32
Klebs-Löffler bacillus, 32
Koch, Robert, 32
Krakatoa, 199
KREEP, 146
Kuiper, Gerald, 184
Kuiper belt objects, 182, 184, 185, 186, 189

La Niña, 160
Labrador, 161
Labrador Current, 160, 161
Lacroix, Alfred, 199
Lahars, 201–02

Lamarck, Jean Baptiste de, 93
Lambda bacteriophage, 35
Lampreys, 16
Lancelets, 13
Lark Quarry Environmental Park, 72
Late Heavy Bombardment, 142, 146, 182
Latin terms, 45
Laurasai, 76
Lava, 198, 201, 202
Laveran, Charles, 32
Law of concealed stratification, 76
Law of original horizontality, 76
Law of superposition, 76
Leaf, 172
Leakey, Louis, 137
Leakey, Mary, 136, 137
Leakey, Richard, 136, 137
Leewenhoek, Anton van, 174
Legends, 200
Legless lizards, 179
Leonardo da Vinci, 76, 112, 151
Leopards, 80
Lepidopterists, 208
Lepidosaurs, 66, 177
Levene, Phoebus Aaron Theodor, 123
Liaoningornis, 42
Libration, 218
Lichens, 9, 119, 142, 175
Life
    Cambrian Explosion and, 9, 10–11
    cladistics and, 48
    description of, 143
    Earth's formation and, 76
    evolution and, 93, 97
    extinction and, 100
    history of, 141–43
    interstellar dust and gas and, 194
    introduction to, 140
    multicellular life, 60, 92, 119, 129, 141
    oceans and, 157
    space and, 144
    Sun and, 195
    timeline of, 140–41
    topic terms and, 143
    vents and, 203
    Web sites for, 213–14
Lifecycles, of fungi, 118, 120
Light year, 218
Lightning, 52, 53, 54, 142
Lignites, 113
Limestone rocks, 78, 94, 114, 136
Limnologists, 208
Lindow, Steve, 32
Linné, Carl von, 45, 92
Lippershey, Hans, 60
Lissamphibians, 3, 4
Lithosphere, 82

Liverworts, 168, 170
Lizards, 177–78, 179
Lobsters, 20
Lock, Richard Alton, 144
Locusts, 19
Löffler, Friedrich, 32
Lomatia tasmanica, 170
Louses, 19, 20
Love, Augustus, 86
Love waves, 86, 90
Lowell, Percival, 144, 185
Lower atmospheric ozone, 6
Lucy, 112, 136
Lunar day, 218
Lunar eclipses, 145, 147, 148
Lunar rocks, 149
Lungfish, 2
Luray Caverns, 78
Luu, Jane X., 185
Lyme disease, 19, 20, 33, 34
Lynx, 80
Lysogenci phase, 35
Lytic phase, 35

McCarthy, Maclyn, 123
MacLeod, Colin, 123
McPhee, John, 130
Madison, James, 52
Magma
    description of, 202–03
    Earth's formation and, 75
    fossils and, 115
    hot spots and, 204
    igneous intrusion and, 77
    lava and, 202
    plate tectonics and, 154
    vents and, 142, 158
    volcanoes and, 198, 199, 200, 201, 203
Magnetism
    birds and, 26–27, 38, 41
    description of, 26–27
    earthquakes and, 85, 86
    extinction and, 104
    layers of Earth and, 81
    sea floor spreading and, 78
    solar cycle and, 194
    volcanoes and, 201
Magnus, Albertus, 45
Maiasaura, 69
Maillet, Benoît de, 93
Makela, Bob, 69
Malaria, 20, 32, 33, 34, 35, 174
Mallet, Robert, 86
Malta, 112
Malthus, Thomas, 93
Mammal theory, 105

Mammals
    birds compared to, 40
    Cenozoic era and, 129
    classification of organisms and, 15–16
    cloning and, 124
    description of, 14–16
    dinosaurs and, 68–69, 100, 104, 105
    estuaries and, 161
    evolution and, 9, 15
    fish and, 107
    hibernation and, 14
    humans and, 133, 134, 139
    Mesozoic era and, 129
    oceans and, 161
    reptiles and, 176, 177, 178
    Species 2000 Programme and, 47
    species and, 98, 133
    as vertebrates, 16
Mammoths, 100
Mandibles, 39, 40, 43
Mannheim Society, 52
Mantell, Gideon Algernon, 67, 68
Mantell, Mary Ann, 67
Mantle
    as layer of Earth, 82
    of Moon, 148
    Moon's formation and, 147
    mountains and, 151, 152
    plate tectonics and, 154, 203
    sea floor spreading and, 78
Mariana Islands, 165
Mariana Trench, 158, 163, 165
Marine biologist, 208
Marine geologists, 209
Mars
    asteroids and, 185, 186
    atmosphere of, 21, 27
    craters and, 104
    description of, 187–88
    discovery of, 183
    life and, 144
    meteoroids and, 187
    solar system and, 182
    as terrestrial planet, 189
    Titius-Bode law and, 185
Marsh, Othniel Charles, 68, 112
Marsupials, 15, 16
Martinique, 199
Maryland darter fish, 102
Mass, 193
Massachusetts, 161
Mastodons, 100
Mathematical physicists, 210
Mather, Cotton, 31
Maury, Matthew, 157–58
Mayer, F., 94
Mayor, Michel, 185
Mean sea level (MSL), 163

Medicine, 169, 208
Mediterranean Sea, 77, 107, 160, 164, 198
Megalosaurus, 67
Meiosis, 62
Mela, Pomponius, 52
Mendel, Gregor Johann, 123, 125
Mercalli Scale, 88–89
Mercury
    atmosphere of, 27, 195
    description of, 187
    discovery of, 183
    lack of satellite, 189
    Late Heavy Bombardment and, 182
    life and, 144
    solar system and, 182
    as terrestrial planet, 189
Mesopause, 25
Mesopotamia, 169
Mesosphere
    as atmospheric layer, 24, 25
    as layer of Earth, 82
Mesozoic era
    birds and, 37
    dinosaurs and, 65, 66, 67, 68, 69, 70–71, 129, 133, 177, 178
    extinction and, 100
    geologic time scale and, 129
    orogenies of, 153
    reptiles and, 177
Metabolic rate, 14, 15
Metamorphic rocks, 153
Metamorphosis
    amphibians and, 3, 4, 5, 6
    arthropods and, 19
    definition of, 218
    description of, 20
    insects and, 18, 20
Metatheria, 15, 16
Metazoa, 11
Meteor Crater, 186
Meteor showers, 187
Meteorites, 187, 200
Meteoroids, 182, 187
Meteorologists, 209
Meteorology, 52–53
Meteors, 187
Methane, 26, 34, 113, 142, 186
Metonic cycle, 163
Mexico, 87, 94, 101, 105
Mexico, Gulf of, 53, 161
Miasmal theory, 31
Mice, 79
Michel, Helen, 69, 94, 101
Michell, John, 85
Michigan, 102, 118
Microbiologists, 208
Microorganisms
    antibiotics and, 118
    bacteria and, 34
    cells and, 61
    discovery of, 31–32, 174
    fossils and, 114
    Species 2000 Programme and, 47
Microsaurs, 2
Microscopes, 35, 60
Mid-Atlantic Ridge, 78, 154, 203
Middle East, 134, 138
Migration, 38, 41
Milky Way galaxy, 193, 194
Miller, Stanley, 142
Millipedes, 79
Milne, John, 86
Minerologists, 209
Minnesota, 6
Mites, 18–19
Mitochondria, 13, 61, 135, 138, 174
Mitosis, 61, 62, 123, 174
Mixed tidal cycles, 164
Modern Synthesis, 96
Mohorovičić discontinuity, 82
Molecular biologists, 208
Molecular physicists, 210
Mollusks, 10, 47, 157, 161
Molting
    arthropods and, 20
    birds and, 40
    frogs and, 5
    snakes and, 179
Monarch butterfly, 96
Mondera, 11
Monerans, 46
Mongolia, 10, 68, 72
Monkeys, 80
Monoplacophorans, 10
Monotremata, 15, 16
Monsoons, 159
Montagu, Mary, 31
Monte Nuovo, 198
Montlivault, Sales-Guyon de, 144
Montreal Protocol of 1987, 27
Moon
    astronomy and, 191
    atmosphere of, 27
    birds and, 38, 41
    description of, 148–49
    Earth's orbit and, 77
    ecliptic path of, 186
    history of, 146–48
    introduction to, 145
    Late Heavy Bombardment and, 182
    life's origins and, 144
    sea level and, 163
    size of, 189
    snowballs above Earth and, 184

timeline of, 145
topic terms and, 148–49
Web sites for, 214
Moonquakes, 147, 149
Moose, 80
Morgan, Thomas Hunt, 123
Morocco, 88
Mosses, 170, 172, 175
Moths, 19, 20, 47, 94
Mountains
description of, 152–53
history of, 151–52
introduction to, 150
Moon and, 146
mountain uplift, 76, 77, 115, 151–52
oceans and, 162
timeline of, 150
topic terms and, 152–54
Web sites for, 214
MSL (mean sea level), 163
Multicellular life
cytology and, 60
eukaryotic cells and, 141
evolution and, 92, 174
fungi and, 119
marine plants and, 166
plants as, 171
Precambrian era and, 129
Multiregionalism theory, 134
Multituberculata, 15
Mushrooms, 117
Mutations, 121
Mutinaite, 63–64
Mycologists, 208
Mycorrhizae, 119

Nariokotome boy, 136, 137
NASA Polar spacecraft, 184
National Center for Atmospheric Research (NCAR), 23
National Endangered Species Programs, 103
National Science Foundation, 23
Native Americans, 33
Natural history
careers in, 207–10
evolution of, xi
fossils and, 114
geologic time scale and, 130
life and, 140
volcanoes and, 198
Web sites for, 211–15
Natural selection, 93, 94, 96, 97
Naturalists, 208
Navigation, 41
NCAR (National Center for Atmospheric Research), 23
Neanderthals, 93, 138
Neap tides, 164
Near-Earth asteroids, 186

Nebulas, 75, 192, 219
Nectar, 171
Nemget Formation, 68
Neogene period, 129
Neognathae, 42
Neornithes, 37, 41–42
Neptune
atmosphere of, 21, 27
description of, 188
as gaseous planet, 186
Kuiper belt objects and, 182, 184, 186
life on, 144
ring system of, 189
satellites of, 189
solar system and, 182
Titius-Bode law and, 185, 189
Neptunism, 77
Nesting, 41
Neurobiologists, 208
New Guinea, 134, 135
New Mexico, 78
New World, 52, 117
New York, 160
Newfoundland, 158, 160, 161
Newts, 2, 3, 5
Nile River, 159
Nitrates, 119, 170
Nitrites, 170
Nitrogen, 23, 24, 27, 34, 124, 125
Nitrogen-14, 114
Nitrous oxides, 26
Nocturnal, 219
Nonflowering plants, 168, 171
Norman, David, 69
North America
caves of, 79
climate and weather and, 52, 54
as continent, 79
forests and, 80
fossils and, 100
mountains and, 151
reptiles and, 178
salamanders and, 5
sea level and, 163
turtles and, 180
North American plate, 77, 159, 204
North Atlantic Current, 160, 161
North Atlantic Drift, 161
North Atlantic gyre, 160
North Carolina, 159
North Equatorial Current, 161
Norway, 158, 161
Norwegian Sea, 164
Nova, 193, 219
Nova Scotia, 163, 165
Novas, Fernando, 43
Nuclear chemists, 208

Nuclear physicists, 210
Nuclear power, 113
Nuclear reactions, 200
Nucleus, 61–62

Oak toad, 5
Observational astronomers, 207
Obsidian, 202
Occluded front, 54
Ocean floor, 158, 162
Ocean margins, 157, 162–63
Oceanic currents, 104
Oceanographers, 209
Oceanography, 155, 157–58
Oceans
    big bombardments and, 184
    Cambrian Explosion and, 11, 13, 130, 142
    caves and, 78
    continental movement and, 77
    crust and, 82
    description of, 162
    early animals and, 9, 13, 22
    earthquakes and, 86
    evolution and, 93
    extinction and, 101
    extraterrestrial impact and, 144
    fish and, 107
    as habitats, 79
    history of, 157–58
    introduction to, 155
    lava and, 202
    life and, 141
    life's origins and, 142
    Moon and, 146
    mountains and, 154
    ozone layer compared to, 168
    Precambrian era and, 129
    sea floor spreading and, 78
    timeline of, 155–56
    topic terms and, 158–65
    vents and, 203
    water and, 28
    water vapor and, 76
    Web sites for, 214
Odontognathae, 42
Offshore breeze, 56
Ogallala Aquifer, 28
Ohio, 117
Oil deposits, 113
Old World python, 179
Oligopeptides, 119
Omnivores, 9, 219
Oncmouse, 124
1992 QB1, 184, 185
Onshore breeze, 56
Ontogeny, 219
Oologists, 208

Oort cloud, 183, 186, 187
Oort, Jan, 186, 187
Open Universe Model, 195
Opossums, 15
Optical physicists, 210
Ordovician period, 100
Organic chemists, 208
Organic evolution, 97
Orientation, 38, 41
Originial horizontality, law of, 76
Orinoco River, 179
Ornithischians, 67, 68, 69, 72
Ornithology, 37–38
Ornithomimid dinosaurs, 43
Ornithopod dinosaurs, 67
Ornitologists, 208
Orogeny, 151, 153–54
Osborne, Thomas, 123
Osteichthyes, 16, 108, 109
Ostriches, 42, 61
Ostrom, John, 38, 42, 68
"Out of Africa" theory, 134, 135, 136
Outer core, 82
Overharvesting, 102
Oviposition, 219
Oviraptor, 43
Owen, Richard, 65, 67–68
Owls, 4, 6, 78–79
Oxygen
    algae and, 175
    amino acids and, 124
    amphibians and, 2
    atmosphere and, 22
    birds and, 39, 40
    DNA and, 125
    fossilization and, 115
    greenhouse effect and, 27
    ionosphere and, 24
    ozone layer and, 24
    photosynthesis and, 9, 171
    plants and, 166, 168, 169, 171
    Precambrian era and, 129
    stromatolites and, 141
    water vapor and, 56
Ozone, 26, 27
Ozone hole, 27–28
Ozone layer
    atmosphere and, 22, 23–24, 141
    as atmospheric layer, 24
    life's origins and, 143
    oxygen and, 168
    ozone and, 27
    supernova theory and, 105

P waves, 90
Pachycephalosaurid dinosaurs, 67
Pacific newt, 5

Pacific Ocean
  continental shelf and, 162
  hurricanes and, 56
  hydrothermal vents and, 142
  oceanography and, 158
  plate tectonics and, 154
  tides and, 164
  trenches and, 165
  volcanoes and, 203
Pacific plate, 77, 78, 159, 204
Pacific rim, 154
Pahoehoe (lava), 202
Pakistan, 88
Palaeognathae, 42
Paleoanthropologists, 209
Paleoanthropology, 134
Paleoecologists, 209
Paleogene period, 129
Paleontologists, 210
Paleontology
  birds and, 36, 42–43
  carbon dating and, 130
  Cuvier and, 112
  definition of, 219
  dinosaurs and, 67, 68, 71
  evolution and, 92
Paleozoic era
  amphibians and, 177
  Cambrian Explosion and, 9, 129, 142, 157
  extinction and, 100, 129
  fungi and, 117
  geologic time scale and, 129
  orogenies of, 153
  reptiles and, 66
Paley, William, 92–93, 94
Palmer Drought Severity Index, 54
Palmieri, Luigi, 85–86
Palms, 47
Panda, 102
Pandaka pygmaea, 109
Pangea, 76, 77
Panopoulos, Nickolas, 32
Parameciums, 14, 174
Parasites
  fungi and, 116, 119
  humans and, 19
  malaria and, 33, 34
  protozoans and, 175
  viruses and, 35
Parasitologists, 208
Particle physicists, 210
Passerines, 40
Pasteur, Louis, 32, 33
Pathologists, 208
Pauling, Linus, 123
Pegasus constellation, 185
Peking Man, 136

Pelagic, 219
Pelee, Mount, 199
Pele's hair, 202
Penguins, 42
Penicillium chrysogenum, 118
Pennsylvanian period, 18, 117
Penumbra, 219
Penumbral lunar eclipse, 148
Perigee, 219
Period, 131
Permian period, 2, 66, 100, 129, 177
Pertussis, 34
Peru, 33, 88, 200
Petals, 171
Petrified wood, 114
Petroleum, 113
Petrologists, 210
Phanerozoic eon, 129
Pharmacists, 208
Pheromones, 119
Philippines, 5, 87, 109, 200
Phobos, 27
Phosphorus, 146
Photoautotrops, 175
Photochemists, 208
Photography, 192
Photosynthesis
  atmosphere and, 22
  carbon dioxide and, 27
  chloroplasts and, 170
  definition of, 219
  description of, 171
  desert plants and, 80
  fungi and, 119
  plants and, 168
  protists and, 174
  stromatolites and, 9, 141
  tropical forests and, 80
Phycomycetes, 119
Phylogenetic systematics, 46
Phylogeny, 2, 219
Physical changes, xi
Physical chemists, 208
Physics, 128, 210
Physiologists, 208
Phytophthora infestans, 117
Piazzi, Guiseppe, 185
Piccard, Auguste, 158
Piccard, Jacques, 158
Pikaia, 13
Pill bugs, 20
Pillow lava, 202
Pinatubo, Mount, 200–201
Pindar, 198
Pinguicula vulgaris, 170
Pinkeye, 34
Pinus longaeva, 170

Piokilothermic, 68
Pisces, 108
Pistil, 171
Pithecanthropus erectus, 136, 137
Placental mammals, 15, 16
Placodonts, 177
Planet X, 185
Planetesimals, 182, 186, 189
Planets
  astronomy and, 191
  description of, 187–88
  ecliptic path of, 186
  solar system and, 181, 182, 183
  solar wind and, 184
Plankton, 20, 28, 107, 109, 113, 161
Plantelogists, 207
Plants
  animals compared to, 11
  atmosphere and, 22
  bacteria and, 34
  boreal forests and, 79
  Cambrian Explosion and, 168
  Cenozoic era and, 129
  classification of organisms and, 45, 46, 169
  coastline and, 158
  cytology and, 58, 60, 62, 63
  Darwin and, 10
  description of, 171–72
  deserts and, 80
  dormancy of, 14
  endangered and threatened species and, 103
  estuaries and, 161
  eukaryotic cells and, 13
  evolution and, 91, 171
  extinction and, 101
  fossil fuels and, 113
  fossils and, 111, 112, 114, 128, 168, 169, 170
  fungi and, 119
  genetics and, 123, 125
  habitat and, 79
  history of, 168–70
  humans and, 9
  introduction to, 166
  land plants, 2, 9, 18, 142, 168, 170
  marine plants, 157, 162, 166
  Mesozoic era and, 129
  mountains and, 151
  oceans and, 155, 157, 159
  ozone layer and, 22–23, 27
  Paleozoic era and, 129
  poison plant theory, 105
  precipitation and, 55
  protists and, 175
  reptiles and, 66
  Silurian period and, 18
  Species 2000 Programme and, 47
  temperate forests and, 80
  timeline of, 166–68
  topic terms and, 170–72
  tropical forests and, 80
  variety of species of, 11
  Web sites for, 214
  winds and, 57
Plasma physicists, 210
Plasmodium, 34, 174
Plate boundaries, 152, 159, 203
Plate tectonics. *See also* Continents
  continental displacement and, 77–78, 86
  description of, 154
  earthquakes and, 84, 86
  mountains and, 151–52
  oceans and, 157
  orogeny and, 151, 153
  Precambian era and, 129
  volcanoes and, 203
Plato, 198
Pleisosaurs, 177
Plesiomorphic, 48
Pliny the Elder, 9
Plovers, 42
Pluto
  atmosphere of, 27
  description of, 188
  discovery of, 185
  Kuiper belt objects and, 182, 184, 186
  Oort cloud and, 187
  satellite of, 189
  as terrestrial planet, 189
  Titius-Bode law and, 189
Plutonism, 77
Pneumatic theory of volcanoes, 198
Poison plant theory, 105
Poisonous snakes, 179
Polar cells, 23, 26
Polar deserts, 81
Polar easterly winds, 26
Polar regions, 79, 81
Polio, 32
Pollination, 169, 171, 172
Pollution, 6, 27–28, 102, 113, 163
Polymer chemists, 209
Polynesia, 200
Pompeii, 198
Pongidae, 139
Portugal, 85, 88
Potassium, 146
Potassium-argon dating technique, 129
Precambrian era
  Cambrian Explosion and, 142
  eukaryotic cells and, 9
  fossils and, 10, 13
  geologic time scale and, 128, 129
  land plants and, 2
  orogenies of, 153
  Sun and, 194

Precipitation, 53, 54–55, 80, 151
Pressure, 219
Pressure gradient force, 56
Primates, 134, 135, 139
Primordial soup, 142
Principle of Superposition, 112
Proconsul, 136
Prokaryotes, 11, 62, 174
Prokaryotic cells, 13
Proteins
    amino acids and, 124, 142
    cells and, 60
    cytology and, 61
    description of, 62, 125–26
    genetics and, 123
    vaccines and, 33
Proterozoic era, 117
Protists
    animals and, 11
    classification of organisms and, 46, 173, 174
    description of, 175
    eukaryotic cells and, 13
    fungi and, 117
    history of, 174
    introduction to, 173
    malaria and, 32, 34
    timeline of, 173
    topic terms and, 174–75
    Web sites for, 214
Proto-galaxies, 193
Proto-sun, 182
Protoctista, 174–75
Prototheria, 16
Protozoans, 60, 141, 173, 174, 175
Pseudomonas syringae, 33
Pseudotsuga menziesii, 170
PSR B1257+12, 185
Psychrophiles, 175
Pterosaurs, 66, 69, 177
Ptolemy, Claudius, 183, 192
Puerto Rico, 159
Puerto Rico Trench, 165
Pumice, 202
Punctuated equilibrium, 96, 97–98
Purkinje, Johannes Evangelista, 60
Pyrenees Mountains, 161
Pyriphlegethon, 198
Pyrite, 63
Pyroclastic flow, 199
Pythagoras, 111

Quagga, 15
Qualitative chemists, 209
Quantitative chemists, 209
Quantum chemists, 209
Quantum physicists, 210
Quantum physics, 192
Quasars, 191
Quaternary period, 37, 129
Queloz, Didier, 185
Quinine, 34

Rabies, 33
Raccoons, 80
Radiation, 24, 219
Radio waves, 24–25
Radioactive decay, 128–29, 131
Radioactive elements, 135
Radioactivity, 114, 200
Radiochemists, 209
Radiometric dating techniques, 101, 114, 128–30
Rain forests
    amphibians and, 3
    arthropods and, 18
    evolution and, 93
    extinction and, 102
    habitats and, 79
    insects and, 20
Rain shadow deserts, 81, 151
Rainier, Mount, 204
Ranapithecus, 136
Rare Earth elements, 146
Ratites, 42
Ray, John, 107, 112
Rayleigh, Lord, 86
Rayleigh waves, 86, 90
Rays, 16, 108
Réaumur, René Antoine Ferchault de, 18
Redwood National Park, 170
Redwoods, 170
Regolith, 146, 148, 149
Reid, H. F., 89–90
Relative humidity, 52
Relative time, 128, 131
Relativity, general theory of, 192
Religion
    astrology as, 192
    Copernican model and, 184
    creationism and, 92–93, 95, 97
    dinosaurs and, 67
    earthquakes and, 84
    epicycle theory of solar system and, 183
    evolution and, 93, 95, 96, 98
    extinction and, 100
    fossils and, 112
    geologic time scale and, 128
    homo sapiens sapiens and, 138
    mountains and, 150, 151
    mushrooms and, 117
    theistic evolution and, 95, 97, 98
    volcanoes and, 198, 200
Reptiles
    amphibians and, 1, 3, 4, 9, 66
    birds and, 38, 40, 42

Reptiles *(continued)*
  description of, 179
  dinosaurs and, 65, 68, 69, 72
  estuaries and, 161
  flowering plants and, 169
  history of, 177–78
  introduction to, 176
  Mesozoic era and, 129
  timeline of, 176
  topic terms and, 178–80
  tropical forests and, 80
  as vertebrates, 16
  Web sites for, 214
Revolution, 219
Rhea, 42
Rhinolophus ferrumequinum, 45
Rhipidistians, 2
Ribonucleic acid. *See* RNA
Richardson, Lewis Fry, 52
Richter, Charles F., 89
Richter Scale, 88, 89
Rigidioporus ulmarius, 118
Ring of Fire, 154, 204
Ring systems, 185, 188–89
Ripples, 165
Rivera Fracture Zone, 158
Rivera, Thomas, 32
Rivers, 159, 163
RNA, 35, 62, 123, 124, 126
Rock
  bacteria and, 142, 143
  coastlines and, 158
  continental movement and, 77
  dating techniques and, 112, 114, 129–30, 134, 135
  early life and, 141, 143
  evolution and, 10
  fossils and, 10, 67
  geologic time scale and, 127, 128
  geological theories and, 77
  groundwater and, 28
  Moon and, 146, 147, 148, 149, 182
  oxygen and, 168
  planetary ring systems and, 188
  relative time and, 128, 131
  self-replication of cells and, 63–64
  superposition and, 76
Rocky Mountains, 81
Rodinia, 76
Roman Empire, 198
Romer, Alfred Sherwood, 2
Roots, 172
Rose, William Cumming, 123
Rose's toad, 5
Ross Ice Shelf, 107
Rossby, Carl Gustav, 52–53
Rostroconchs, 10
Rotation, 219

Rubber boa, 179
Ruben, John, 42
Rubidium-strontium dating technique, 129
Russia, 52, 200
Rutherford, Ernest, 128

S waves, 90
Sabin, Albert, 32
Sachs, Julius von, 169
Saffir-Simpson Damage Potential Scale for Hurricanes, 56
Sagan, Carl Edward, 193
Sahara Desert, 80, 81
Salamanders, 2, 3, 4, 5, 14
Salientia, 4, 5
Salk, Jonas Edward, 32
San Andreas fault, 77–78, 86, 90, 154
San Francisco earthquake of 1906, 90
Sandstones, 113
Saprobes, 119
Satellites, 181, 182, 185, 189
Satellites, artificial, 25, 53, 193, 194
Saturation point, 56
Saturn
  atmosphere of, 21, 27
  description of, 188
  discovery of, 183
  as gaseous planet, 186
  life on, 144
  rings of, 184, 185
  solar system and, 182
Sauer, Franz, 38
Saurischians, 67, 68, 69, 72
Sauropods, 67
Scandinavia, 79, 163
Scanning electron microscope, 61
Schiaparelli, Giovanni, 144
Schimper, Karl, 77
Schleiden, Matthias Jakob, 60
Schoetensack, Otto, 136
Schwann, Theodor, 60
SCN (suprachiasmatic nucleus), 14
Scopes, John, 95–96
Scorpions, 9, 17, 18, 19
Scot, Michael, 45
Scotland, 157
Scripps Laboratory, 157
Sea anemones, 161
Sea floor spreading, 78, 86, 199
Sea gulls, 37
Sea levels, 157, 159, 163, 165
Seas, 157, 164. *See also* Oceans
Seawater, 157, 158, 159, 162, 163–64
Sediment
  coastal plains and, 159
  continental rise and, 163
  continental shelf and, 162
  definition of, 219

erosion and, 55
fossilization and, 114, 115, 178
oceans and, 105, 162
stromatolites and, 143
tides and, 165
turbidity currents and, 158, 165
winds and, 57
Sedimentary rock
carbon dating techniques and, 114
dinosaur fossil locations and, 72
fossils and, 13, 115
geologic time scale and, 129
Leonardo da Vinci and, 76
Precambrian era and, 10
self-replication of cells and, 64
stromatolites and, 143
Sedimentologists, 210
Seed plants, 171–72
Seeds, 172
Seeley, Harry, 68, 72
Seismic tomography, 90
Seismic waves, 86, 88, 89, 90
Seismicity, 201
Seismographs, 86, 89, 90
Seismology, 86
Seismometers, 85, 147
Seismoscopes, 85
Self-replication, 60, 62–64
Semi-diurnal tides, 164
Semmelweiss, Ignaz, 31
Sepals, 171
Sequoia sempevirens, 170
Sequoiadendron giganteum, 170
Sesquiterpines, 119
Sex cells, 62
Seysenegg, Erich Tschermak von, 123, 125
Shale, 113
Shallow-water waves, 165
Sharks, 16, 107, 108
Shasta daisy, 170
Shasta, Mount, 204
Shergotty-Nakhla-Chassignys (SNGs), 187
Shi-Chen, Li, 9
Shih Shen, 191
Shoemaker-Levy 9, 184
Short-horned lizards, 179
Shrimp, 20, 79
Siberia, 79
Siebold, Karl von, 174
Sierra Nevada mountains, 151
Silica, 143, 202–03
Silicon, 143, 194
Silurian period, 2, 9, 18, 117
Sinanthropus pekinensis, 136
Singapore, 87
Sinosauropteryx, 42
Six-Kingdom Classification, 48

Skates, 108
Skeletal systems, 40–41, 66
Skin cancer, 28
Slime molds, 175
Slime nets, 175
Small shelly fossils (SSFs), 10
Smallpox, 31, 33, 34, 35
Smith, J. L. B., 108
Snakes, 4, 6, 14, 177, 178, 179
SNGs (Shergotty-Nakhla-Chassignys), 187
Snowballs above Earth, 184
Society Islands, 20
Sociobiologists, 208
Soddy, Frederic, 128
SOHO (Solar and Heliospheric Observatory), 193
Solar astronomers, 207
Solar cycle, 193, 194
Solar eclipses, 145, 147, 148
Solar flares, 194
Solar and Heliospheric Observatory (SOHO), 193
Solar nebula, 22, 182, 186, 189
Solar power, 113
Solar storms, 193, 194
Solar system
astronomy and, 192
evolution and, 91
history of, 182–85
introduction to, 181
timeline of, 181–82
topic terms and, 185–89
Web sites for, 214–15
Solar wind, 105, 184, 194
Solid-state chemists, 209
Solid-state physicists, 210
Sonar, 158
Songs, of birds, 41
Sound, 219
South African clawed frog, 124
South America
climate and weather and, 54
coastlines and, 159
as continent, 79
continental movement and, 77
currents and, 160
forests and, 80
fungi and, 117
habitats and, 79
reptiles and, 178
snakes and, 179
volcanoes and, 203
South American tinamous, 42
South China Sea, 164
Soviet Union, 10
Space Weather Program, 193
Space-time, 192, 193
Spain, 43, 138
Sparrows, 41

Spawning, 219
Species 2000 Annual Checklist, 47
Species 2000 Programme, 47
Species
    adaptations of, 11
    amphibians and, 3
    birds and, 37, 38–39, 42
    cells and, 61
    crustaceans and, 20
    definition of, 220
    description of, 98
    dinosaurs and, 66
    endangered and threatened species, 99, 101, 103
    evolution and, 92
    extinction of, 99, 100, 101, 102
    fish and, 107–08
    fungi and, 119
    habitats and, 79–81
    humans and, 134–35, 136, 137, 138
    insects and, 20
    mammals and, 15, 98, 133
    origin of, 10
    puntuated equilibrium and, 96, 97
    religion and, 67
    species introduction and, 102
Speed of light, 220
Sphenondon, 178
Spiders, 9, 17, 18, 19, 79
Sponges, 9, 10, 11, 157
Spores, 120, 171
Sprigg, R. C., 10
Spring tides, 164
Squirrels, 80
SSFs (small shelly fossils), 10
Stabiae, 198
Stamen, 171
Stars
    big bang theory and, 191
    black holes and, 193
    constellations and, 185, 191–92
    description of, 194
    extrasolar planets and, 186
    movement of, 191–92
    positional measurements of, 183
Stationary front, 54
Steady-State Model, 195
Stegosaurid dinosaurs, 67
Steller astronomers, 207
Stem, 172
Steno, Nicolaus, 76, 112
Steno's law, 112
Stereochemists, 209
Sternberg, Charles H., 68
Sternberg, Charles M., 68
Sternberg, George, 68
Sternberg, Levi, 68
Stoichiometrists, 209

Stone Age, 111
Stone tools, 137, 138
Strabo, 198
Strasburger, Eduard Adolf, 60–61
Stratigraphers, 210
Stratopause, 24, 25
Stratosphere, 24
Strep throat, 34
Strike-slip faults, 77–78, 87
Striped newts, 5
Stromatolites, 9, 141, 142, 143
Structural geologists, 210
Subage, 131
Subduction, 154, 165, 201, 203–04
Submarine, 158
Subtropical deserts, 81
Sudbury crater, 104
Suess, Edward, 77
Sulfides, 113
Sulfur dioxide, 201
Sun
    asteroids and, 185
    astronomy and, 191, 193
    atmosphere and, 22, 23
    birds and, 38, 41
    cells and, 26
    comets and, 186
    currents and, 159, 160
    description of, 194–95
    ecliptic path of, 186
    greenhouse effect and, 27
    life and, 141
    lizards and, 179
    magnetism and, 26
    Moon and, 145
    ozone and, 27
    ozone layer and, 24, 27
    plants and, 171
    solar system and, 181
    solar wind, 184
    thermoclines and, 160
    tides and, 164
    volcanoes and, 198, 199
Sun spots, 194
Sunset Crater, 78
Supernova, 191, 193, 194, 195
Supernova theory, 105
Superposition, law of, 76
Superposition, Principle of, 112
Suprachiasmatic nucleus (SCN), 14
Surface currents, 159, 160
Surface waves, 90
Surinam Toad, 6
Swammerdam, Jan, 18
Sweat peas, 49
Swedenborg, Emmanuel, 192
Swimming, and fish, 107, 109

Synapsids, 15, 66, 177, 178
Synthetic chemists, 209
Syphilis, 32
Syria, 88
Systematics, 46, 48

T-cells, 34
Tadpoles, 2, 3, 4, 5, 6
Tagmata, 19
Tambora, 199
Tanzania, 136, 137
Taphonomy, 114–15
Tarantulas, 19
Tasmania, 103
Tasmanian devils, 15
Taung Baby, 136
Taxodium distichum, 170
Taxonomists, 208
Taxonomy, 46, 47, 48–49
Telegraph, 52
Telescopes, 183, 184, 192
Television Infrared Observation Satellite-1 (TIROS-1), 53
Temnospondyls, 2
Temperate forests, 80
Temperature, 24, 143, 217, 220
Tennessee, 95–96
Tephra, 201, 203
Termites, 19, 20
Terrestrial planets, 75, 182, 189
Tertiary period, 37, 69, 72, 129
Tetrapods, 2, 15
Thales, 85
Thecodontosaurus, 67
Thecodonts, 66–67, 177
Theistic evolution, 95, 97, 98
Theophrastus, 169
Theoretical astronomers, 207
Thermal radiation, 220
Thermochemists, 209
Thermoclines, 160
Thermodynamicists, 210
Thermoluminescence, 114, 135
Thermometer, 184
Thermosphere, 25
Theropod dinosaurs, 37, 42, 67
Thomson, Charles Wyville, 157
Thomson, William, 128
Thorium, 128, 200
Thorium-230, 114, 136
Thrinaxodon, 15
Thymine, 125
Tibetan Plateau, 154
Ticks, 18–19
Tides
    coastlines and, 159
    description of, 164–65
    estuaries and, 161
    Moon and, 145, 147, 164
    oceanography and, 157
    oceans and, 155
    sea level and, 163
Tigers, 49
Time units, 130, 131
Tinamous, 42
TIROS-1 (Television Infrared Observation Satellite), 53
Titan, 27, 144
Titius, Johann, 184
Titius-Bode law, 184–85, 189
Toads, 2, 3, 4, 5–6
Tobacco mosaic virus, 35
Tombaugh, Clyde William, 185
Tonga Trench, 165
Topography, 220
Tornadic-type winds, 53, 55
Tornadoes, 55
Toro, 147–48
Torricelli, Evangelista, 52
Tortoises, 177, 178, 180
Trace fossils, 112, 115
Trackways, of dinosaurs, 72
Trade winds, 23, 26, 28
Transduction, 33
Transform fault, 77–78
Transformation, 3, 18, 19. *See also* Metamorphosis
Transgenics, 125
Tree frogs, 80
Trenches, 158, 162, 165
Triassic period
    crocodiles and, 178
    dinosaurs and, 65, 66, 67, 70, 177
    extinction and, 100
    reptiles and, 177–78
Trilobites, 13
Trieste, 158
Triton, 189
Troglodytes, 78
Tropical cyclones, 53, 55–56
Tropical depressions, 55–56
Tropical forests, 2, 80. *See also* Rain forests
Tropical storms, 56
Tropopause, 24
Troposphere, 23, 24, 26, 27
Tsunamis, 199, 204, 220
Tuberculosis, 32, 33, 34
Tunicates, 13
Turbidity currents, 158, 163, 165
Turkey, 87
Turtles, 177, 178, 179–80
Tyndall, John, 23
Typhoid bacillus, 32
Typhoons, 56
Typhus, 20
Tyrrell, Joseph Burr, 68

Umbellifers, 47
Umbra, 220
Umbral eclipse, 148
Unenlagia comaheunsis, 43
UNEP (United Nations Environment Programme), 47
Uniformitarianism, 76–77
United Kingdom. *See* England
United Nations Convention on Biological Diversity, 47
United Nations Environment Programme (UNEP), 47
United States
    amphibians and, 5, 6
    arthropods and, 19
    bacteria and viruses and, 33
    classification of organisms and, 46
    climate and weather and, 52, 53, 55
    coastlines and, 158, 159
    currents and, 160
    dinosaurs and, 68, 72
    endangered and threatened species and, 101, 103
    evolution and, 95–96
    extinction and, 102
    forests and, 80
    fungi and, 117, 118
    liverworts and, 168
    magma and, 202
    oceanography and, 157, 158
    sea level and, 163
    snakes and, 179
    volcanoes and, 199, 204
Universe
    description of, 195
    history of, 191–93
    introduction to, 190
    timeline of, 190–91
    topic terms and, 193–95
    Web sites for, 215
Upper atmospheric ozone, 6
Upper Carboniferous period, 18
Upwellings, 160
Uranium, 128, 136, 200
Uranium-lead dating method, 129
Uranium-series dating, 114, 136
Uranium-thorium-lead dating techniques, 129
Uranus
    atmosphere of, 21, 27
    description of, 188
    as gaseous planet, 186
    life on, 144
    ring system of, 189
    solar system and, 182
    Titius-Bode law and, 185
Urey, Harold, 142
Ussher, James, 128
Utricularia vulgaris, 170

Vaccination, 31, 32, 33, 34–35
Vacuum, 184
Valley Fever, 118
Vapor pressure, 52
Vascular plants, 2, 9, 168
Vendian period, 10, 13, 142
Vents
    coastal margin and, 163
    description of, 203
    Galapagos Islands and, 158, 201
    life's origins and, 142–43
    ocean floor and, 162
Venus
    atmosphere of, 21, 27, 195
    description of, 187
    discovery of, 183
    greenhouse effect and, 27
    lack of satellite, 189
    life and, 144
    phases of, 184
    solar system and, 182
    as terrestrial planet, 189
Venus flytrap, 170
Vertebrates
    animals and, 7
    Aristotle and, 9
    Cambrian period and, 142
    Chordata and, 13
    cloning and, 124
    definition of, 220
    description of, 16
    evolution of, 11
    fish and, 108
    humans and, 134
    oceans and, 161
    variety of species of, 11
Vesuvius, Mount, 85, 198
Viceroy butterfly, 96
Vine, Fred, 78
Virginia, 78, 161
Virgo constellation, 185
Virologists, 208
Virology, 31, 32
Viruses
    definition of life and, 143
    description of, 35
    history of, 31–33
    introduction to, 29
    protists and, 174
    size of, 141
    Species 2000 Programme and, 47
    timeline of, 29–31
    topic terms and, 34–35
    Web sites for, 212
Vision, of fish, 107, 109
Viviani, Vincenzo, 52
Volcanic theory, 105
Volcano types, 201

Volcanoes
    Aristotle and, 85
    atmosphere and, 27
    catastrophism and, 94–95
    caves and, 78
    cytology and, 60
    description of, 203–05
    Earth's formation and, 76
    extinction and, 67, 72, 101, 104, 105
    fossils and, 115
    geologic time scale and, 129
    history of, 198–201
    hydrothermal vents and, 142
    introduction to, 196
    iridium anomaly and, 101
    Late Heavy Bombardment and, 182
    Moon and, 146
    mountains and, 152
    oceans and, 157, 162
    oxygen and, 23
    ozone and, 27
    sea floor spreading and, 78
    timeline of, 196–97
    topic terms and, 201–05
    water and, 144
    Web sites for, 215
Volcanologists, 210
Volcanology, 198–201, 203
Von Middendorff, Alfred, 38
Voyager spacecraft, 185
Vredefort crater, 104
Vulcanists, 77

Wales, 118
Walker, Alan, 136, 137
Wallace, Alfred Russel, 18, 93
Walsh, Donald, 158
Warm front, 54
Washington, George, 52
Washington (state), 118
Wasps, 19, 20
Water
    amphibians and, 2, 3
    atmosphere and, 23
    comets and, 186
    description of, 28
    fungi and, 117, 119
    interstellar dust and gas and, 194
    life's origins and, 143, 144
    magma and, 202–03
    Mars and, 144
    mountains and, 151
    oceans and, 157, 162
    pollination and, 172
    precipitation and, 55
    protozoans and, 175
    reptiles and, 177, 178
    turtles and, 179

Water cycle, 28
Water fleas, 20
Water molds, 174
Water vapor
    atmosphere and, 27, 75–76
    climate and weather and, 52
    description of, 56
    earth's formation and, 75–76
    greenhouse effect and, 26
    life's origins and, 142
    precipitation and, 54–55
    snowballs above Earth and, 184
    volcanoes and, 198
Waterspouts, 55
Watson, James, 61, 123, 125
Waves, 157, 158, 163, 165
Weather. *See* Climate and weather
Wegener, Alfred, 77, 78, 86, 199
Weissman, August, 93
Werner, Abraham, 77
Westerly winds, 26
Wetlands, 28, 159, 161
Whale shark, 109
Wheeler, Mount, 170
Whiston, William, 76
White, Gilbert, 46
White Mountains, 170
Whooping cough, 34
Wilkins, Maurice, 123
Willughby, Francis, 107
Willy-willys, 56
Wilson, J. Tuzo, 77, 86, 204
Winds
    climate and weather and, 53
    currents and, 159, 160, 161
    description of, 56–57
    mountains and, 151
    oceans and, 158
    pollination and, 169, 172
    solar wind, 184
    tornadic-type winds, 53, 55
    trade winds, 23, 26, 28
Wolszczan, Alex, 185
Wolverine, 80
Woodlice, 20
Woodpeckers, 41
Woods Hole Oceanographic Institution, 157
Woolly bear caterpillars, 14
Worm lizards, 179

Xenophanes, 111

Yellow fever, 20
Yellowstone National Park, 204
Yosemite frogs, 6
Yosemite National Park, 6

Zeolite, 63
Zinjanthropus boisei, 136
Zircon, 129
Zodiac, 186
Zodiacal light, 184

Zoologists, 208
Zoology, 9–10, 45, 48, 107
Zygomycetes, 119
Zygotes, 11

Not to be taken from the Library